UNIVERSITY OF STRATHCLYDE
30125 0059319
KV-575-691

Intelligent Sensors

HANDBOOK OF SENSORS AND ACTUATORS

Series Editor: S. Middelhoek, Delft University of Technology, The Netherlands

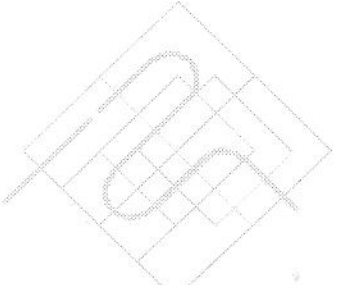

Volume 1 Thick Film Sensors (edited by M. Prudenziati)
Volume 2 Solid State Magnetic Sensors (by C.S. Roumenin)
Volume 3 Intelligent Sensors (edited by H. Yamasaki)

HANDBOOK OF SENSORS AND ACTUATORS 3

Intelligent Sensors

Edited by
Hiro Yamasaki
Yokogawa Research Institute Corporation
Tokyo, Japan

1996
ELSEVIER
Amsterdam - Lausanne - New York - Oxford - Shannon - Tokyo

ELSEVIER SCIENCE B.V.
Sara Burgerhartstraat 25
P.O. Box 211, 1000 AE Amsterdam,
The Netherlands

ISBN: 0 444 89515 9

This book is printed on acid-free paper.

Printed in The Netherlands

Introduction to the Series

The arrival of integrated circuits with very good performance/price ratios and relatively low-cost microprocessors and memories has had a profound influence on many areas of technical endeavour. Also in the measurement and control field, modern electronic circuits were introduced on a large scale leading to very sophisticated systems and novel solutions. However, in these measurement and control systems, quite often sensors and actuators were applied that were conceived many decades ago. Consequently, it became necessary to improve these devices in such a way that their performance/price ratios would approach that of modern electronic circuits.

This demand for new devices initiated worldwide research and development programs in the field of "sensors and actuators". Many generic sensor technologies were examined, from which the thin- and thick-film, glass fiber, metal oxides, polymers, quartz and silicon technologies are the most prominent.

A growing number of publications on this topic started to appear in a wide variety of scientific journals until, in 1981, the scientific journal Sensors and Actuators was initiated. Since then, it has become the main journal in this field.

When the development of a scientific field expands, the need for handbooks arises, wherein the information that appeared earlier in journals and conference proceedings is systematically and selectively presented. The sensor and actuator field is now in this position. For this reason, Elsevier Science took the initiative to develop a series of handbooks with the name "Handbook of Sensors and Actuators" which will contain the most meaningful background material that is important for the sensor and actuator field. Titles like Fundamentals of Transducers, Thick Film Sensors, Magnetic Sensors, Micromachining, Piezoelectric Crystal Sensors, Robot Sensors and Intelligent Sensors will be part of this series.

The series will contain handbooks compiled by only one author, and handbooks written by many authors, where one or more editors bear the responsibility for bringing together topics and authors. Great care was given to the selection of these authors and editors. They are all well known scientists in the field of sensors and actuators and all have impressive international reputations.

Elsevier Science and I, as Editor of the series, hope that these handbooks will receive a positive response from the sensor and actuator community and we expect that the series will be of great use to the many scientists and engineers working in this exciting field.

Simon Middelhoek

Preface

Sensors are the front–end devices for information acquisition from the natural and/or artificial world. Higher performance of advanced sensing systems is realized by using various types of machine intelligence. Intelligent sensors are smart devices with signal processing functions shared by distributed machine intelligence.

When I agreed to be the editor of this book, I recollected typical examples of intelligent sensors: the receptors and dedicated signal processing systems of the human sensory systems. The most important job of information processing in the sensory system is to extract necessary information from the receptors signals and transmit the useful information to the brain. This dedicated information processing is carried out in a distributed manner to reduce the work load of the brain. This processing also lightens the load of signal transmission through the neural network, the capacity of which is limited.

The performance of the receptors in our human sensory system is not always ideal and is frequently inferior to that of man–made sensors. Nevertheless, the total performance is usually far superior to those of our technical sensing systems. The weak points of human receptors are masked by the information processing. This processing makes our sensory system adaptable to the environment and optimizes system performance.

The basic idea of this book, which contains new computing paradigms, is that the most advanced intelligent sensing system is the human sensory system. The book was designed and suitable authors were selected with this idea in mind.

Section I reviews the technologies of intelligent sensors and discusses how they developed. Typical approaches for the realization of intelligent sensors emphasizing the architecture of intelligent sensing systems are also described. In section II, fundamental technologies for the fabrication of intelligent sensors and actuators are presented. Integration and micro–miniaturization techniques are emphasized. In section III, advanced technologies approaching human sensory systems are presented. These technologies are not directly aimed at practical applications, but introduce the readers to the development of engineering models of sensory systems. The only exception is the echo location system of bats. The signal processing in a bat's system is reasonable and may be useful for practical applications. Section IV introduces technologies of integrated intelligent sensors, which will be in practical use soon. These sensors are fabricated on a silicon chip using monolithic IC technology. In section V, examples are given of intelligent sensing systems which are used in industrial installations. Hardware for machine intelligence is not integrated at present, but can soon be implemented in the monolithic integrated structure. Without this machine intelligence, new functions, for example, self diagnosis or defects identification, cannot be realized. This section also demonstrates the potential of intelligent sensors in industry. Section VI introduces two interesting topics which are closely related to intelligent sensing systems. The first one is multisensor fusion. It is expected to be one of the fundamental and powerful technologies for realizing an advanced intelligent sensing system. The second is visualizing technology of the sensed states for easy comprehension of the dynamic multi–dimensional state. This is useful for intelligent man–machine interfaces.

As Editor of this book, I am very grateful to the authors for their contributions and to the staffs of Elsevier Science for their support and cooperation. I sincerely hope this book will be widely accepted as fruitful R & D results and recognized by the readers as a milestone in the rapid progress of intelligent sensors.

Hiro Yamasaki

TABLE OF CONTENTS

Intelligent Sensors
H. Yamasaki (Editor)

What are the Intelligent Sensors

Hiro YAMASAKI

Yokogawa Research Institute Corporation, Musashino, Tokyo, Japan

(Professor Emeritus of The University of Tokyo)

1. INTRODUCTION

Sensors incorporated with dedicated signal processing functions are called intelligent sensors or smart sensors. Progress in the development of the intelligent sensors is a typical example of sensing technology innovation. The roles of the dedicated signal processing functions are to enhance design flexibility of sensing devices and realize new sensing functions. Additional roles are to reduce loads on central processing units and signal transmission lines by distributing information processing in the sensing systems [1–5].

Rapid progress in measurement and control technology is widely recognized. Sensors and machine intelligence enhance and improve the functions of automatic control systems. Development of new sensor devices and related information processing supports this progress.

Application fields of measurement and control are being rapidly expanded. Typical examples of newly developed fields are robotics, environmental measurement and biomedical areas. The expanded application of intelligent robots requires advanced sensors as a replacement of the sensory systems of human operators. Biomedical sensing and diagnostic systems are also promising areas in which technological innovation has been triggered by modern measurement and automatic control techniques.

Environmental instrumentation using remote sensing systems, on the other hand, informs us of an environmental crisis on the earth.

I believe that in these new fields attractive functions and advanced features are mostly realized by intelligent sensors. Another powerful approach to advanced performance is the systematization of sensors for the enhancement of sensing functions.

In this chapter the recent development of intelligent sensors and their background is described.

2. WHY INTELLIGENT SENSORS?

2.1 General information flow through three different worlds

Measurements, control and man–machine communications

Let us consider the technical background of the intelligent sensors. First we will analyze the general information flow relating to measurement and control systems. This information flow includes not only communications between objects and measurement systems, but also man–machine communications. It can be depicted as communication between three different worlds, (Fig.1). It is easily understood that a smooth and efficient information flow is essential to make the system useful and friendly to man [6].

The three different worlds shown in Figure 1 are as follows:

(1) Physical world. It represents the measured and controlled object, sensors and actuators. Natural laws rule this world in which the causality is strictly established. Information is transmitted as a physical signal.

(2) Logical world. It represents an information processing system for measurement and control. Logic rules this world in which information is described by logical codes.

(3) Human intellectual world. It is the internal mental world of the human brain. Information is translated to knowledge and concepts in this world. What kinds of laws rule this world? The author cannot answer this question at the moment. This is the one of the old problems left unsolved.

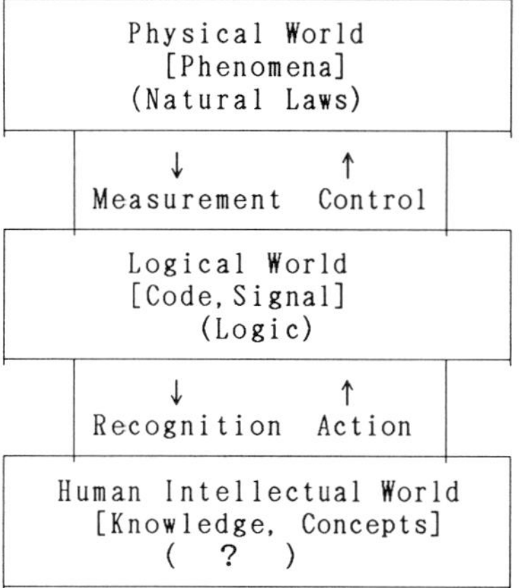

Figure 1 Information flow between three different worlds

Information in the physical world is extracted by sensing or measurement techniques and is transferred into logical world. The information is processed according to the objectives of the measurement and control systems in the logical world, and it is recognized by man through man–machine interfaces that display the measured and processed results. Thus, the information is transferred into the human intellectual world and then constructed as a part of the knowledge and concepts in the human world. Masses of obtained information segments are first structured and then become a part of knowledge. Assembled knowledge constitutes concepts. Groups of concepts constitute a field of science and technology.

Human behavior is generally based on obtained information and knowledge. Intentions are expressed as commands and transmitted to the system by actions toward man–machine interfaces. The logical control system regulates its objects through actuators in the physical world based on the command. We can paraphrase "measurement and control" as the information flow process: measurement, recognition, action and control in three different worlds in which human desire is realized in the physical world.

Between the physical world and the logical one, information is communicated through measurement systems and control systems. There, sensors and actuators act as the interfaces. Sensors are the front–end elements of the sensing or measuring systems, and they extract information from the objects in the physical world and transmit the physical signal.

Between the logical world and the human intellectual world, the communications are carried out through a man–machine interface.

An intelligent sensor is one which shifts the border of the logical world to to the physical world. Part of the information processing of the logical world is replaced by information processing within the sensors.

An intelligent man–machine interface comprises a system by which the border between the human world and the logical world is shifted to the human side. Part of human intelligence for recognition is replaced by machine intelligence of the man–machine interfaces.

2.2 Needs for intelligent sensors

Neede for present sensor technology

Present sensor technology has been developed to meet the various past needs of sensors. Sensor technology is essentially needs–oriented. Therefore, we can forecast the future of sensing technology by discussing the present needs of sensors.

The present needs for the development of intelligent sensors and sensing systems can be described as follows: [6,7]

(1) We can sense the physical parameters of objects at normal states with high accuracy and sensitivity. However, detection of abnormalities and malfunctions are poorly developed. Needs for fault detection and prediction are widely recognized.

(2) Present sensing technologies can accurately sense physical or chemical quantities at single point. However, sensing of multidimensional states is difficult. We can imagine the environment as an example of a multidimensional state. It is a widely spread state of characteristic parameters are spatially dependent and time–dependent.

(3) Well–defined physical quantities can be sensed with accuracy and sensitivity. However, quantities that are recognized only by human sensory systems and are not clearly defined cannot be sensed by simple present sensors. Typical examples of such quantities are olfactory odors and tastes.

Difficulties in the above–described items share a common feature: the ambiguity in definition of the quantity or difficulty in simple model building of the objects. If we can define the object clearly and build its precise model, we can learn the characteristic parameter that describes the object's state definitely. Then, we can select the most suitable sensor for the characteristic parameter. In the above three cases, this approach is not possible.

Human expertise can identify the abnormal state of an object. Sometimes man can forecast the trouble utilizing his knowledge of the object. Man and animals know whether their environment is safe and comfortable because of their multimodal sensory information. We can detect or identify the difference in complicated odors or tastes in foods. Man senses these items without the precise model of the object by combining multimodal sensory information and knowledge. His intelligence combines this multiple sensory information with his knowledge.

Accordingly in the world of physical sensors, we can solve these problems by signal processing and knowledge processing. Signal processing combines or integrates the outputs of multiple sensors or different kinds of sensors utilizing knowledge of the object. This processing is called sensor signal fusion or integration and is a powerful approach in the intelligent sensing system. An advanced information processing for sensor signal fusion or integration should be developed and is discussed later in another chapter.

3. ARCHITECTURE OF INTELLIGENT SENSING SYSTEMS

3.1 Hierarchy of machine intelligence [8]

Before we describe intelligent sensors, we will discuss the architecture of intelligent sensing systems in which intelligent sensors are used as the basic components. Flexibility and adaptability are essential features in an intelligent sensing system. We can consider the human sensory system as a highly advanced example of an intelligent sensing system. Its sensitivity and selectivity are adjustable corresponding to the objects and environment.

The human sensory system has advanced sensing functions and a hierarchical structure. This hierarchical structure is suitable architecture for a complex system that executes advanced functions.

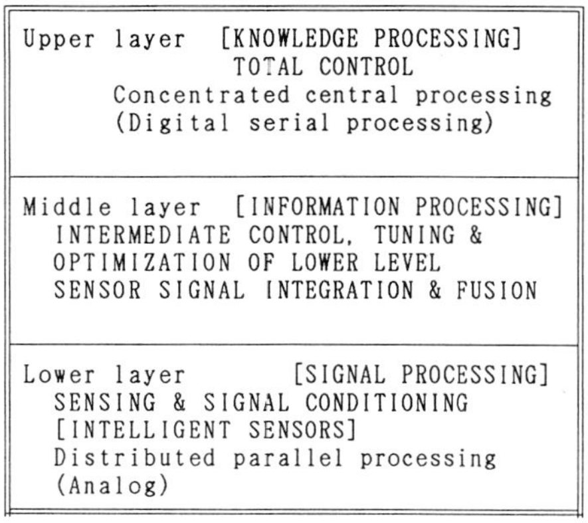

Upper layer [KNOWLEDGE PROCESSING] TOTAL CONTROL Concentrated central processing (Digital serial processing)
Middle layer [INFORMATION PROCESSING] INTERMEDIATE CONTROL, TUNING & OPTIMIZATION OF LOWER LEVEL SENSOR SIGNAL INTEGRATION & FUSION
Lower layer [SIGNAL PROCESSING] SENSING & SIGNAL CONDITIONING [INTELLIGENT SENSORS] Distributed parallel processing (Analog)

Figure 2 Hierarchical structure of intelligent sensing system

One example of the hierarchical architecture of intelligent sensing systems has a multilayer as shown in Figure 2.

The most highly intelligent information processing occurs in the top layer. Functions of the processing are centralized, like in the human brain. Processed information is abstract and independent of the operating principle and physical structure of sensors.

On the other hand, various groups of sensors in the bottom layer collect information from external objects, like our distributed sensory organs. Signal processing of these sensors is conducted in a distributed and parallel manner. Processed information is strongly dependent on the sensor's principles and structures

Sensors incorporated with dedicated signal processing functions are called intelligent sensors or smart sensors. The main roles of dedicated signal processing are to enhance design flexibility and realize new sensing functions. Additional roles are to reduce loads on central processing units and signal transmission lines by distributing information processing in the lower layer of the system.

There are intermediate signal processing functions in the middle layer. One of the intermediate signal processing functions is the integration of signals from multiple sensors in the lower layer. When the signals come from different types of sensors, the function is referred to as sensor signal fusion. Tuning of sensor parameters to optimize the total system performance is another function.

In general, the property of information processing done in each layer is more directly oriented to hardware structure in the lower layer and less hardware–oriented in the upper layer. For the same reason, algorithms of information processing are more flexible and need more knowledge in the upper layer and are less flexible and need less knowledge in the lower layer.

We can characterize processing in each layer as follows: signal processing in the lower layer, information processing in the middle layer, and knowledge processing in the upper layer.

3.2 Machine intelligence in man–machine interfaces

Intelligent sensing systems have two different types of interfaces. The first is an object interface including sensors and actuators. The second is a man–machine interface. The man–machine interface has a hierarchical structure like the one as shown in Figure 2. Intelligent transducers are in the lower layer, while an advanced function for integration and fusion is in the middle layer. Machine intelligence in the middle layer harmonizes the differences in information structure between man and machine to reduce human mental load. It modifies the display format and describes an abnormal state by voice alarm. It can also check error in human operation logically, which makes the system more user friendly.

3.3 Role of intelligence in sensors

Sensor intelligence performs a distributed signal processing in the lower layer of the sensing system hierarchy. The role of the signal processing function in intelligent sensors can be summarized as follows:

1) reinforcement of inherent characteristics of the sensor device and
2) signal enhancement for the extraction of useful features of the objects.

1) Reinforcement of inherent characteristics of sensor devices

The most popular operation of reinforcement is compensation of characteristics, the suppression of the influence of undesired variables on the measurand. This operation of compensation is described below. We can depict the output signal of a sensor device by the eqation

$$F(x_1,\ x_2,\ldots x_n),$$

where x_1 is the quantity to be sensed or measured and $x_2, \ldots x_n$ are the quantities which affect the sensor output or sensor performances. The desirable sensor output is dependent only on x_1 and t and independent of any change in $x_2, \ldots x_n$. Compensation is done by other sensors that are sensitive to change in $x_2, \ldots x_n$. A typical compensation operation can be described mathematically for the case of a single influential variable for simplicity. The increment of x_1 and x_2 are represented by Δx_1 and Δx_2, espectively. The resultant output y is as follows:

$$y = F(x_1+\Delta x_1,\ x_2+\Delta x_2) - F^*(x_2+\Delta x_2) \tag{1}$$

$$\begin{aligned} y \approx\ & F(x_1,\ x_2) + (\partial F/\partial x_1)\Delta x_1 + (\partial F/\partial x_2)\Delta x_2 \\ & + \tfrac{1}{2}\{(\partial^2 F/\partial x_1^2)(\Delta x_1)^2 + 2(\partial^2 F/\partial x_1 \partial x_2)\Delta x_1 \Delta x_2 + (\partial^2 F/\partial x_2^2)(\Delta x_2)^2\} \\ & - F^*(x_1,\ x_2) - (\partial F^*/\partial x_2)\Delta x_2 - \tfrac{1}{2}(\partial^2 F^*/\partial x_2^2)(\Delta x_2)^2 \end{aligned} \tag{1)'$$

If we can select function F^* to satisfy the conditions in eq. (3) within variable range of x_2, we obtain eq. (2), and can thus reduce the influence of change in x_2, as shown in Figure 3.

$$y \approx (\partial F/\partial x_1)\Delta x_1 + \tfrac{1}{2}(\partial^2 F/\partial x_1^2)(\Delta x_1)^2 + (\partial^2 F/\partial x_1 \partial x_2)\Delta x_1 \Delta x_2 \qquad (2)$$

$$\begin{aligned} F(x_1, x_2) &= F^*(x_1, x_2) \\ (\partial F/\partial x_2) &= (\partial F^*/\partial x_2) \\ (\partial^2 F/\partial x_2^2) &= (\partial^2 F^*/\partial x_2^2) \end{aligned} \qquad (3)$$

However, since eq. (2) contains a second–order cross–term including x_2, the compensation is not perfect. In this case, the output at the zero point cannot be influenced by x_2, but the output at the full scale point of x_1 is influenced by x_2. When $F(x_1, x_2)$ is expressed as a linear combination of single variable functions, as shown in eq. (4), then the cross–term is zero and the compensation is perfect.

$$F(x_1, x_2) = \alpha F_1(x_1) + \beta F_2(x_2) \qquad (4)$$

When $F(x_1, x_2)$ is expressed as a product of single variable functions, as shown in eq. (5), we can realize the perfect compensation by dividing $F(x_1, x_2)$ by the suitable function $F_2(x_2)$.

$$F(x_1, x_2) = \gamma \cdot F_1(x_1) \cdot F_2(x_2) \qquad (5)$$

Conventional analog compensation is usually represented by eq. (2); thus perfect compensation is not realized easily. If we can use the computer for the operation of compensation, we can obtain much more freedom to realize perfect compensation. We can suppress the influence of undesired variables using table–lookup method as long as the undesired variables can be measured, even if the relationship between input and output is complicated. Therefore, the new sensor device can be used. But if the sensitivity for the undesired variables cannot be measured, the device cannot be used.

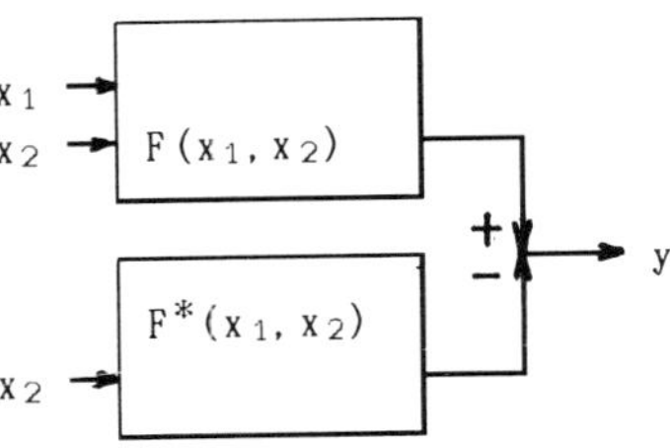

Figure 3 Fundamental structure of compensation for undesired variables

The compensation technique described above is considered to be a static approach to the selectivity improvement of the signal. This approach takes advantage of the difference in the static performance of the sensor device for the signal and the undesired variables or noise. Compensation is the most fundamental role for machine intelligence in relation to intelligent sensors.

2) Signal enhancement for useful feature extraction

Various signal processing is done for useful feature extraction. The goals of signal processing are to eliminate noise and to make the feature clear.

Signal–processing techniques mostly utilize the differences in dynamic responses to signal and noise, and are divided into two different types, frequency–domain processing and time–domain processing. We consider them a dynamic approach for selectivity improvement of the signal (Fig. 4).

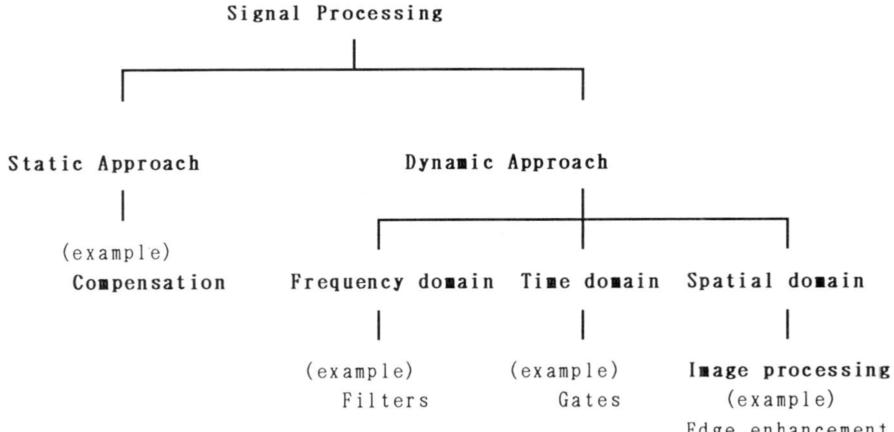

Figure 4 A classification of signal processing relating to intelligent sensors

The typical signal processing used in intelligent sensors is real–time filtering of noise. They are low–pass, high–pass and band–pass filtering in frequency–domain processing. The filtering is done mostly in analog circuits.

Another important area of signal processing in intelligent sensors is the spatial domain. The most popular spatial domain processing is image processing for visual signal. Noise suppression, averaging, and edge enhancement are functions of processing, and primary features of the target image are extracted.

More advanced image processing, for example for pattern recognition, is usually done in a higher hierarchical layer. However, information quantity of the visual signal from multipixel image sensors is clearly enormous, and primary image processing is very useful for the reduction of load for signal transmission line and processors in higher layers.

We can summarize the roles of intelligence in sensors as follows:

The most important role of sensor intelligence is to improve signal selectivity of individual sensor devices in the physical world. This includes simple operations of output from multiple sensor devices for primary feature extraction, such as primary image processing. However, this does not include optimization of device parameters or signal integration from multiple sensor devices, because this requires knowledge of the sensor devices and their object.

3.4 Roles of intelligence in middle layer

The role of intelligence in the middle layer is to organize multiple output from the lower layer and to generate intermediate output. The most important role is to extract the essential feature of the object. In the processing system in the middle layer, the output signals from multiple sensors are combined or integrated. The extracted features are then utilized by upper layer intelligence to recognize the situation. This processing is conducted in the logical world.

Sensor signal fusion and integration [8]

We can use sensor signal integration and sensor signal fusion as the basic architecture to design an adaptive intelligent sensing system. Signals from the sensor for different measurands are combined in the middle layer and the results give us new useful information. Ambiguity or imperfection in the signal of a measurand can be compensated for by another measurand. This processing creates a new phase of information.

Optimization

Another important function of the middle layer is parameter tuning of the sensors to optimize the total system performance. Optimization is done based on the extracted feature and knowledge of target signal. The knowledge comes from the higher layer as a form of optimization algorithm.

Example of architecture of intelligent sensing system:"Intelligent microphone"

Takahashi and Yamasaki [9] have developed an intelligent adaptive sound-sensing system (Fig. 5), called the "Intelligent Microphone", this is an example of sensor fusion of auditory and visual signals. The system receives the necessary sound from a signal source in various noisy environments with improved S/N ratio. The location of the target sound source is not given, but some feature of the target is a cue for discriminating the signal from the noise. The cue signal is based on the visual signal relating to the movement of the sound-generating target. The system consists of an audio subsystem and a visual subsystem. The audio subsystem consists of one set of multiple microphones, a multiple input linear filter and a self-learning system for adaptive filtering. The adaptive sensing system has a three-layered structure, as shown in Figure 6. The lower layer includes multiple microphones and A/D converters. The middle layer has a multiple input linear filter, which combines multiple sensor signals and generates an integrated output. A computer in the upper layer tunes and controls the performance of the filter. The middle layer filter functions as an advanced signal processor for the middle layer, as described previously. The computer in the upper layer has a self-learning function and optimizes filter performance on the basis of knowledge. The knowledge is given by an external source (in this case, the source is man).

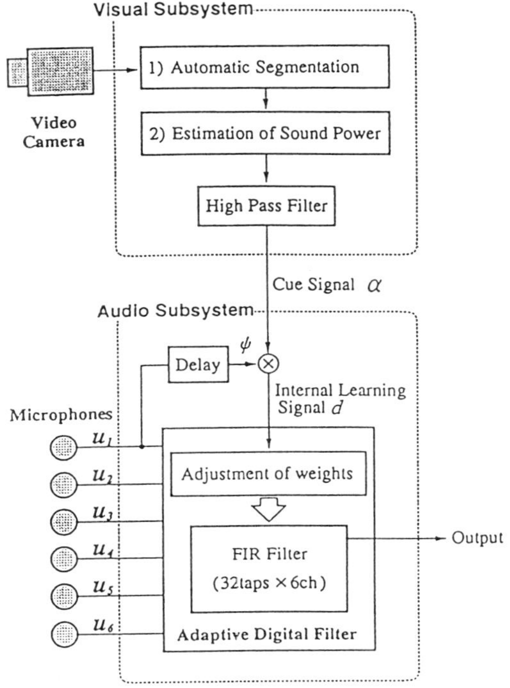

Figure 5 Block diagram of the intelligent sound-sensing system consisting of an audio subsystem and a visual subsystem

The visual subsystem consists of a visual sensor and image-sensing system. It also has three layers in its hierarchical structure. It extracts the object's movement and generates the cue signal. The output from the optimized filter is the final output of the system. Max S/N improvement of 18 dB has been obtained experimentally.

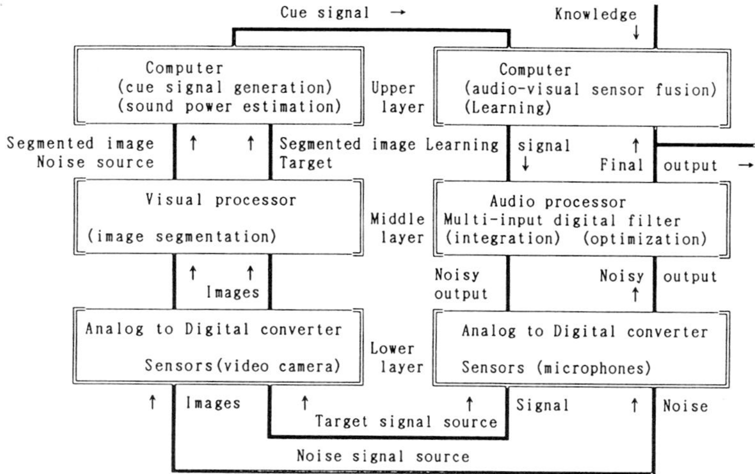

Figure 6 Intelligent adaptive sound sensing system.
Both subsystems have a 3–layer hierarchical structure

Possible applications of the system are the detection of human voice or abnormal sounds from failed machines in a noisy environment.

4. APPROACHES TO SENSOR INTELLIGENCE REALIZATION

In the author's opinion, there are three different approaches to realize sensor intelligence ([10] and Fig. 7).

1) Integration with computers
   (Intelligent integration)
2) Use of specific functional materials
   (Intelligent materials)
3) Use of functional geometrical structure
   (Intelligent structure)

Integration with Signal Perocessors (Intelligent Integration)
Use of Specific Materials (Intelligent Materials)
Use of Functional Geometrical Structure (Intelligent Structure)

Figure 7 Three different approaches to realize sensor intelligence

The first approach is the most popular. A typical example of intelligent integration is integration of the sensing function with the information processing function. It is usually imaged as the integration of sensor devices and microprocessors. The algorithm is programmable and can be changed even after the design has been made.

Information processing is optimization, simple feature extraction, etc. Some operation is done in real time and in parallel with analog circuitry or network. We call such integration a computational sensor [11].

In the second and third approaches, signal processing is analog signal discrimination. Only the useful signal is selected and noise or undesirable effects are suppressed by the specific materials or specific mechanical structure. Thus, signal selectivity is enhanced.

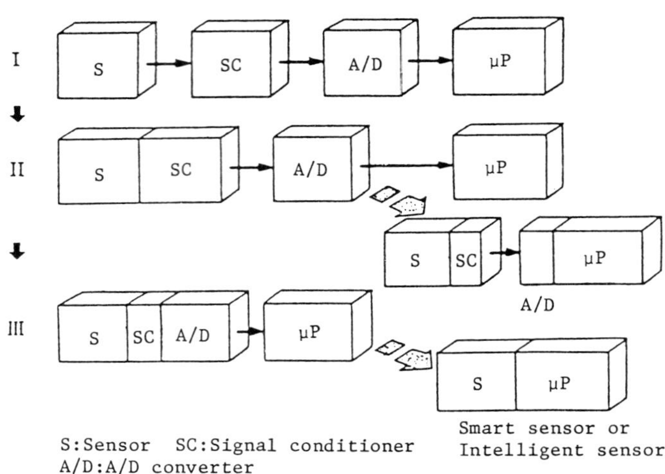

Figure 8 Trends in integration with microprocessors and sensors

In the second approach, the unique combination of object material and sensor material contributes to realization of an almost ideal signal selectivity. Typical examples of sensor materials are enzymes fixed on the tip of biosensors.

In the third approach, the signal processing function is realized through the geometrical or mechanical structure of the sensor devices. Propagation of optical and acoustic waves can be controlled by the specific shape of the boundary between the different media. Diffraction and reflection of the waves are controlled by the surface shape of the reflector. A lens or a concave mirror is a simple example. Only light emitted from a certain point in the object space can be concentrated at a certain point in the image space. As a result the effect of stray light can be rejected on the image plane.

The hardware for these analog processes is relatively simple and reliable and processing time is very short due to the intrinsically complete parallel processing. However, the algorithm of analog processing is usually not programmable and is difficult to modify once it is fabricated.

Typical examples of these three technical approaches are described in the following sections. They range from single-chip sensing devices integrated with microprocessors to big sensor arrays utilizing synthetic aperture techniques, and from two-dimensional functional materials to a complex sensor network system.

4.1 Approach using integration with computer

The most popular image of an intelligent sensor is an integrated monolithic device combining sensor with microcomputer within one chip. The development process toward such intelligent sensors is illustrated in Figure 8 [12]. Four separate functional blocks (sensor, signal conditioner, A/D converter and microprocessor) are gradually coupled on a single chip, then turned into a direct coupling of sensor and microprocessor. However, such a final device is not yet in practical use.

We are in the second or third stage of Figure 8. Many different examples are introduced by the other authors in this book. Below some of the examples in the stages will be discussed; for example, two samples of pressure sensors that are in practical use for process instrumentation. Figure 9 shows a single-chip device including a silicon diaphragm differential pressure sensor, an absolute pressure sensor, and a temperature sensor. The chip is fabricated according to a micromachining technique. The output signals from these three sensors are applied to a microprocessor via an A/D converter on a separate chip. The processor calculates the output and at the same time compensates for the effects of absolute pressure and temperature numerically. The compensation data of each sensor chip is measured in the manufacturing process and loaded in the ROM of the processor, respectively. Thus, the compen–sation is precise and an accuracy of 0.1% is obtained as a differential pressure transmitter [13].

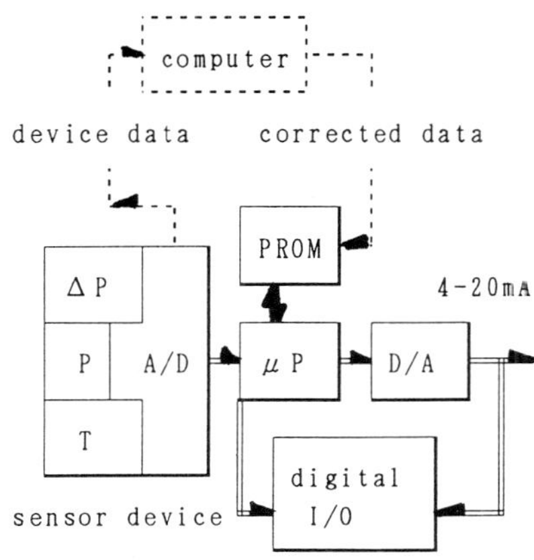

Figure 9 A smart differential pressure transmitter with digital signal compensation

Figure 10 shows another pressure sensor, which has a silicon diaphragm and a strain–sensitive resonant structure also made by micromachining technology. Two "H" shape vibrators are fabricated on a silicon pressure–sensing diaphragm as shown in the figure, and they vibrate at their natural frequencies, incorporating oscillating circuits. When differential pressure is applied across the diaphragm, the oscillation frequency of one unit is increased and the other one is decreased due to mechanical deformation of the diaphragm. The difference frequency between two oscillators is proportional to the differential pressure. The mechanical vibrators are sealed inside a local vacuum shell to realize a high resonant Q factor. The effects of change in temperature and static pressure are automatically canceled by the differential construction. Sensors with a frequency output are advantageous in interfacing with microprocessors.

These differential pressure transmitters have a pulse communication ability that is superposed on the analog signal line by a digital communication interface. Remote adjustment of span and zero, remote diagnosis and other maintenance functions can be performed by digital communication means.

The range of analog output signal is the IEC standard of 4–20 mA DC. Therefore, total circuits, including the microprocessor, should work within 4 mA. The problem can be overcome by a CMOS circuit approach [14].

4.2 Approach using specific functional materials

Enzymes and microbes have a high selectivity for a specified substance. They can even recognize a specific molecule. Therefore, we can minimize the time for signal processing by rejecting the effects of coexisting chemical components.

One example of an enzyme sensor is the glucose sensor. Glucose–oxidase oxidizes glucose exclusively and produces gluconic acid and hydrogen peroxide (H_2O_2). An electrode detecting H_2O_2 generates an electric current that is proportional to the glucose concentration. The enzyme is immobilized on the sensor tip. In this way, we can develop a variety of biosensors

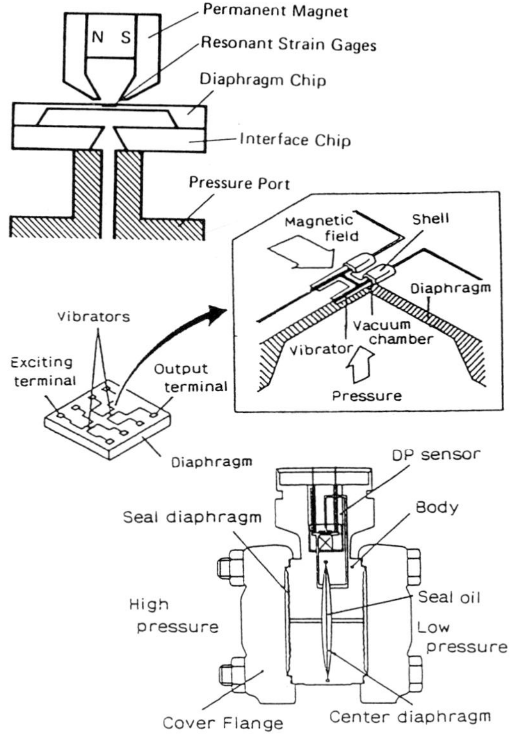

Figure 10 A differential pressure sensor having a strain–sensitive resonant structure fabricated by micromachining

utilizing various enzymes and microbes. If we immobilize the antigen or antibody on the sensor tip, we can achieve a high sensitivity and selectivity of the immuno assay. This type of sensor is called a biosensor because it uses the biological function of enzymes. Since biosensors have almost perfect selectivity, to detect many kinds of species the same amounts of different biosensors are necessary.

Another approach to the chemical intelligent sensor for various species has been proposed. It uses multiple sensors with different characteristics and imperfect selectivity. Kaneyasu describes an example of the approach in the chapter on the olfactory system. Six thick film gas sensors are made of different sensing material, which have different sensitivities for the various object gases. They are mounted on a common substrate and the sensitivity patterns of the six sensors for the various gases are recognized by microcomputer. Several examples of sensitivity (conductivity) patterns for organic and inorganic gases have been found. Typical patterns are memorized and identified by dedicated microprocessors using similarity analysis for pattern recognition. The microprocessor identifies the species of the object gas and then calculates the concentration by the magnitude of sensor output [15].

A more general expression of this approach has been proposed and is shown in Figure 11. Multiple sensors S_1,S_2,. . . S_i have been designated for different gas species X_1, X_2,. . . . X_N. Matrix Q_{ij} describes multiple sensor sensitivities for gas species. It expresses selectivity and cross–sensitivity of the sensors. If all sensors have a unique selectivity for a certain species, all matrix components except diagonal ones are zero. However, gas sensors have imperfect selectivity and cross–sensitivity for multiple gas species; therefore, a microprocessor identifies an object species using the pattern recognition algorithm [16]. In this approach, imperfect selectivity of sensor materials is enhanced by electronic signal processing of microprocessor.

4.3 Approach using functional mechanical structure

If the signal processing function is implemented in the geometrical or mechanical structure of the sensors themselves, processing of the signal is simplified and a rapid response can be expected.

An optical information processing relating to image processing is discussed as an example.

Figure 12 shows a configuration for a coherent image processing system. Three convex lenses (focal length:f) are installed in coaxial arrangement. This system performs processing of Fourier transform and correlation calculation for spatial pattern recognition or image feature extraction.

Coherent light from the point source is collimated by lens L_1. The input image to be processed is inserted as a space varying amplitude transmittance $g(x_1, y_1)$ in plane P_1. Lens L_2 transforms g, producing an amplitude distribution $k_1G(x_2/\lambda f, y_2/\lambda f)$ across P_2, where G is the Fourier transform of g, and k_1 is a complex constant and λ is wave length of the light. A filter h is inserted in this plane P_2 to modulate the amplitude and phase of the spectrum G. If H represents the Fourier transform of h, then the amplitude transmittance of spatial frequency filter τ should be

$$\tau(x_2, y_2) = k_2H(x_2/\lambda f, y_2/\lambda f) \qquad (6)$$

The amplitude distribution behind the filter is proportional to GH. Finally, the lens L_3 transforms this amplitude distribution to yield an intensity distribution $I(x_3, y_3)$ on P_3.

$$I(x_3, y_3) = K|\iint_D g(\varepsilon, \eta)\bullet h(x_3-\varepsilon, y_3-\eta)d\varepsilon d\eta|^2 \qquad (7)$$

Reversal of the co-ordinate system on the plane P_3 is introduced to remove the sign reversal due to two Fourier transformations [17].

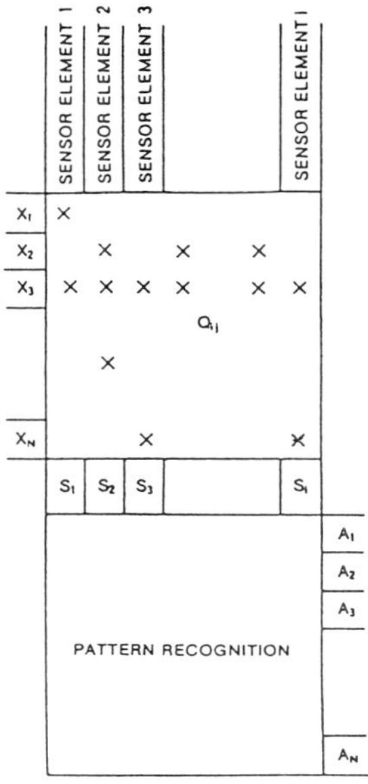

Figure 11 Gas analysis using multiple gas sensors and pattern recognition

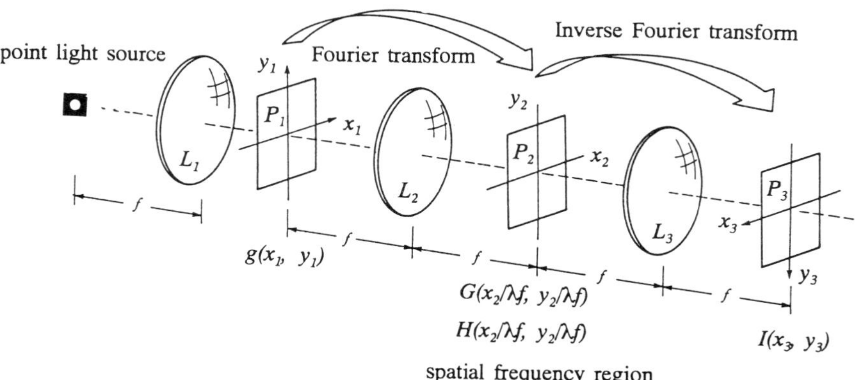

Figure 12 An optical configuration of intelligent mechanical structure

If we use filter h with suitable characteristics, we can extract useful feature from input image g by the filtering in spatial frequency region. As seen eq. (7) the output image on P_3 is proportional to two dimensional correlation of g and h. Computing of correlation between two figures can be carried out instantaneously utilizing Fourier transforming properties of lenses. This intelligent structure can be applicable for pattern recognition.

Another example is man's ability to tell three-dimensional directions of sound sources with two ears. We can also identify the direction of sources even in the median plane.
The identification seems to be made based on the directional dependency of pinna tragus responses. Obtained impulse responses are shown in Figure 13. The signals are picked up by a small electret microphone inserted in the external ear canal. Sparks of electric discharge are used as the impulse sound source. Differences in responses can be easily observed.
Usually at least three sensors are necessary for the identification of three-dimensional localization. Therefore, pinnae are supposed to act as a kind of signal processing hardware with inherent special shapes. We are studying this mechanism using synthesized sounds which are made by convolutions of impulse responses and natural sound and noise [18].
Not only human ear systems, but also sensory systems of man and animals are good examples of intelligent sensing systems with functional structure.
The most important feature of such intelligent sensing systems is integration of multiple functions: sensing and signal processing, sensing and actuating, signal processing and signal transmission. Our fingers are typical examples of the integration of sensors and actuators. Signal processing for noise rejection, such as lateral inhibition, is carried out in the signal transmission process in the neural network.

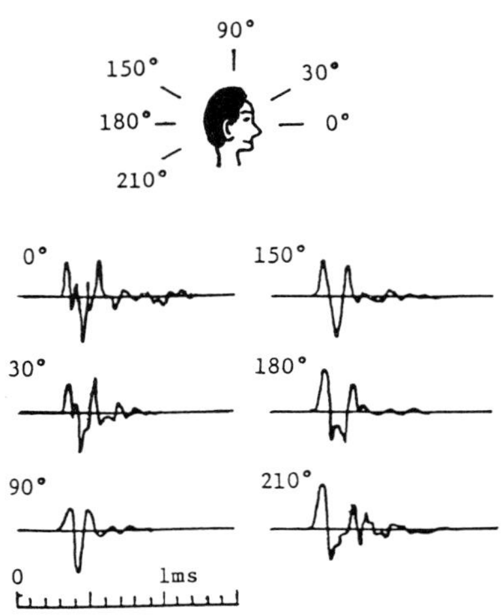

Figure 13 Pinna impulse responses for sound source in the median plane

4.4 Future image of intelligent sensors [7]

Intelligent sensing system on a chip
The rapid progress in LSI circuit technologies has shown a highly integrated advanced sensing system with learning ability. Neural networks and optical parallel processing approaches will be the most reasonable to overcome adaptability and wiring limitations.
An optical learning neurochip integrated on a single chip device has been proposed. The device is developed for pattern recognition and has a multilayer neural network structure with an optical connection between the layers. A number of photoemitter devices are arrayed on the top layer, while the signal transmission layer is built in the second layer. A template for an object pattern or special pattern memories is in the third, computing devices are in the fourth, and power supplies are in the bottom layer.

Image processing, such as feature extraction and edge enhancement, can be performed in the three-dimensional multifunctional structure. The pattern recognition function is formed by the neural network configuration and parallel signal processing. As previously described in this paper, the important feature of sensing systems of man and animals is adaptability. The feature is formed by such integration of multiple functions and distributed signal processing.

It is important to note that the approach to the future image is not single but three fold, each of which should be equally considered. In this book, these approaches are discussed with typical examples for various purposes. It is also important to note that a hierarchical structure is essential for advanced information processing and different roles should be reasonably allocated to each layer.

5. FUTURE TASKS OF INTELLIGENT SENSORS

This section will discuss what is expected of the intelligence of sensors in future. The author considers the following three tasks to be the most important examples to meet social needs.

5.1 Fault detection and forecast using machine intelligence

High performance or a highly efficient system is very useful if it works correctly, while it is very dangerous, if it does not. It is important to detect or forecast trouble before it becomes serious.

As I previously pointed out in the beginning of this chapter, a well-defined model of the abnormal state has not yet been realized. The present sensor technology is weak with regard to abnormal detection. Sensor fusion is a possible countermeasure to overcome the difficulty. At the moment the problem has been left unsolved for improved machine intelligence combining sensed information and knowledge.

5.2 Remote sensing of the target composition analysis

Chemical composition analysis is mostly carried out on the basis of sampled substance. In some cases, sampling of object material itself is difficult. For example, ozone concentration in the stratosphere is space- and time-dependent, but it is also a very important item to know for our environmental safety.

Remote sensing is indispensable for monitoring the concentration and distributions. Spectrometry combined with radar or lider technologies may be a possible approach for the remote sensing of ozone concentration.

Another important area of remote composition analysis is noninvasive biomedical analysis. In the clinical or diagnostic analysis, noninvasive composition analysis of human blood is very useful, if the analysis is reliable.

Composition analysis without sampling is easily influenced by various noise or media in between the sensing system and the object ingredient. Machine intelligence of the sensing system is expected to solve the problems.

5.3 Sensor intelligence for efficient recycle of resources

Modern production systems have realized an efficient and automatized production from source material to products. However, when the product is not in use or abandoned, the recycling process is not efficient nor automatized. The costs for recycling reusable resources are expensive and paid for by society. If the recycling of reusable resources is done efficiently and automatically, it can prevent contamination of our environment and a shortage of resources. Life cycle resource management will then be realized, as shown in Figure 14.

A sensing system for target material or components is indispensable for an automatic and efficient recycling process.

For this sensing system machine intelligence is necessary to identify the target ingredient or a certain component.

This is very important task of the intelligent sensing system.

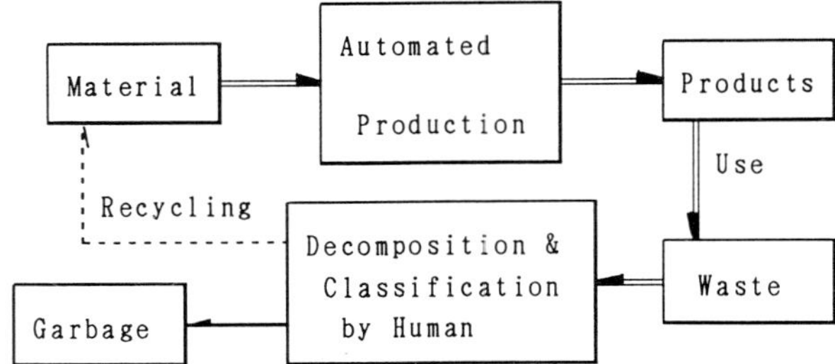

(a) Present: Only production is automated.

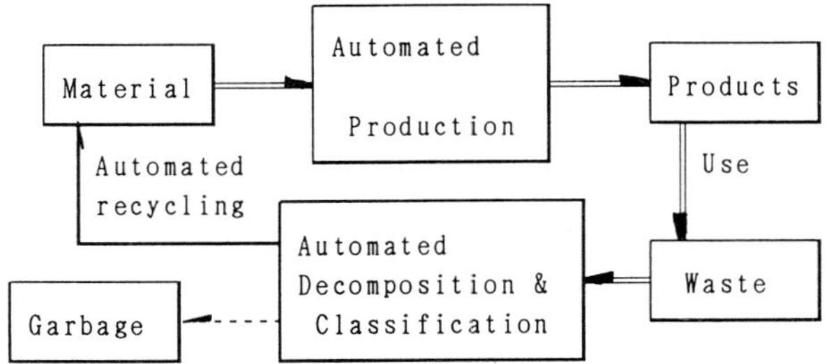

(b) Future: Life cycle resource management using automated decomposition and classification.

Figure 14 Advanced resource recycling process using intelligent sensing systems

REFERENCES

1 W.H. Ko & C.D. Fung, VLSI and Intelligent Transducers, Sensors & Actuators Vol.2(1982) pp.239.

2 S. Middelhoek & A.C. Hoogerwerf, Smart Sensors: When and Where?. Sensors & Actuators Vol.8(1985) pp.39.

3 J.M. Giachino, Smart Sensors, Sensors & Actuators Vol.10 (1986) pp.239.

4 K.D. Wise, Integrated Sensors; Interfacing Electronics to a Non electronic World, Sensors & Actuators Vol.2 (1982) pp.229.

5 J.E. Brignell, Sensors in Distributed Instrumentaion Systems, Sensors & Actuators, Vol.10 (1986) pp.249.

6 H. Yamasaki, Sensors and Intelligent systems, Y. Hashimoto and W. Day ed. Mathematical and Control Applications in Agriculture and Horticulture (Proc. IFAC/ISHS Workshop Matsuyama 1991) pp.349, Pergamon Press.

7 H. Yamasaki, Future Tasks for Measurement Technologies, Journal of The Society of Instrument and Control Engineers (SICE) Vol.31 (1992) pp.925 (in Japanese).

8 H. Yamasaki and K. Takahashi, An Intelligent Sound Sensing System Using Sensor Fusion, Digest of Technical Papers Transducers'93 pp 2 1993.

9 K. Takahashi and H. Yamasaki: Self-adapting Microphone System, S. Middelhoek ed. Proc. of Transducers'89 Vol.2 pp.610 Elsevier Sequoia S.A. 1990.

10 H. Yamasaki: Approaches to Intelligent Sensors, Proc. of the 4th Sensor Symposium, pp.69,(1984).

11 J. Van der Spiegel, Computational Sensors of the 21st Century, Extended Abstracts of International Symposium on Sensors in the 21st Century (1992 Tokyo) pp.51.

12 Mackintosh International, Sensors, Vol.24 (1981).

13 Digital smart transmitter DSTJ 3000, Journal of SICE 22(12) pp.1054 (1983)(in Japanese).

14 K. Ikeda et al., Silicon Pressure Sensor with Resonant Strain Gauges Built into Diaphragm, Technical Digest of the 7th Sensor Symposium, Tokyo,(1988), pp.55.

15 A. Ikegami and M. Kaneyasu: Olfactory Detection Using Integrated Sensor, Digest of Technical papers for Transducers'85 pp.136 (1985).

16 R. Muller and E. Lange, Multidimensional Sensor for Gas Analysis, Sensor and Actuators Vol.9 (1986), pp.39.

17 J. Goodman, Introduction To Fourier Optics, McGraw-Hill Physical and Quantum Electronics Series (1968).

18 Y. Hiranaka & H. Yamasaki:Envelope Representations of Pinna Impulse Responses Relating to 3-dimensional Localization of Sound Sources, J. Acoustical Society of America, Vol.73(1983), pp.291.

Intelligent Sensors
H. Yamasaki (Editor)

Computational Sensors
The basis for truly intelligent machines

Jan Van der Spiegel

The Moore School of Electrical Engineering, Center for Sensor Technologies
University of Pennsylvania, Philadelphia, PA 19104-6390, U.S.A.

1. INTRODUCTION

The second half of the 20th century witnessed the birth of the information age. It is fair to say that one of the principle driving forces has been the silicon integrated circuit. Basic and applied research, together with the most sophisticated manufacturing methods of modern history, have led to impressive reductions in the minimum feature size of semiconductor devices and have brought us chips with a staggering number of transistors (Figure 1). As can be seen from the trend line in Figure 1, chips with over one billion elements will be available before the end of this century[1]. One of the beneficiaries of this technological development has been the digital computer, in particular microprocessors, whose computational power has increased by several orders of magnitude over the last couple of decades.

Digital computers have reached a level of sophistication at such low prices that they are currently being used at an ever-increasing rate in areas such as modeling of complex systems,

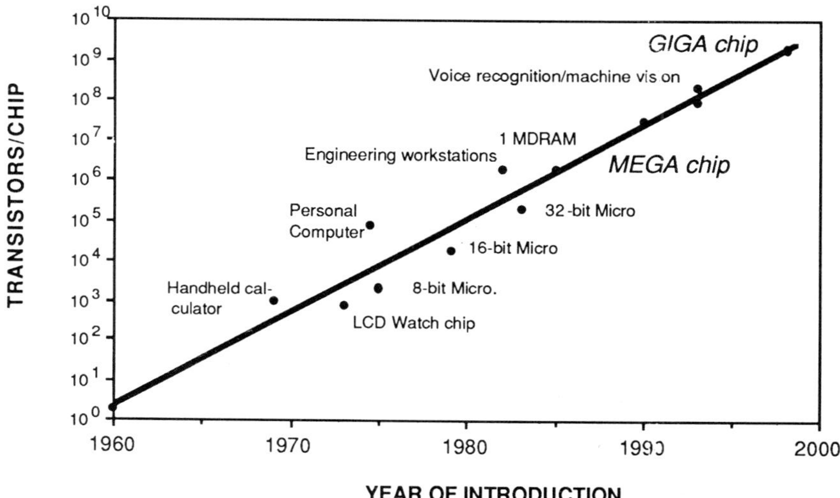

Figure 1: Evolution of chip complexity and application areas since the invention of the integrated circuit (Source: Siemens AG, IBM Corp., ISSCC).

manufacturing, control systems, office automation, automotive systems, communications, consumer products, and health care systems, just to name a few. Continued technological improvements will give rise to more powerful computers that will not only make computationally intensive tasks easier and faster, but also carry out tasks which are traditionally performed by humans. These tasks will become increasingly more sophisticated and extend well beyond the simple repetitive jobs that robots currently do, for instance, on assembly lines. To be successful, these machines will need to be equipped with advanced vision and other sensors to aid in the recognition process. In addition, there is one other important consequence of the technological change: the increasing capability to manipulate, store, display and communicate large amounts of information. This is causing a shift in the computing paradigm from *"computational speed"* to *"information-oriented"* computing[2]. These applications will require improved human-machine interfaces that will incorporate sensors for speech, vision, touch, etc. in order to become really useful. The input and output of these machines will not be just a keyboard, mouse, display, or printer, but will be a range of elaborate sensors and actuators that act as the computer's "*eyes*" and "*ears*". Also, the interface may become intelligent enough so that it is capable of responding to our thought patterns[3] As an example, when one talks to such a machine in Japanese to retrieve some documents, it should recognize that one wants the information in Japanese. A lot of the tasks that these machines will perform can be grouped under "*pattern recognition*" and include optical pattern recognition (OCR), image processing, speech and sonar recognition, and electronic image data management. In order to make this happen, these input systems will require a wide variety of sensors coupled to neural networks and massively parallel computers for sensory data processing. As a result, a new breed of "*user-friendly*" and "*intelligent*" machines is expected to emerge in the next couple of decades.

The goal of this chapter is to discuss what the above mentioned systems will look like and to forecast what type of technologies will realize these systems, with special emphasis on the sensory acquisition and processing stages. In the first section, a discussion of computational sensors will be presented describing what they are and what their role will be. This will be followed by a broad overview of the current state-of-the-art computational sensors with a few current examples. In the next section, the role of neural processing will be briefly discussed since neural networks, with their cognitive capabilities, are expected to play an important role in information processing systems. Finally, we offer visions of exciting uses for these systems as the third millennium begins.

2. THE BIOLOGICAL SENSORY SYSTEM AS A PARADIGM FOR BUILDING INTELLIGENT ACQUISITION AND INFORMATION PROCESSING SYSTEMS

For the majority of sensing and control problems, a small number of sensors and actuators with moderate computational complexity are quite adequate. A home-heating system or an environmental conditioning system in a car are typical examples. These systems are well-described in the literature but are not the topic of this chapter. The purpose of this chapter is to look into the future and anticipate what types of new information processing systems are likely to emerge.

Sensor and computer technologies have begun to merge in a much more profound way than has been the case. Future systems will typically consist of several stages of processing, with front ends consisting of sensors tightly coupled to conditioning and feature extraction circuitry, while the subsequent stages will employ massively parallel digital processors. The input to these systems will consist of real-world data, such as visual and tactile images, sounds or odors. The complex input patterns will be processed in real time. Such machines do not exist and are currently still beyond the capabilities of available technologies. This may be surprising considering the fact that very powerful computers, made of picosecond switching devices, and sensor and actuator systems with on-chip electronics exist today[4-6]. The problem is not the speed of the individual elements but rather the very architecture of current processing systems that are serial and synchronous in nature. This architecture is extremely useful in dealing with computational problems that can be algorithmically solved, such as numerical calculations. Understandably, it is tempting to use these "powerful" machines to solve other types of problems. However, one should be aware of the inherent limitations of this approach, which has been referred to as the "von Neumann" bottleneck (Figure 2).

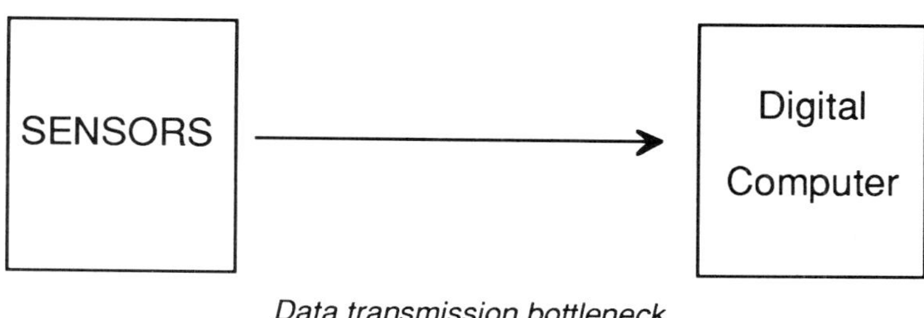

Figure 2: A conventional sensor and data processing system suffering from bandwidth limitations and serial nature of the processor: von Neumann bottleneck.

The von Neumann bottleneck refers to the serial operation of the processor with its separation of processing and memory units. A similar communication bottleneck exists between the sensor and the processor. Systems using this approach are generally incapable of dealing in real time with applications that involve huge amounts of data, as is the case for instance in vision. One can prove[7] that 10^{13} parallel processors are required to inspect the entire visual field of the retina and recognize it within 100 ms. Not even the biological system has this number of neural processors available. This mind-boggling number tells us that brute computational force on the raw image will be highly inadequate and that specialized computations on segments of the image are needed. It is this type of real-world application that is becoming important and will pose the greatest challenge for the development of sensory processing systems in the years ahead.

It is really fascinating to see how elegantly biology has succeeded in solving complex lower and higher level sensory processing tasks such as speech and vision. It decomposes the acoustic patterns or visual images into primitive features and processes these features to do real-time pattern recognition. It is even more impressive when one realizes that the brain has a weight of about 1-1.5 kg, consumes about 8 watts of power, and consists of slow processing elements (ms response time) when compared to those used in digital computers. This raises the important question: what makes the biological system so powerful for sensory information processing?

Without going into detail on the anatomy of the nervous system, one can recognize several major operational differences between biological information processors and conventional computer and sensor systems. The nervous system is highly parallel, consisting of many nonlinear processing elements (neurons) massively interconnected by synapses[8-9]. The values of the spatially distributed synaptic weights represent the stored information and constitute the memory of the system. The network processes information in a fully parallel fashion and evolves to a stable state within a few neural time constants[10]. The synaptic time constant allows it to represent time as an independent variable that enables the network to do spatiotemporal processing. This corresponds to solving a large set of coupled nonlinear differential equations -- a task that is very computationally intensive for conventional digital serial computers.

The unique architecture of the nervous system is only one of two aspects that gives biology its ability to do sensory information processing so effectively. The nature of the *sensors and the sensory data preprocessing* and representation at the cerebral cortex are equally important. Let us focus again on the visual system because it best illustrates the complexity of the task at hand. From progress made in understanding the biological vision system, one knows that many early vision tasks occur in parallel at the sensor site without input from the brain. In other words, the retina is more than just an array of photo detectors. The sheet of neural tissue located at the back of the eye carries out some form of image analysis so that certain features of the impinging visual input are enhanced while others are down-played[11]. This is accomplished by five layers of nerve cells in which information flows both vertically and horizontally. The intricate coupling of photocells with the processing layers is believed to be essential for the formation of visual images and the reason the biological system is so efficient, notwithstanding the slow response of the individual nerve cells.

The power of the biological nervous and sensory system is so striking that it cannot be ignored. As has been eloquently expressed by Mead and Mahowalt, the biological system provides us with an "alternative engineering paradigm that has the potential of dealing with real-world problems in real time"[12] A sensory information processing system modeled after this scheme is shown in Figure 3. It consists of sensors that serve not only as input devices but also as distributed local processing elements linked to a neural network that serves as the "brain" behind the sensors and which is further connected to a digital processor[13]. All three of these system blocks will be essential in order to build machines that can deal efficiently with real-world problems.

The role of computational sensors is not merely to transduce the physical world and generate signals, but also to transform the data in order to eliminate redundant information and to extract simple features[14]. This is usually done in parallel and in real time before the data goes into the processor. These sensors were coined *"Computational Sensors"*[15] and were the topic of a DARPA workshop held on this topic at the University of Pennsylvania[16]. After the initial "computational sensing", the next stage of processing will be done by a neural network that integrates sensory data from different sensor modalities[17-18] and performs higher level, more global processing. The network must be trained to combine features extracted by the sensor such as edges, line orientations, end stops, etc., to recognize, for instance, letters, faces or other patterns. Once the neural

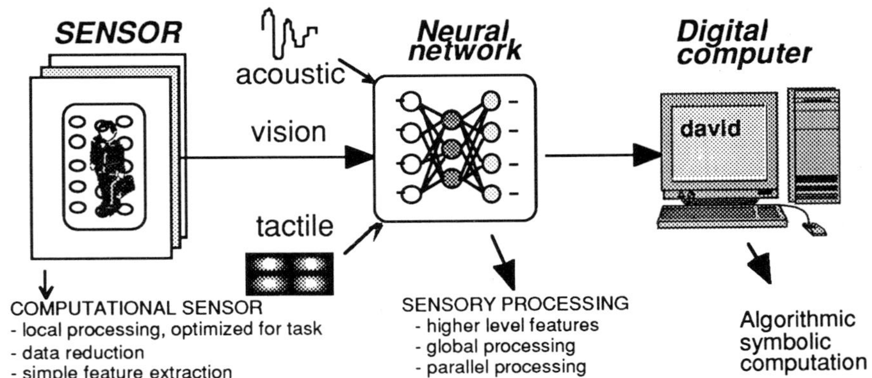

Figure 3: Schematic diagram of a sensory information processing channel consisting of a front-end computational sensor followed by neural network and a digital processor.

network has processed the data, it will be transferred for further processing. The digital system with its flexibility of programming and implementing algorithmic computations will now work at a higher level of abstraction and symbols. The overall strategy is one of "Divide and Conquer" in which one processes the information both locally and in a hierarchical fashion instead of dealing with all the data at once, a situation that can be overwhelming for even the most powerful super computers.

3. COMPUTATIONAL SENSORS - WHERE ARE WE NOW?

Although we concentrated on visual sensors in the previous section, computational sensors span the whole spectrum of sensing applications, including visual, acoustic, mechanical (tactile), chemical, magnetic sensors, etc. Vision has received the most attention so far because conventional methods have simply been inadequate in real-world situations. This is due, to a large extent, to the band width required at the sensing and processing stages. However, early vision processing functions such as edge detection, contrast and motion can be done locally using simple processing elements. These functions are ideally done at the sensing site itself by merging the sensor and the preprocessing function on the same substrate. Current VLSI technology allows us to realize such functions and indeed, several prototypes of such "computational sensors" have been developed[19-20]. Different approaches are discussed in this section[16].

3.1 Uni-plane computational sensors

In "uni-plane" computational sensors, sensors and processing elements are distributed over the sensing plane. The processor is usually relatively simple, dedicated to a specific task and connected to its neighbors. Preferably, the processing is done in the analog domain in order to take full advantage of the analog nature of the sensor signals and to allow truly parallel and simultaneous operation. In the case of optical sensors, the uni-plane is usually referred to as the focal plane.

A popular approach to uni-plane processing, pioneered by C. Mead, makes use of elementary computational primitives that are the direct consequence of the fundamental laws of device physics[21]. The proper choice of primitives is important for the efficient implementation of computational functions. This approach has led to several successful implementations of artificial retinas that respond to light over several orders of magnitude and have a spatial and temporal response similar to the biological retina. This approach has also been used to detect motion and compute optical flow[22-28].

One such example, a motion detector, is described next[24]. It consists of three stages: the photoreceptors, contrast and edge detection, and motion detection as is schematically shown in Figure 4. The photodetectors are bipolar transistors operating in the conductive mode. This has the advantage of continuous, asynchronous operation. In addition, by using an MOS transistor in the subthreshold as active load, one can achieve a logarithmic response similar to the Fechner-Weber law observed in biological systems. However, the sensor does not have adaptation capabilities.

The second stage implements an approximation of a "Difference of Gaussian Operator" (DOG) to detect edges. This is realized using a resistive grid. Operational amplifiers (not shwown) produce an output that corresponds to an approximated difference in Gaussian (DOG) function. The third stage, the motion detector, detects motion when an object disappears from one pixel and reappears at the nearest neighboring pixel. The approach is based on a combination of the Reichardt and Ullman-Marr motion schemes[29-30]. The implementation of this motion scheme is shown in Figure 5.

The velocity of an object can be extracted using a sequence of thresholding and correlation techniques at each pixel. Experimental results and the timing diagram have been reported somewhere else[24,31].

Position and orientation detectors have been reported in the literature as well. One such implementation makes use of resistive grids to determine the first and second moments of the object in order to calculate the centroid and axis of least inertia[32]. Still other approaches have been reported in the literature. Among these is a vision sensor whose first layer consists of photodetectors and analog circuit elements that implement an approximation of a first order Gabor function to detect local orientation of lines and edges. The next stage of processing consists of feedforward parallel networks that recognize more complex patterns in groups of these oriented line directions[33-34].

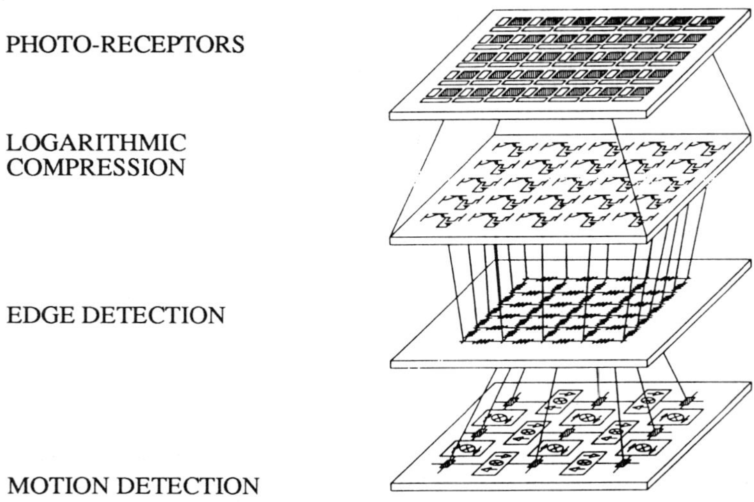

Figure 4: Image sensor with motion detection: photo detectors interconnected through a resistive grid with lateral inhibition that implements an approximation of a difference-of-Gaussian edge-detector (after ref. 31).

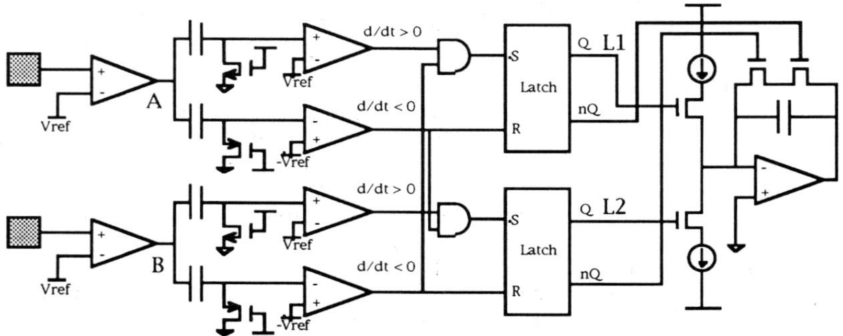

Figure 5: Schematic circuit of a motion detector, consisting of a zero-crossing locator, time derivatives, correlators, latches and motion integration pulse (after ref. 24)

Another interesting application is a range detector based on light stripe triangulation[35]. This allows the 3 -dimensional (3-D) location of imaged object points. It makes use of a light strip that is continuously swept across the scene. From the known geometry of the projected light stripe as a function of time and the line-of-sight geometry of the cells, one can find the 3-D location of the imaged object through triangulation.

The technology of choice for the implementation of the computational sensors is Complementary Metal Oxide Semiconductor (CMOS) technology. CMOS is widely available at a relatively low cost and provides devices with effectively infinite input impedance. The technology enables one to design compact, low-power circuits and to incorporate processing element at the sensor site (computational sensors or active pixel sensors)[36]. An alternative technology employs charge-coupled devices (CCD). This technology has been the choice for high-performance video cameras and is, in general, more complicated and expensive than CMOS technology. However, several CCD computational sensors have been developed, such as a CCD-based programmable image processor[37-38]. More recently, both CCD and CMOS technology have become available in the same process, which has allowed the fabrication of a CCD imager with a CMOS focal plane signal processor[39].

The same principles used in focal plane computational sensors may be employed in mechanical and magnetic sensors. Tactile sensors that have a pressure-sensitive layer based on resistance or capacitance changes have been developed[40-42]. Also, magnetic sensor arrays with on-chip signal processing capabilities have been described[43].

Although uni-plane processing has many benefits, there are some disadvantages as well. The processing element that is merged between the pixels reduces the resolution of the sensor. This is because the technology used to fabricate these devices is planar and does not allow the stacking of layers on top of each other, as is the case in the biological visual system. As a result, the pixel pitch increases from 5 μm as in conventional CCD cameras to 20-100 μm. One solution is to use 3-D integration. This technology had been suggested for the realization of a visionary 3-D retina sensor[44].

More recently, a 3-D sensor for pattern recognition was reported[45]. It consists of four layers and employs template matching with an associative ROM, as schematically shown in Figure 6. It is made with silicon-on-insulator (SOI) technology. The first layer consists of photodiodes that operate in the conductive mode. The second layer digitizes the image and makes a majority decision. The bottom two layers are used for matching and consists of queue and mask registers and an associative read-only memory. A test chip with 5040 pixels has been fabricated and is able to detect 12 characters at the same time, taking about 3 μs to identify each one. It can recognize up to 64 characters in both upper and lower case.

It is worthwhile to discuss 3-D integration technology. It provides a natural method for implementing sensory functions and allows both lateral and vertical flow of data that is often required in vision tasks. Three-dimensional stacking also allows a high packing density. As a result, 3-D integration permits the efficient implementation of architectures with a high degree of parallelism. Some of the disadvantages of 3-D integration are that it is an expensive technology, it lags behind state-of-the-art ICs by several generations, the yield is usually lower, and power dissipation may be a problem for high-speed applications. For computational sensors, 3-D integration will undoubtedly be an important technology. It is really one of the few methods, besides optical means, to provide a solution for massively parallel interconnections found in visual sensing and processing systems. The same 3-D techniques are being investigated for implementing high-density DRAMs, a large market that can ensure that the technology will mature and become more widely available at a lower cost than today.

An alternative way to obtain 3-D functionality is to use separate, although tightly coupled, processing and sensing modules. Flip-chip technology with solder bumps is a good way of ensuring tight coupling and preserving a large band width. Advanced packaging techniques will play an important role in the success of this approach[46]. Three-dimensional architectures for a parallel processing photoarray have been described that make use of the

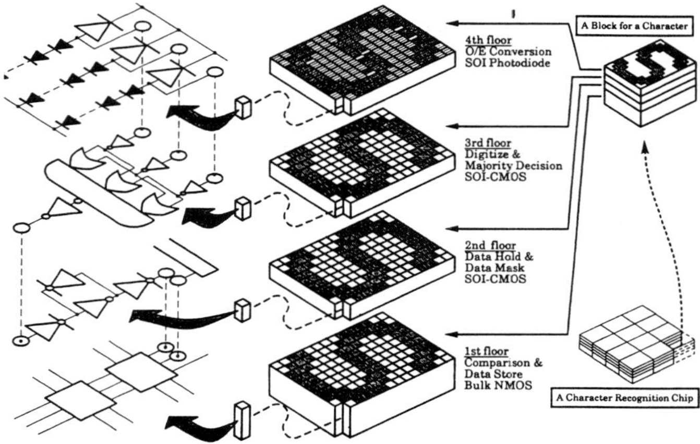

Figure 6: A schematic diagram of a 3-D character recognition chip consisting of 4 layers made with SOI technology (K. Choi, et al., J. Sol.-St. Crct., Vol. 27, 1992 ©IEEE; Ref. 45).

thermo migration of aluminum pads. Thermo migration was applied earlier for the fabrication of chemical sensors and to make backside contacts through the wafer[47]. However, when applied to sensor arrays, this method requires a relatively large pixel separation due to the random walk of the thermo migrated regions. More recently, multi-chip modules and 3-D packaging technologies have allowed stacking of chips on top of each other with interconnections either through the chips or allong the faces of an expoxy cube that encapsultes the chips [48].

In vision, the bottleneck is not so much data acquisition but rather the processing of large amounts of data. This suggests that one can, under certain circumstances, use a standard CCD camera. The data can then be read by a fast processor such as a neural network or a massively parallel computer. This approach has been successfully demonstrated. A real-time image velocity extraction system that performs light adaptation, spatial contrast enhancement, and image feature velocity extraction on images of 256 x 256 by 8-bits at a frame rate of 30 per second has been implemented on a parallel computer[49]. The processor is an 8-stage multi-instruction multi data (MIMD) machine performing over one billion 8-bit operations per second. Another example is the pyramid chip for multi resolution image analysis[50]. It is a digital VLSI chip that processes image samples sequentially, in a pipelined fashion. The pyramid processing is based on filtering and resampling of the image. It supports low-pass, bandpass and sub-band pyramids. The chip is capable of constructing Gaussian and Laplacian pyramids from a 512 x 480 pixel image at rates of 44 frames per second.

3.2 Spatio-geometric computational sensors

Another approach to computational sensors is through the functional geometric and mechanical structure of the sensors. This method was demonstrated several years ago for responses associated with the external ear[6]. Another example is the functional structure of the retina whose sampling grid has a space-variant layout. Spatio-geometrical sensors allow new architectures where computing is now assisted by the geometry of the sensor.

One of the challenges for machine vision is the simultaneous need for a wide field of view to maximize the perception spans, and for high resolution in order to improve accuracy of observation[51]. Research on the anatomy of the human visual system revealed that the biological vision system has solved this problem by distributing the photoreceptors nonuniformly over the retina. The cone density shows a peak in the center of the visual field and decreases towards the periphery[52-53]. If the biological system is to serve as an example for building artificial sensors, this would argue for a nonuniform sampling grid. The question then is, what kind of distribution to use? In the 3-D world of robot vision, perspective projection introduces range scaling to every image. As a result, magnification transformations are very important. The only coordinate system that embodies magnification invariance while preserving the isotropy of local neighborhoods is the log-polar grid[54]. In addition, the log-polar sampling structure provides a good compromise between high resolution and a wide field of view as is illustrated in Figure 7. The figure on the left corresponds to the image plane, while the one on the right was obtained by taking the logarithm of the first one. Rotation and magnification now become simple translations in the computation plane. Indeed, a point in the image plane can be described by its polar coordinates r and θ,

$$z = r\, e^{j\theta}.$$

Mapping the retinal image into the computation plane gives the new coordinates

$$w = \ln(z) = \ln(r) + j\theta = u + jv.$$

After applying a magnification and rotation in the image plane, the new coordinates become

$$z' = ar\, e^{j(\theta+\theta')}, \qquad \text{and} \qquad w' = \ln(r) + \ln(a) + j(\theta + \theta') = u' + jv',$$

which corresponds to simple translation along the u and v axes in the computational plane (Figure 8).

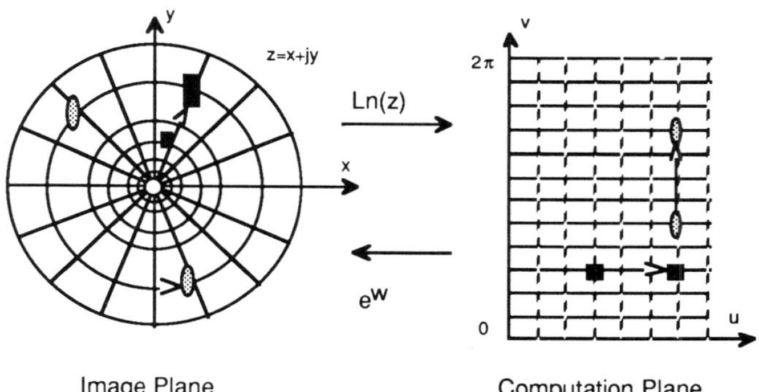

Figure 7: Schematic of a log-polar sampling structure. The left picture gives the actual geometry of the pixels on the image plane while the right picture gives the transformed image (computation plane). Rotation and scale invariances are illustrated by the ellipses and rectangles.

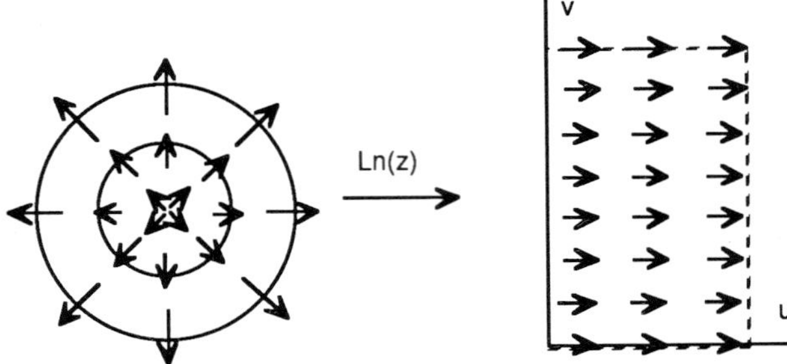

Figure 8: The use of a log-polar structure simplifies time-to-impact and optical flow calculations.

Besides the invariances to rotation and scale, one has compressed the data in the periphery. This will considerably reduce the burden on the computer, thus speeding up the vision computations. The geometry of the sensor is also natural for the detection of optic flow and time-to-impact[55], as is illustrated in Figure 9.

Several groups have studied the use of the nonuniform sampling scheme and nonrectangular pixel geometry to build computational functions into hardware structures[56-57]. The first log-polar image sensor modeled after this scheme was built a few years ago using CCD technology[58-59]. The sensor consisted of 30 concentric circles whose diameters increased exponentially with 64 pixels per circle, and a constant high resolution central area called the fovea. A photograph of the sensor is shown in Figure 9.

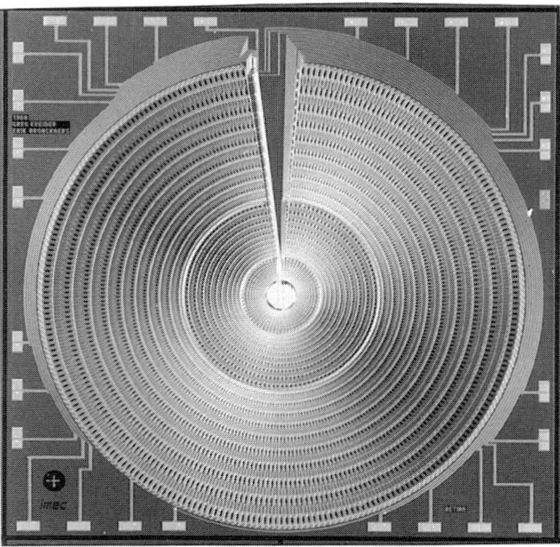

Figure 9: Photograph of a foveated log-polar CCD sensor, consisting of 30 circles and 64 pixels per circle; the chip size is 11mm by 11mm.

We found from experiments that computational, log-polar sensors outperform the conventional (television) imager in two areas. The geometrical properties simplify several calculations, such as line detection and time-to-impact prediction. The transform also leads to a substantial reduction in the quantity of sampled data which naturally speeds the analysis of such images. As a result, practical image processing can now approach TV frame rates on general-purpose engineering workstations. The sensor has been demonstrated to have advantages for edge detection, line detection, adaptation, lens control (auto-focusing), time-to-impact prediction, and simple tracking. The log-polar algorithms run on the order of 50 times faster than equivalent programs on conventional images[60].

3.3 Binary optics

As pointed out previously, one of the main limitations of electronic implementation of computational sensors is the interconnection bottleneck due to the two-dimensional nature of planar technology. Optics, because it is inherently 3-D, provides us with the means for generating massively parallel and high-speed interconnections. The drawback is that optics is bulky, fragile, and expensive. However, advances in electronic circuit-patterning techniques with nanoscale dimensions now allow the mass fabrication of low-cost, high-quality components. These optical components show promise in solving the interconnection bottleneck in optical sensors between processors and displays[61]. They are called "binary" optical components and are based on diffractive optics in contrast to the conventional refractive optics. By combining diffractive and refractive optics, one can reduce the dispersive nature of diffractive elements. This provides an additional degree of freedom for designing new and improved optical systems.

By employing proper lithographic techniques, for instance, one can fabricate close-packed arrays of lenses with a pixel pitch in the order of tens of microns. These lenses can be used to focus light on tiny photodetectors surrounded by processing circuitry, thus increasing the fill factor to about 100%. It is even feasible to integrate these components on the chips themselves, giving rise to ultracompact sensing systems as schematically shown in Figure 10. Large arrays of coherent lenslets of 20,000 elements per square cm with f/1 speed can be fabricated in Si, GaAs, CdTe, quartz, and other materials.

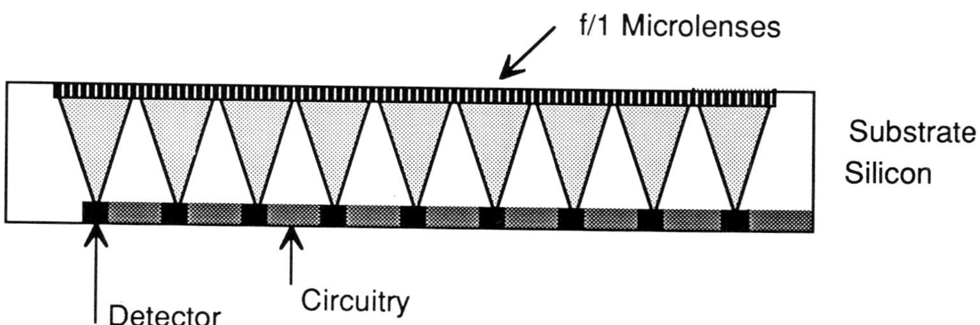

Figure 10: Schematic of a microlens array etched in silicon to focus light on the photo-detectors between the processing elements in order to increase the field factor of the detector array.

One can also create binary optical components that implement a variety of transformations of an optical wave front. Even more far-reaching is the use of integrated photo-emitters and micro-optics to pass the output of one layer to another. The ability to interconnect different image-processing layers of cells that perform low-level functions such as those found in the mammalian retina (motion detection, edge enhancement, etc.) may lead to systems that perform the elaborate functions performed in the visual cortex[62-63]. Such stacked optical signal processing systems have been coined "Amacronics". An example of such a multilayer sensory processing system is shown in Figure 11. Going from right to left, one finds telescopic mixed optics that collect the light. A beam-steering system consists of a positive and a negative lens array. The next stage is multiplexing arrays that split and share light, followed by a microlens array that projects the image on the electronic detector array that performs local processing on the pixels. The next stage, which is not shown, is a neural network or massively parallel processor.

As illustrated in the above example, binary optics applications range from conventional imaging to light manipulation between processors and machine vision tasks. It is a technology that will play a key role in computational sensors in the next century.

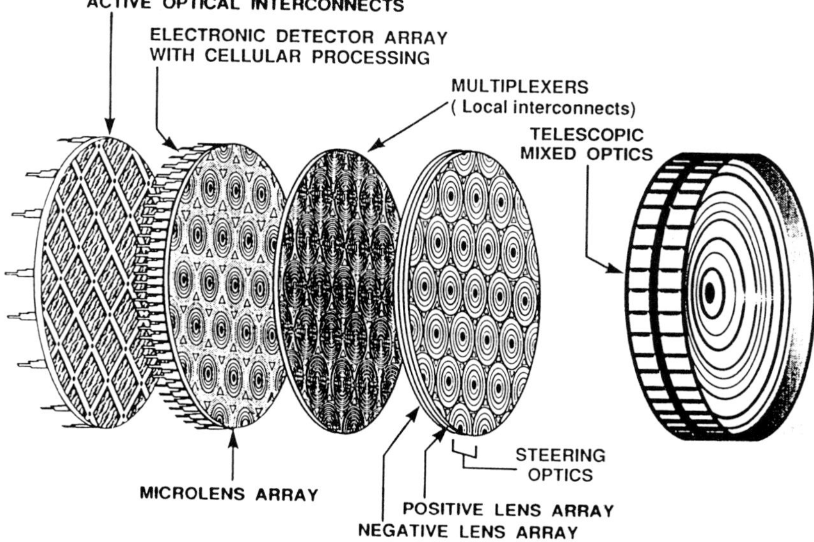

Figure 11: An opto-electronic sensory processing system built with binary optical components (after W. Veldkamp).

4. THE BRAINS BEHIND THE SENSORS: NEURAL NETWORKS AND MASSIVELY PARALLEL PROCESSORS

While computational sensors are likely to become the front-end of large sensory information processing systems, as illustrated above, they will only perform dedicated and limited but specific functions. The higher level processing will be done by a neural computer. This network will integrate data from different sensor modalities and allow flexibility in terms of architecture and processing operations. The goal of the neural network is to mimic some of the cognitive functions found in the human nervous system. Several of these neural computers are currently being developed at major research laboratories[64]. A prototype of an analog neural computer for real-time, real-world computations of sensory data was built in the author's laboratory[65]. It consists of VLSI custom-designed modules that model the neural, axodendritic, synaptic, and synaptic temporal functions of the cerebral cortex. Its architecture as well as the neurons, synapses, and time constants are modifiable. Among its unique features is temporal processing, which is essential for dynamic sensory applications such as speech, optical, and sonar pattern recognition.

A schematic block diagram of the overall system is shown in Figure 12a. Each block consists of a collection of neurons, synapses, time constants, and switches. The modules have only short connections to their neighbors except for a couple of global control lines. The insert on the left side of the picture shows the signal flow in the network. Neuron outputs exit north and south and are routed via analog switches (axodendrites) anywhere in the network. Inputs to the neuron are in the east-west direction. The switch positions, characteristic of neurons, and the synaptic weights and time constants are set by the digital host shown to the right of the picture. The configuration data is stored in a local memory on the individual modules. The output of each neuron is sampled and displayed on the digital host computer. This allows us to follow the activity of the neurons and to use it for implementing learning algorithms. However, the network runs independently of the host and operates in both parallel and analog mode. A block diagram of the software organization which resides on the digital host is shown in Fig. 12b. The digital host takes care of downloading the network configuration, of displaying the neuron activity, and learning. The network can be described using a user-friendly graphical interface at the logical or at the physical level [67].

The above network has been used for pattern recognition, winner-take-all circuits, associative memory, and processing of auditory signals[66-67]. For speech recognition application, the network decomposes the sounds to its acoustic primitives. The input to the network is an artificial cochlea that consists of high-Q bandpass filters. The first units calculate the local temporal rise and decay of sound frequencies. The next layer neurons are configured to compute changes in the frequency maxima and the direction of the change. They are, in essence, motion detectors. The network basically decomposes speech into its primitives, which are functions of energy, space, and time. These primitives are then used as inputs into a pattern recognition network in order to recognize individual phonemes.

The next stage of the information channel is a digital computer (Figure 3). The current trend in computers is parallelism. This development will be beneficial for sensory processing problems because these machines can handle large amounts of data in parallel. Depending on the complexity of the task, one will be able to use either custom-built boards with arrays of DSPs or massively parallel processors (MPP). The architecture of the MPP is based on the interconnection of thousands of microprocessor-based nodes rather than a few but more powerful and expensive processors, as is the case in current vector-based supercomputers. By the end of this decade, one expects that MPP with a processing power of teraFLOPS (10^{12} operations per second) will be available[68]. Currently, supercomputers can deliver about 20 gigaFLOPS.

By the end of this century, the computational power of computers will be staggering. Even more mind-boggling will be the computers of the 21st century. They may consist of nanoscale elements made of quantum electronics, optical, and organic devices, and even biological cells[69-71]. Free-space optical addressing of individual organic molecules that store bits of information are a distinct possibility. The result may be computers based on quantum-mechanical principles rather than conventional logic gates. These devices would lend themselves well to 3-D connections and parallel information processing. Sensory information processing is a natural application for these advanced quantum mechanical or bioelectronic devices.

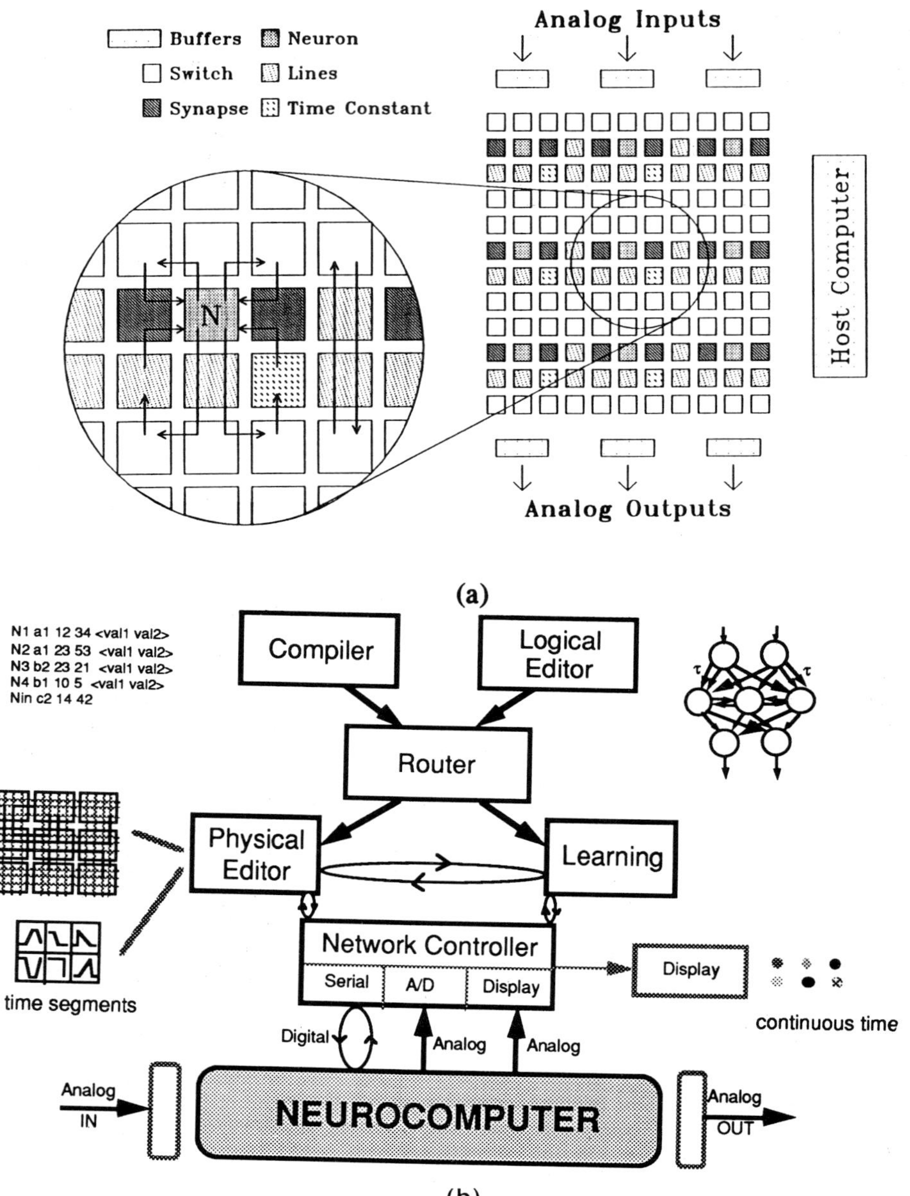

(b)

Figure 12: (a)A schematic diagram of the system architecture of an analog neural computer designed for dynamic sensory computations. The network architecture, as well as the neuron, synaptic weight and time constants are modifiable. Inputs and outputs are analog signals. A digital host is used to display the state of the neurons and to download the network configuration; (b) Schematic overview of the software organization, residing on the digital host.

5. USE OF COMPUTATIONAL SENSORS

Most of the above examples exist today in the research stage. They serve as a modest illustration of what the future may bring. By extrapolating technological progress into the next century, we can assume that truly intelligent machines will become a reality. This raises an important question: "how to best use this computational and technological power?" The guiding principle should be improvement in the quality of life. This includes all aspects, from lifestyle to health care, education, entertainment, communication and the environment[72].

Let us look at some of the application areas and the need for sensors and actuators. The health care system will be dramatically different from what it is today. There is likely to be a changing emphasis from today's large impersonal hospitals to the more human dimension of a doctor's office or the patient's home. This will demand superfast access to large central databases and the exchange of diagnostic information such as X-rays and other pictures between the doctor and medical centers in order to improve care. Central to this new approach will be smart sensors and actuators that not only help doctors perform the diagnosis, but also monitor the condition of the elderly who are living at home. Smart sensors and actuators will remind patients to take their medication when their condition warrants it or even control the release of medication. In the case of emergencies, the doctor will be informed automatically and will be able to request sensor readings remotely. Cost containment is another important driving force behind changes in health care. Intelligent, low-cost sensing systems are needed to replace expensive medical instruments with large operating costs. Such systems will play a key role in this evolution. Diagnostic tools, such as the I-STAT^{TM} sensor, are precious examples of the direction in which the health care system is moving[73]. In addition, sensors and actuators that control artificial limbs will gain in importance. Aid for the blind that allows them to navigate or recognize objects at a primitive level will become high in demand.

Another area that will benefit from computational sensors and intelligent machines is the "smart house" which not only provides remote health monitoring and care, but is more energy-efficient and has voice-activated appliances, robots with learning capabilities, and sophisticated personal security systems[74]. These "smart houses" will play an important role in providing the fast-growing population of aging people with the means to remain in their own homes. It is clear that sensors, coupled to intelligent machines, will play a key role in creating this new "living space".

Since 1950 the world population has doubled to 5.3 billion and will grow to 8.5 billion by the year 2021. Depletion of natural resources and pollution will place high demands on the earth's life-support system if one realizes that the earth's capacity is somewhere between 8 and 12 billion people[75-76]. This brings the interdependencies of lifestyle, environmental change, and technology into sharper focus because our very survival will be at stake. Environmental management is already a topic of importance and is going to become much more so in the next century. There will be an acute need for sensors to monitor the environment and develop electronic computing systems that help solve our environmental problems.

Related to the environmental issue is energy production. Clean, non-fusel, and low temperature energy sources, such as photovoltaic or wind power generation, will, by necessity, gain importance. Successful utilization of these resources will require a much more pervasive use of sensors and electronics to learn the atmospheric conditions and to allow energy-producing equipment to optimize energy output. The use of more energy-efficient appliances, transportation systems, and industrial processes is the other side of the same coin.

Finally, in the 21st century the ancient dream of a person being someplace he is not - at a time he is not - will become a "virtual reality". Virtual Worlds Technologies (VWT) or tele-existence systems will emerge rapidly. This emerging field depends on smart sensors and actuators that will provide the individual sensation of being present at the center of the virtual display. In addition, tele-existence will act as human interfaces to computers, robots, and other machines[77].

6. CONCLUSIONS

The trend towards improved human-machine interfaces and intelligent machines will increase the need for sensors that will do more than just acquire and condition the data. These sensors will perform feature extraction and preprocessing of the incoming signals. A sensory processing channel consists of "computational sensors" which may in part be modeled after the biological system, linked to a neural network which extracts higher level features and performs some cognitive functions. These sensory systems have the potential to revolutionize a whole range of applications such as robotics, entertainment, media, health care, communications, the workplace and industrial inspection, just to name a few [78].

Computational sensors will make use of a combination of micro electronics, micro-optics, and micro mechanical technologies to achieve its full power. The major challenges are the resources required to make it happen, especially human resources. There is a growing need for scientists and engineers from different disciplines to work together and cross their traditional boundaries.

7. ACKNOWLEDGMENTS

The author would like to thank Professors Jay Zemel, Kenneth Laker and Ralph Etienne-Cummings for reviewing the manuscript and providing insightful input. Also, Drs. P. Mueller, W. O. Camp, D. Blackman, G. Kreider, C. Donham, Messrs. S. Fernando, Z. Kalayjian, J. Kinner and Ms. M. Xiao for their valuable contributions. Finally, support from the ONR, NSF, NASA Lyndon B. Johnson Space Center, the Ben Franklin Partnership of Southeastern Pennsylvania and Corticon, Inc. is greatly appreciated.

8. REFERENCES

1 J. D. Meindl, "Prospects for Gigascale Integration (GSI) Beyond 2003," Digest Tech. Papers, Intl. Solid-State Circuits Conf., Vol. 36, pp. 94-95, 1993.

2 P. Chatterjee, "ULSI-Market Opportunities and Manufacturing Challenges," Tech. Digest of the IEEE Intl. Electron Device Meeting, pp. 1.3.1-1.3.7, 1991.

3 M. Morris, "Human-Machine Interface," in 2021 AD: Visions of the Future, The ComForum Summary and Research Report, Natl. Engineering Consortium, J. Sheth and R. M. Janowiak, eds., Chicago, IL, 1991.

4 G. Zorpette, "Teraflops Galore-Special Report on Supercomputers," IEEE Spectrum, pp. 26-27, Sept. 1992.

5 K. D. Wise, "Integrated Microinstrumentation Systems: Smart Peripherals for Distributed Sensing and Control," Digest Tech. Papers, Intl. Solid-State Circuits Conf., Vol. 36, pp.126-127, 1993.

6 H. Yamasaki, "Multi-Dimensional Intelligent Sensing Systems Using Sensor Arrays," Dig. Intl. Conf. Solid-State Sensors and Actuators, pp. 316-321, 1991.

7 J. Tsotos, D. Fleet and A. Jepson, "Towards a Theory of Motion Understanding in Man and Machine," in Motion Understanding, Eds. W. Martin and J. Aggarwal, Kluwer Academic Publishers, Boston, 1988.

8 S. Deutch and A. Deutch, *Understanding the Nervous System - an Engineering Perspective,* IEEE Press, New York, 1993.

9 S.W. Kuffler, J.G. Nicholls and A.R. Martin, *From Neurons to Brain*, 2nd Edition, Sinauer Assoc. Inc. Publ., Sunderland, MA, 1984.

10 E. Sanchez-Sinencio and C. Lau, eds. *Artificial Neural Networks*, IEEE Press, New York, 1992.

11 R. H. Masland, "The Functional Architecture of the Retina," Sc. American, pp. 102-110, Dec. 1986.
12 M. A. Mahowald and C. Mead, "The Silicon Retina," Sc. American, pp. 76-82, May, 1991.
13 J. Van der Spiegel, "Computational Sensors of the 21st Century, Ext. Abstracts. Intl. Symp. on Sensors in the 21st Century, pp. 51-59, Jap. Electr. Development Assoc., Tokyo, 1992.
14 M. Ishikawa, "Parallel Processing for Sensory Information," Electronics and Communications in Japan, Vol. 75, Part 2, No. 2, pp. 28-43, 1992.
15 J. Clark, Introduction to Special Issue of the Intl. J. of Computer Vision, Intl. J. Comp. Vision, Vol. 8:3, pp. 175-176, 1992.
16 T. Kanade and R. Bajcsy, DARPA workshop on Computational Sensors, University of Pennsylvania, Philadelphia, May 11-12, 1992.
17 M. Ishikawa, "Sensor Fusion: The State of the Art," J. of Robotics and Mechatronics, Vol. 2, No. 4, pp. 19-28, 1989.
18 H. Yamasaki and K. Takahashi, "An Intelligent Sound Sensing System Using Sensor Fusion," Dig. Tech. Papers Transducers'93, pp. 217, Yokohama, June 7-10, 1993.
19 B.P. Mathur and C. Koch (eds.), *Visual Information Processing: From Neurons to Chips*, SPIE Proc., Vol. 1473, Bellingham, WA, 1991.
20 C. Koch and H. Li (eds.), *VLSI Chip: Implementing Vision Algorithms with Analog VLSI Circuits*, Selected Reprint Volume, IEEE Press, New York, NY, 1995.
21 C. Mead, *Analog VLSI and Neural Systems*, Addison Wesley Publ., Reading, MA, 1988.
22 W. Bair, C. Koch, "Real-Time Motion Detection Using an Analog VLSI Zero-Crossing Chip," SPIE Proc., Vol. 1473, on *Visual Information Processing: From Neurons to Chips*, pp. 59-65, 1991.
23 C. Mead and M. Ismail, eds., Analog VLSI Implementation of Neural Systems, Kluwer Publ., Boston, MA, 1989.
24 R. Etienne-Cummings, S. Fernando, J. Van der Spiegel and P. Mueller, "Real-Time 2-D Analog Motion Detector VLSI Circuit," Proc. IJCNN92, Vol. IV, pp. 426-431, Baltimore, MD, June 7-11, 1992.
25 A. G. Andreou, K. Strohbehn and R. E. Jenkins, "Silicon Retina for Motion Computation," Proc. IEEE Intl. Symp. on Circuits and Systems, Vol. 5, pp. 1373-1376, 1991.
26 K. A. Boahen and A. G. Andreou, "A Contrast Sensitive Silicon Retina with Reciprocal Synapses," in *Adv. in Neural Information Processing Systems 4*, eds. J. Moody, S. Hanson and R. Lippmann, M. Kaufman, San Mateo, CA, 1992.
27 T. Horiuchi, W. Bair, B. Bishofberger, A. Moore, C. Koch and J. Lazzaro, "Computing Motion Using Analog VLSI Vision Chips: An Experimental Comparison Among Different Approaches," Intl. J. Computer Vision, Vol. 8:3, pp. 203-216, 1992.
28 T. Delbruck, "Silicon Retina with Correlation-Based Velocity-Tuned Pixels," IEEE Trans. Neural Networks, Vol. 4, pp. 529-541, 1993.
29 W. Reichardt, "Autocorrelation, A Principle for the Evaluation of Sensory Information by the Central Nervous System," *Sensory Communication*, J. Wiley Publ., New York, 1961.
30 D. Marr, *Vision*, Freeman Publ. Co., New York, 1982.
31 R. Etienne-Cummings, *Biologically Motivated VLSI Systems for Optomotor Tasks*, Ph.D. Thesis, Univ. of Pennsylvania, Dept. of Electrical Engineering, Philadelphia, PA, 1994.
32 D. Stanley, "An Object Position and Orientation IC with Embedded Imager," IEEE J. Solid-State Circuits, Vol. 26, pp. 1853-1860, 1991.
33 W. O. Camp, J. Van der Spiegel and M. Xiao, "A Silicon VLSI Optical Sensor Based on Mammalian Vision," Proc. Conf. on Parallel Problem Solving from Nature, Brussels, Sept. 28-30, 1992.

34 W. O. Camp and J. Van der Spiegel, "A Silicon VLSI Optical Sensor for Pattern Recognition", Proc. of the 7th Intl. Conf. on Solid-State Sensors and Actuators (Transducers '93), Yokohama, pp. 1066-1069, June, 1993.
35 A. Gruss, R. R. Carley and T. Kanade, "Integrated Sensor and Range-Finding Analog Signal Processor," IEEE J. Solid-State Circuits, Vol. 26, pp. 184-191, 1991.
36 S.K. Mendis, S.E. Kemeny and E.R. Fossum, "A 128 x 128 CMOS Active Pixel Image Sensor for Highly Integrated Imaging Systems," Tech. Digest, IEDM, pp. 583-586, 1993.
37 W. Yang, "Analog CCD Processor for Image Filtering," SPIE Proc. Vol. 1473, on *Visual Information Processing: From Neurons to Chips,* pp. 114-127, 1991.
38 A. M. Chiang and M. L. Chuang, " A CCD Programmable Image Processor and Its Neural Network Applications," IEEE J. Solid-State Circuit, Vol. 26, pp. 1894-1901, 1991.3
39 C. L. Keast, C. G. Sodini, "A CCD/CMOS-Based Imager with Integrated Focal Plane Signal Processing", IEEE J. Solid-State Circuits, Vol. 28, pp. 431-437, 1993.
40 K. Suzuki, K. Najafi and K. Wise, "Process Alternatives and Scaling Limits for High-Density Silicon Tactile Imagers," Sensors and Actuators, Vol. A21-23, pp. 915-918, 1990.
41 K. Chun and K. D. Wise, "A High-Performance Silicon Tactile Imager Based on a Capacitive Cell," IEEE Trans. Electr. Dev., ED-32, pp. 1196-1201, 1985.
42 S. Sugiyama, K. Kawahata, M. Abe, H. Funabashi, I. Igarashi, "High Resolution Silicon Pressure Imager with CMOS Processing Circuits," Proc. 4th Intl. Conf. Solid-State Sensors and Actuators, Tokyo, Japan, pp. 444-447, 1987.
43 J. Clark, "CMOS Magnetic Sensor Arrays," Abstract, Transducers and Actuators Workshop, Hilton Head, pp. 72-75, 1988.
44 S. Kataoka, "An Attempt Towards an Artificial Retina: 3-D IC Technology for an Intelligent Image Sensor," Proc. Intl. Conf. 1985 Solid-State Sensors and Actuators, Philadelphia, PA, pp. 440-442, 1985.
45 K. Kioi, T. Shinozaki, K. Shrakawa, K. Ohtake and S. Tsuchimoto, "Design and Implementation of a 3D-LSI Image Sensing Processor," IEEE J. Solid-State Circuits, Vol. 27, pp. 1130-1140, 1992.
46 A. S. Laskar and S. Blythe, "Epoxy multichip modules: A solution to the problem of packaging and interconnection of sensors and signal processing chips," Sensors and Actuators, Vol. A36, pp. 1-27, 1993.
47 C. Wen, T. Chen and J. Zemel, "Gate-Controlled Diodes for Ionic Concentration Measurement," IEEE Trans. Electron Devices, Vol. ED-26, pp. 1945-1961, 1979.
48 J.M. Stern, et al., "An Ultra Compact, Low-Cost, Complete Image Processing System," Digest Tech. Papers, IEEE Intl. Solid-State Circuits Conf., pp. 230-231, 1995.
49 D. Fay and A. Waxman, "Neurodynamics of Real-Time Image Velocity Extraction," *Neural Networks for Vision and Image Processing,* Ed. G. Carpenter and S. Grossberg, MIT Press, 1992.
50 G. S. Van der Wal and P. J. Burt, "A VLSI Pyramid Chip for Multiresolution Image Analysis," Intl. J. Computer Vision, Vol. 8:3, pp. 177-189, 1992.
51 C. F. Weiman, Space Variant Sensing, DARPA Workshop on Computational Sensors, University of Pennsylvania, Philadelphia, PA, May 11-12, 1992.
52 A. I. Cohen, "The Retina," in *Adler's Physiology of the Eye*, 8th edition, R. A. Moses and W. M. Hart, eds., C.V. Mosby Co., St. Louis, MO, 1987.
53 E. Schwartz, "Computational Anatomy and Functional Architecture of Striate Cortex: A Spatial Mapping Approach to Perceptual Coding," Vision Res., Vol. 20, pp. 645-669, 1980.
54 C. F. Weiman and G. Chaikin, "Logarithmic Spiral Grids for Image Processing and Display," Computer Graphics and Image Processing, Vol. 11, pp. 197-226, 1979.
55 M. Testarelli and G. Sandini, "On the Advantages of Polar and Log-Polar Mapping for Direct Estimation of Time-to-Impact from Optical Flow", PAMI, March/April 1993.

56 C. F. Weiman, "3-D Sensing with Exponential Sensor Array," Proc. SPIE Conf. on Pattern Recognition and Signal Processing, Vol. 938, 1988.

57 G. Sandini and V. Tagliasco, "An Anthropomorphic Retina-Like Structure for Scene Analysis," Comp. Graphics and Image Proc., Vol. 14, pp. 365-372, 1980.

58 J. Van der Spiegel, G. Kreider, et al., "A Foveated Retina-Like Sensor Using CCD Technology," in Analog VLSI Implementation of Neural Systems, pp. 189-210, Eds. C. Mead and M. Ismail, Kluwer Publ., Boston, MA, 1989.

59 I. Debusschere, et al., "A Retinal CCD Sensor for Fast 20D Shape, Recognition and Tracking," *Sensors and Actuators A*, Vol. 21, pp. 456-460, 1990.

60 G. Kreider, "A Treatise on Log-Polar Imaging Using a Custom Computational Sensor" Ph.D. Thesis, Univ. of Pennsylvania, Dept. Electrical Engr., Philadelphia, PA, 1993.

61 W. Veldkamp and T. McHugh, "Binary Optics," Sc. American, pp. 92-97, May 1992.

62 Y. A. Carts, "Microelectronic Methods Push Binary Optics Frontiers," Laser Focus World, p. 87, Feb. 1992.

63 W. Veldkamp,"Wireless Focal Planes - On the Road to Amocronic Sensors," IEEE J. Quantum Electr., Vol. 29, pp. 801-813, 1993.

64 Y. Hirai, "VLSI Neural Networks," IMC 1992 Proc., pp. 12-19, Yokohama, June 3-5, 1992.

65 J. Van der Spiegel, P. Mueller, D. Blackman, P. Chance, C. Donham, R. Etienne-Cummings and P. Kinget, "An Analog Neural Computer with Modular Architecture for Real-Time Dynamic Computations," IEEE J. Solid-State Circuits, Vol. 27, pp. 82-92, 1992.

66 P. Mueller, J. Van der Spiegel, D. Blackman, C. Donham, "Design and Performance of a Prototype Analog Neural Computer," J. Neurocomputing, Vol. 4, pp. 311-324, 1992.

67 C.D. Donham, *The Design and Implementation of a General Purpose Analog Neural Computer and Its Application to Speech Recognition*, Ph.D. Thesis, Univ. of Pennsylvania, Dept. of Electrical Engineering, Philadelphia, PA, 1995.

68 G. Zorpette, "The Power of Parallelism," IEEE Spectrum, pp. 28-33, 1992.

69 T. Bell, "Beyond Today's Supercomputers," IEEE Spectrum, pp. 72-75, Sept. 1992.

70 S. Isoda, "Bioelectronic Devices Based on Biological Electron Transport Systems Using Flavin-Porphyrin Molecular Heterojunction," Future Electron Devices, Vol. 2, Suppl. pp. 59-65, 1992.

71 D. DiVincenzo, "Principles of Quantum Computing," Digest Tech. Papers, IEEE Intl. Solid-State Circuits Conf., pp. 312-313, 1995.

72 J. Sheth and R. M. Janowiak, eds. "2021 AD: Visions of the Future, The ComForum Summary and Research Report," Natl. Engineering Consortium, Chicago, IL, 1991.

73 I-STAT™, Portable Clinical Analyzer, Princeton, NJ.

74 S. Tachi, "Sensors and Sensing Systems in Advanced Robotics," Digest of the 1991 Intl. Conf. on Sensors and Actuators, pp. 601-606, 1991.

75 F. G.Splitt, "Environment 2021: Will we be able to make or use anything," *2021 AD: Visions of the Future , The ComForum Summary and Research Report,* J. Sheth and R. M. Janowiak, ed., Natl. Engineering Consortium, Chicago, IL, 1991.

76 United Nations and World Bank data.

77 S. Tachi, "Virtual Reality as a Human Interface in the 21st Century," Extended Abstracts Intl. Symp. on Sensors in the 21st Century, pp. 121-128, Jap. Electr. Development Assoc., Tokyo, 1992.

78 J. Van der Spiegel, "New Information Technologies and Changes in Work," in *The Changing Nature of Work,* Ann Howard (ed.), Jossey-Bass, San Francisco, CA, 1995.

Intelligent Sensors
H. Yamasaki (Editor)

Intelligent Materials

Hiroaki YANAGIDA

Department of Applied Chemistry, Faculty of Engineering, University of Tokyo, 7-3-1 Hongo, Bunkyo-ku, Tokyo, 113 JAPAN

1. Introduction: the background of intelligent materials development

We often hear the term "intelligent material". Some international conferences on intelligent materials have already been held. While intelligent materials have become popular, the present author is afraid that very complicated materials are being developed because of the term "intelligent materials". He proposes that we keep in mind the original aim of R & D for intelligent materials [1], which is to solve the unfriendliness of technology to the public. Some recent advanced technology has led to technostress due to much too complicated control systems. Intelligent mechanisms involve self-diagnosis, self-adjustment, self-recovery, tuning (including recycling) capability, etc. These functions may be achieved by installing sensors and actuators within materials. If the structure is complicated, the material is no longer intelligent. One may believe the idea that the more complicated something is, the more advanced it is. This false recognition is sometimes called the "spaghetti syndrome", i.e. many electric wires tangling the object. Once technologies suffer these syndromes, they become increasingly complicated. We need intelligent materials to save complicated and unstable circuits. One of the important objectives for intelligent materials is to cure this syndrome.

Another objective for intelligent materials is to develop technology that is friendly to the environment. Materials for technology must be designed to meet certain requirements: reasonable fabrication cost, high reliability during use, and capability for recycling. Improvement of the reliability has been tried by making the structure thicker. This, however uses many resources and energy and produces much waste. Even though one may make materials stronger and more durable, fractures may occur unnoticed. Increase in strength makes recycling more difficult. The materials break when undesired and do not break when desired. If we can obtain a certain signal prior to the fatal fracture, we may be able to avoid accidents. One may put strain gauges on materials or structures. Many gauges are needed to cover a large or long specimen. One may of course apply an acoustic emission method to detect the generation of cracks. The signal obtained, however, is usually very subtle and it is not always practical to install an acoustic emission apparatus in every application. These methods are complicated and may lead to the "spaghetti syndrome". The monitoring method must be very simple with distinct signals. One of the most promising ways to assure the reliability of materials is to install a self-diagnosis mechanism, where signals that warn of fractures must be very distinct, while the materials still resist stress. If materials could be made with such a self-diagnostic mechanism, we could take action to avert disasters caused by material's failure. We could also avoid the unnecessary sacrifice of materials by replacing them prematurely lest they should fail.

2. The measure of intelligence -- Technological Wisdom Index

The author proposes the following index to measure the merits of intelligence.

$$M_I = \text{number of merit} \; / \; (\text{number of component})^n \qquad (n>1)$$

Simpler structures with less components are considered to be more advanced. When there is addition of a sensor for an additional merit, M_I decreases. The way to achieve higher values is integration of functional and structural materials. The following examples explain the index.

2.1. No necessity of electrical assistance

Since most functional materials need computers and integrated circuits now, it may seem difficult to get a useful function without the assistance of electricity. The present author, however, considers the humidity adjustment of wood in old Japanese log storehouses an original intelligent mechanism. In comparison, the "spaghetti syndrome" needs humidity sensors, dehumidifiers, humidifiers, temperature sensors, heaters, coolers, many circuits, and so on. In addition, emergency dynamos and circuits are necessary to provide against accidents. The system becomes complicated and consumes much energy.

One recent example is photochromic glass, where optical transmittance is automatically adjusted without electricity. This character may be achieved by installing an optical sensor and polarizer on ferroelectric ceramics. The degree of intelligence is much higher in photochromic glass, even though the response is slower than the ferroelectric device. Another example is the vibration damper already adopted for buildings which works even during electricity outage. Of course there are proposed dynamic damping mechanisms. We cannot expect that they work perfectly during very strong earthquakes. Electricity may eventually become out-of-date.

2.2. No additional sensors

The present author has discovered the self-diagnosis function in CFGFRP, carbon fiber and glass fiber reinforced plastics, which are used as replacements for iron bars in concrete structures [2]. A typical example of the load-strain change of electrical resistance characteristic of the materials is shown in Figure 1. We can see a tremendous change in electrical resistance at the point corresponding with one where carbon fibers are broken before fatal fracture as shown around (A) in Figure 1. We can note a not-yet fatal fracture with a very distinct signal. This corresponds with a health examination. Residual increase of resistance observed when unloaded, around (B) in Figure 1, shows a history of the specimens which have suffered damage close to fatal fracture. This method, therefore, also tells of past disease. Maintenance of large-scale structures, architecture or aircraft, is expected to be easier with this method. The material, CFGFRP, has intelligence to not fracture suddenly as structural material, to diagnose present and past stress, and to foresee fatal fracture as functional material, a sensor. There are many merits without the addition of complicated sensor systems. The method applied to check the damage of materials is very simple. Only a very conventional tester is needed. The number of components is not large because of integration of structural and functional materials. It is clear that the value of the wisdom index is high.

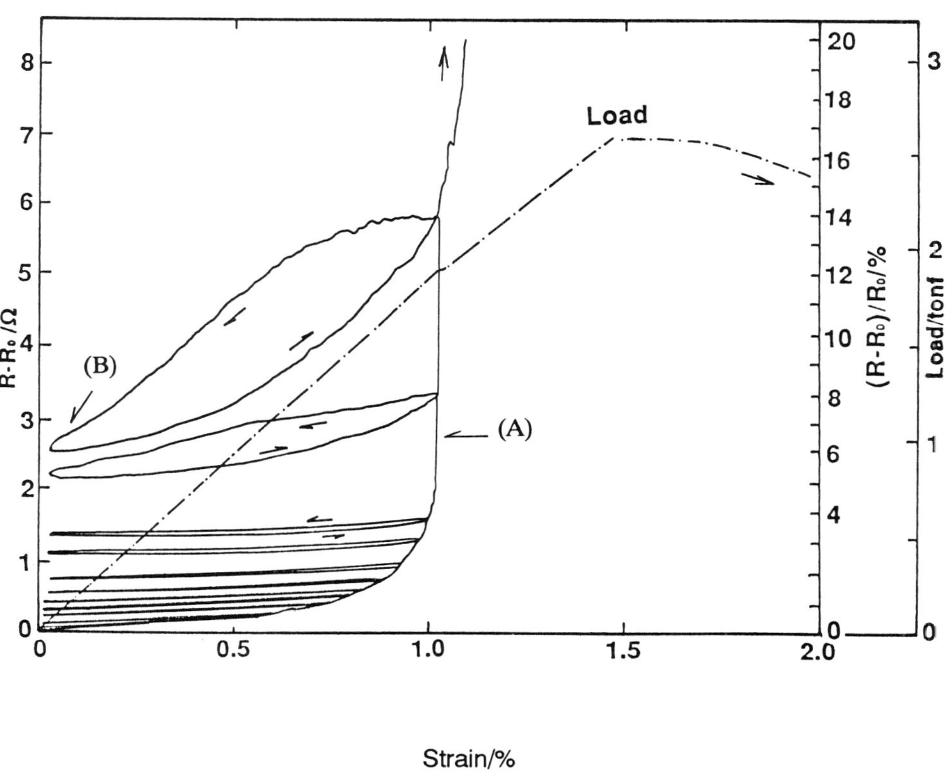

Figure 1. Load - Strain - ΔR -Characteristic of CFGFRP (C= PAN-HMCF, R_0=40Ω).

2.3. Simplicity of Structures

2.3.1. Thermistors

A PTC (positive temperature coefficient) thermistor is a good example of intelligent material with a simple structure. From the viewpoint of heater performance, the PTC thermistor is more intelligent than the NTC (negative temperature coefficient) one. Figure 2 shows a comparison between a PTC thermistor and an NTC one. The resistivity of NTC thermistor decreases with temperature. An increase in temperature gives rise to an increase in electric current. This is positive feedback and difficult to control. When an NTC thermistor is used as a heating element, we have to add a temperature-control circuit. On the other hand, the PTC thermistor is a heater below the Curie Temperature (T_C), it is a critical temperature sensor, and it is a switch since above the temperature the resistivity increases tremendously. It is multifunctional material. The structure is very simple and it is intelligent not because of the multifunctions but because of the self-adjusting mechanism. The temperature vs. resistivity characteristics give rise to a strong negative-feedback mechanism. We do not have to equip any electric circuits to control temperature.

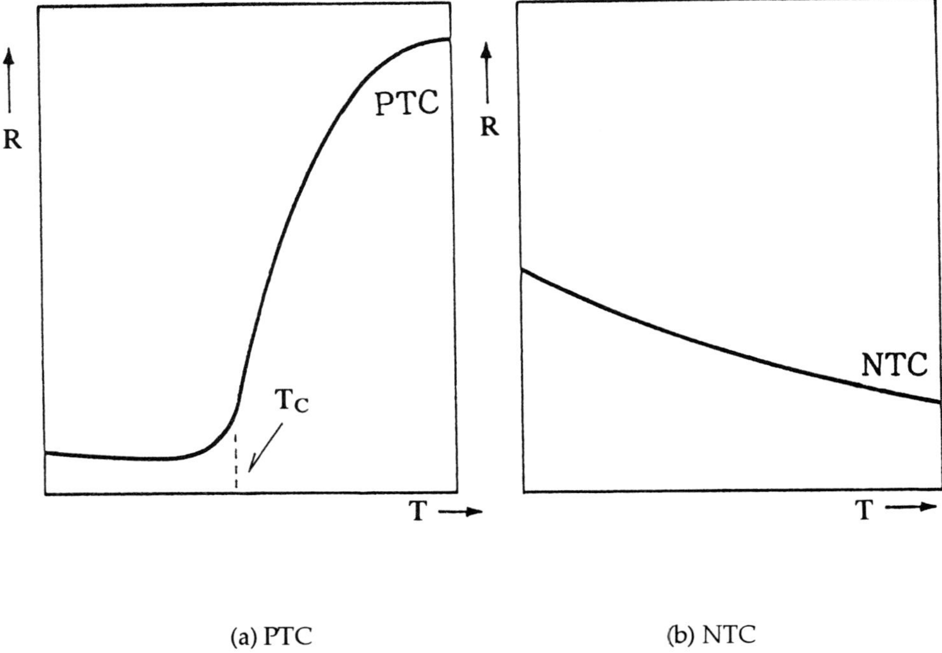

(a) PTC (b) NTC

Figure 2. Two types of ceramic thermistors.

2.3.2. Humidity sensor

Some porous ceramic semiconductors work as a humidity sensor. The conductivity of the material increases with the adsorption of water vapor under humid conditions and decreases with the removal of adsorbed water molecules under dry conditions. Removal is, however, very slow at room temperature. In order that the water molecule adsorption sites remain fully active, one must frequently carry out a so-called cleaning operation to remove the adsorbed water molecules at high temperature. During such cleaning treatments, the sensor cannot supply any information about current humidity. This is a disadvantage arising from the lack of a self-recovery mechanism.

The author's group has constructed a new type of ceramic humidity sensor [3]. This sensor, made of p/n contact, CuO p-type and ZnO n-type semiconductors, measures current across the interface required for electrolyzing water adsorbed around the contact points [4]. The adsorbed water is electrolyzed into gases. This means automatic cleaning. This has saved a circuit or treatment to remove water molecules adsorbed around the contact points, while the humidity sensor made of porous ceramic semiconductors cannot be operated without the treatment. The voltage-current characteristic changes with humidity, as shown in Figure 3.

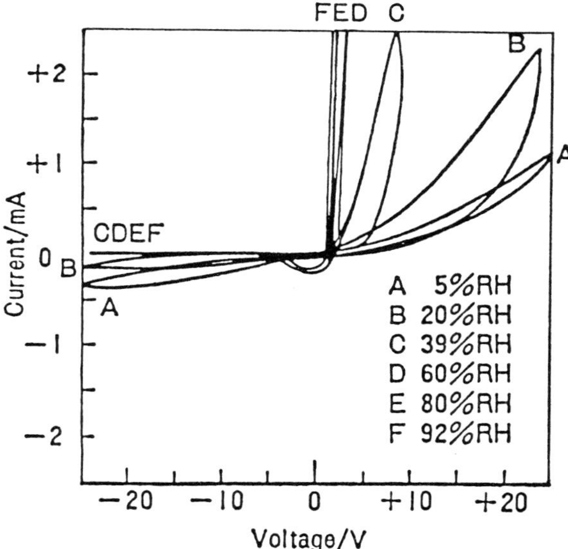

Figure 3. Humidity-dependent voltage-current characteristic changes of an intelligent humidity sensor made by heterocontact between copper oxide and zinc oxide (All measurements were carried out at room temperature, RH = relative humidity).

In the CuO/ ZnO heterocontact type of humidity sensor, the amount of water adsorption in the vicinity of the heterocontacts changes with humidity. This is also found for porous humidity sensors of the usual type with homophase contacts. Electron holes are injected by the p-type semiconductor electrode into the adsorbed water molecules, giving rise to protons in the adsorbed water. The positive charges from the protons are neutralized with negative charges injected by the n-type semiconductor electrode to liberate hydrogen gas. As a consequence, the adsorbed water is electrolyzed. During measurements, the cleaning process is always working, since the cleaning or self-recovery treatment is itself a working mechanism. If we look at the CuO/ ZnO heterocontact from the viewpoint of humidity sensor performance compared with the porous ceramic type humidity sensor, we see more intelligence in the former.

2.4. Tuning capability

There is the porous ceramic semiconductor type gas sensor using zinc or tin oxide. It is, however, very difficult to distinguish one gaseous species from another, for example, carbon monoxide from hydrogen. It has been proposed to use computers to judge different kinds of gaseous species. The computers memorizing how sensors react to each gaseous species evaluate the data from the sensors and give judgments. They need many sensors.

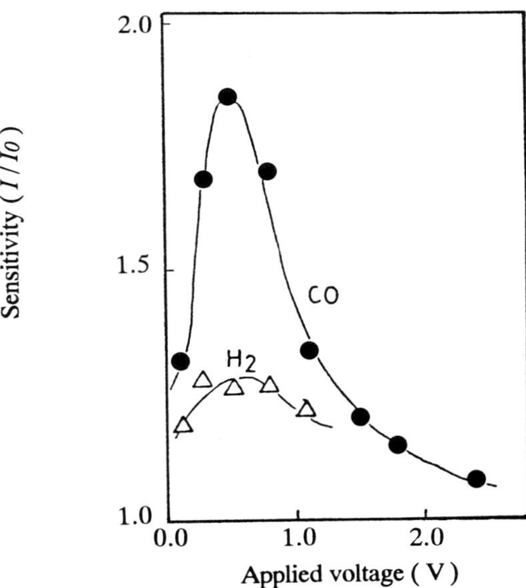

Figure 4. The relation between gas sensitivity and the forward applied voltage in CuO/ ZnO heterocontact. Gas concentrations are 8000 ppm. Measuring temperature is 260 ℃.

The p/n contact humidity sensor also works as a very selective carbon monoxide gas sensor [5]. The sensitivity changes with the bias applied across the p/n contact, as shown in Figure 4. Selectivity and sensitivity for carbon monoxide gas is maximum when the CuO side is positively biased at about 0.5 volt. Although sensitivity decreases above the bias, selectivity becomes uniform for all kinds of flammable gaseous species. This effect constitutes a "tuning".

Sometimes we need to change the behavior or characteristics of materials from the outside. We may call this "tuning" the device. It avoids having to employ a large number of different components. The material with tuning capability achieves a high value in wisdom index.

The present author proposes that the intelligent mechanisms are classified into two levels. Self-diagnosis, self-recovery and self-adjustment belong to primary intelligence. Tuning capability to control these primary intelligent mechanisms as we want should be called advanced intelligence.

3. Conclusion

Intelligent materials are required to make technology friendly for the environment and people. The intelligent mechanisms, such as self-diagnosis, self-recovery, self-adjustment and tuning capability for saving complexity and/or recycling must be achieved by simple methodology. The author proposes the technological wisdom index to measure the merits of intelligence and has shown typical cases.

References

1. K. Yamayoshi and H. Yanagida, Chemistry Today, 1992 [5] (1992) 13; H. Yanagida, Intelligent Materials, 2 [1] (1992) 6, 1st International Conference on Intelligent Materials, March 23-25, 1992 Oiso, Japan; H. Yanagida, Symposium of University of Tokyo, Jan. 28, 1992; H. Yanagida, Electronic Ceramics, 22 [5] (1991) 5; H. Yanagida, Lecture. U. S. - India - Japan Joint Symposium upon Electronic Ceramics, Jan. 16, 1989, Pune, India; H. Yanagida, Angewandte Chemie, 100 [10] (1988) 1443.
2. N. Muto, H. Yanagida, M. Miyayama, T. Nakatsuji, M. Sugita and Y. Ohtsuka, J. Ceram. Soc. Jpn., 100 [4] (1992) 585.
3. K. Kawakami and H. Yanagida, Yogyo Kyokaishi, 87 (1979) 112.
4. Y. Nakamura, M. Ikejiri, M. Miyayama, K. Koumoto and H. Yanagida, Nippon Kagaku Kaishi, 1985 (1985) 1154.
5. Y. Nakamura, T, Tsurutani, M. Miyayama, O. Okada, K. Koumoto and H. Yanagida, Nippon Kagaku Kaishi, 1987 (1987) 477.

Intelligent Sensors
H. Yamasaki (Editor)

Micromachining

S. Shoji and M. Esashi

Department of Mechatronics & Precision Engineering,
Faculty of Engineering, Tohoku University,
Aza Aoba, Aramaki, Aoba-ku, Sendai, 980, Japan

Abstract

In order to fabricate circuit integrated microsensors, micro fabrication techniques based on LSI technologies, that micromachining, play important roles. In this section useful micromachining techniques for fabricating integrated sensors are described. The micropackaged structure for integrated sensors is also introduced.

1. INTRODUCTION

Various kinds of silicon-based microsensors have been studied and some of them are already in commercial use. Especially, microsensors using three-dimensional silicon structures, for example pressure sensors, accelerometers, have been developed for practical applications. Micromachining based on microfabrication technologies in LSI, i.e. photolithography, etching, deposition, etc., have been applied to the fabrication of such sensors(1).

One of the recent trends in the development of microsensor is the integration of the sensors and circuits for readout and interface on the same silicon chip, the so-called integrated sensors. This is also leading to the advanced monolithic intelligent sensors. For realizing integrated sensors, IC-compatible sensor processing techniques must be taken into account.

The technique of encapsulating only the part comprising the sensor but also the circuit is very important in practical use. Micromachining can also be applied to the packaging(1). For the packaging of integrated sensors, bonded silicon and Pyrex glass structures are very useful, having the advantages of small size and low cost etc. In order to realize such packaging, the bonding technique and the micromachining for Pyrex glass are very important.

This paper describes important techniques of micromachining such as etching and bonding. Then micropackaging structure for integrated sensors is introduced.

2. SILICON ETCHING

Silicon etching techniques are generally classified into two categories. One is called isotropic etching and the other is anisotropic etching. Each can be carried out by wet etching and dry etching. A nitric acid, fluoric acid and acetic acid system is generally used for the isotropic wet etching. Since this etching technique shows large undercut and poor controllability of etching properties, it is difficult to apply when the precise geometrical control is necessary. The dry etching techniques of plasma etching and reactive ion etching have not been used so often for deep balk etching because of the small etching rate. High etching rate anisotropic dry etching techniques have recently been developed but this is a method for the next generation of micromachining techniques (2).

2.1 Anisotropic wet etching

Anisotropic wet etching plays a very important role in micromachining. Fine three-dimensional micro structures, i.e. cantilevers, bridges, diaphragms etc., are fabricated as shown in Fig.1. Various alkaline hydroxide and organic aqueous solutions have been used as etchants. Potassium hydroxide, ethylendiamine and hydrazine are the typical etchants. The important properties of this etching method are anisotropy, selectivity, handling, non-toxicity and process compatibility. Ammonium hydroxide (AH) and quaternary ammonium hydroxide (QAH) solutions have recently been developed to obtain integrated circuit compatibility. The characteristics of these etchants are summarized in Table 1.

2.1.1 Alkaline metal hydroxide

Purely inorganic alkaline solutions like KOH, NaOH and LiOH have been used for the etchant(3,4). The effect of different anions are minor if solutions of the same molarity are used. CsOH has been developed and has the advantages of a large (110)/(111) etching ratio and small SiO_2 etching rate(5). KOH solutions are the most commonly used, and their etching procedures have been studied in detail(3).

Uniform and constant etching can be realized with highly

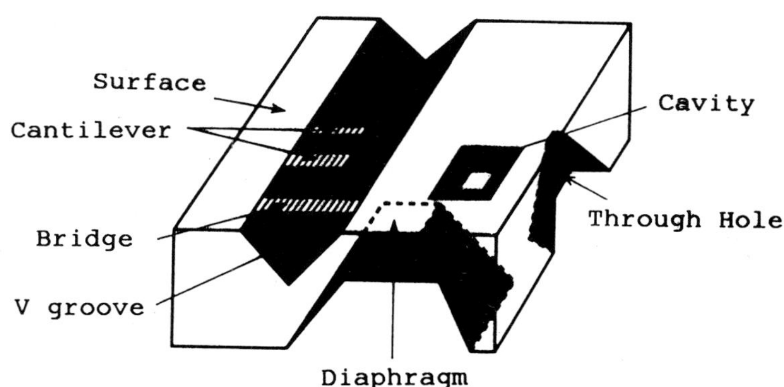

Fig.1 Microstructures fabricated by wet aniostropic etching (1).

Table.1 Chracteristics of Silicon Anisotropic Etching

Etchant	Etching Rate (μm/min)	(100)/(111) Etching Ratio	(110)/(111) Etching Ratio	Etching Rate SiO_2(Å/min)
KOH[2] Water (35wt%)	1.0 (80°C)	400	600	14
CsOH[5] Water (60wt%)	0.15 (50°C)	-	200	0.25
EPW[2] +Pyrazine	1.25 (115°C)	35	-	2
Hydrazine[6] Water(1:1)	3.0 (100°C)	16	9	1.7
AHW[7] (4wt%)	0.7 (50°C)	-	200	8
TMAHW[9] (22wt%)	1.0 (90°C)	33	-	24

* EPW : Ethylenediamine-Pyrocatechol-Water
** AHW : Ammonium hydroxide-Water
*** TMAHW : Tetramethyl ammonium hydroxide-Water

concentrated KOH solutions of above 20 wt% at temperatures higher than 60°C. These etchants are less toxic. The etching rate of silicon dioxide is, however, large compared to that of the other etchants.

2.1.2 Organic aqueous solutions

Ethylenediamine pyrocatechol water (EDP) and Hydrazine water are the typical etchants for this category (3,6). The basic etchant of the EDP system consists of ethylenediamine and water. An addition of pyrocatechol causes a large increase in the etching rate.

When the etchant is exposed to oxygen, 1,4-benzoquinone and other products are formed, leading to an increase in the etching rate. This effect can be avoided by purging the etching chamber with an inert gas. An unwanted increase in the etching rate is also caused by the presence of trace amounts of pyrazine. The addition of enough pyrazine to the solution can avoid this problem. This etchant has a very high selectivity between silicon and SiO_2, which is one of its remarkable advantages. The etching rate ratios of (100)/(111) and (110)/(111) are small compared to those of the pure alkaline solutions.

Hydrazine and water, normally a 1:1 solution is used, has similar etching properties to those of the EDP solution: for example a small etching rate for SiO_2, and small orientation dependency of the etching rate. The etching rate is about two times larger than that of EPW. Pyramidally shaped etch pits are sometimes observed in this etching. Ag, Au, Ti, Ta are resistant to this etching solution, while Al, Cu, Zn are attacked. EDP and hydrazine are carcinogens(6).

2.1.3 Ammonium hydroxide and QAH solutions

For integrated sensors, integrated-circuit compatible etching techniques are required. Ammonium hydroxide-water (AHW) solutions and quaternaly ammonium hydroxide-water (QAHW) solutions have been developed for this purpose(7-10).

AHW shows similar orientation selectivity to those of EDP etchants. The (100) etching front is covered with etch pits. These etch pits can be suppressed by adding small amounts of H_2O_2(7). Aluminum is not attacked by silicon-doped AHW.

TEAHW and TMAHW have similar etching characteristics. The maximum of the (100) etching rate appeared at 20 wt% for TEAH and at 5 wt% for TMAH. Etch pits were observed at low QAH concentration, appearing below 20 wt% in TEAHW and below 15 wt% in TMAHW. The selectivity of SiO_2 to (100) and (111) of QAH is high and increases with decreasing temperature.

2.2 Etching stop (selective etching)

Some kinds of doping-selective, bias-controlled etching techniques have been developed. In anisotropic etching, these selective etching techniques are simply classified into two categories as shown in Fig.2. One uses highly boron doped silicon and the other uses a biased p-n junction.

2.2.1 p+ etching stop

All alkaline anisotropic etchants exhibit a strong reduction in their etching rate at high boron concentrations(4). The group of purely alkaline hydroxide solutions, like KOH, and NaOH and the other group of aqueous solution of organic substances, like EDP and hydrazine, show different properties. The former group shows high alkaline-concentration dependency of the etching rate reduction on the boron concentration. The

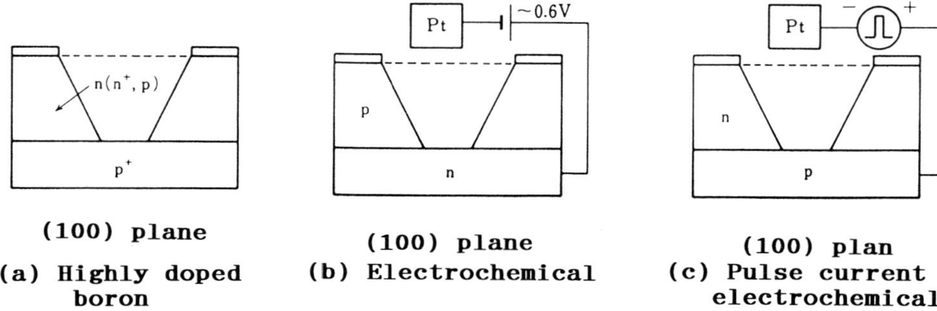

(100) plane
(a) Highly doped boron

(100) plane
(b) Electrochemical

(100) plan
(c) Pulse current electrochemical

Fig.2 Etching stop methods in anisotropic etching.

latter group generally shows a smaller etching rate for the boron-doped layer than the former group.

2.2.2 p-n junction etching stop

There are some methods of p-n junction etching stop using 2 electrodes, 3 electrodes and 4 electrodes(11,12). A set-up for a typical 4-electrode method, which has a reference electrode, a counter electrode, n-type and p-type silicon (12). The etching-stop is caused by the well-known anodic oxidation which passivates the etching. The reverse-biased p-n junction can provide large selectivity of p-type silicon over n-type in anisotropic etching solutions. The negative voltage is applied to prevent oxidation of the surface of the p-type silicon, while the positive voltage is applied to the n-type silicon to be anodically oxidized when the p-type silicon is etched completely. In the 3 electrode-system the p-type silicon is floating while the etching proceeds. The etching stop is obtained in both alkaline hydroxide solutions and organic aqueous solutions.

The opposite dopant selective etching can be realized in a KOH etching solution using pulsed anodization. It passivates p- type silicon while n-type silicon is etched.(13)

3. ETCHING AND DRILLING OF PYREX GLASS

In order to fabricate sophisticated sensors the micromachining techniques for Pyrex glass substrate are also necessary. Pyrex glass can be etched by concentrated HF using Cr-Au as a mask (14). The etching profile is of course isotropic. Electrodischarge drilling, a traditional ceramic machining technique, is useful for boring very fine through-holes (14). A schematic diagram of the apparatus and the principle of the drilling technique are shown in Fig.3. The glass is placed in an alkaline hydroxide solution. A fine needle is put on to the glass and a negative voltage of around 40 V is applied. A

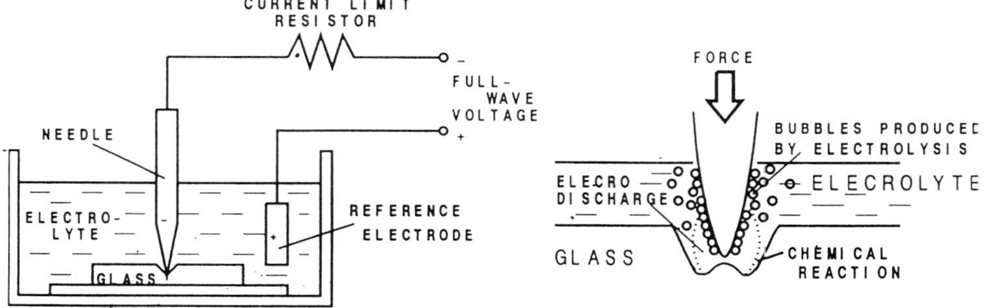

(a) Schematic diagram of the apparatus

(b) Principle of the drilling technique

Fig.3 Schematic diagram of the set-up and the principle of the technique of electro-chemical discharge drilling

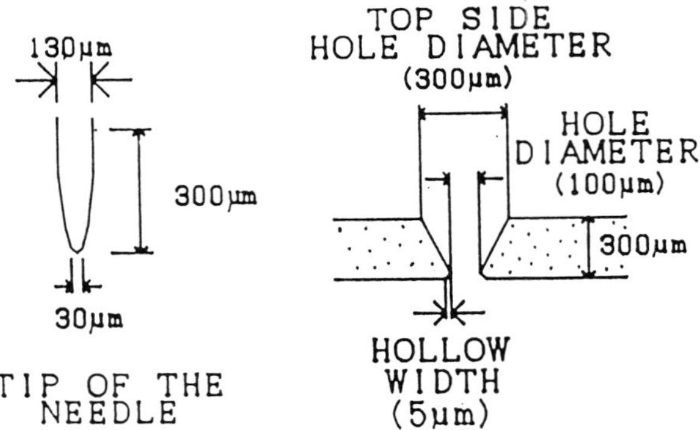

Fig.4 A cross-sectional view of an example of a needle and a hole

discharge occurs at the tip of the needle; this increases the temperature locally at the tip. The glass at the tip is removed by thermally accelerated chemical etching. Fig.4 illustrates an example of a needle and the hole it produces. A smooth etching surface is achieved when a highly concentrated NaOH solution of larger than 35 wt% is used.

4. BONDING

The bonding of one substrate to another has become an important technology in micromachining. Silicon to silicon bonding techniques and silicon to Pyrex glass bonding techniques have been developed.

4.1 Direct bonding

The silicon to silicon direct bonding technique is very attractive for microsensor fabrication (15). It also offers the new fabrication concept of bonding silicon on insulator (SOI) materials. A high process temperature of 700°C to 1100°C is necessary to achieve high bond strength. Since the maximum temperature for bonding integrated circuits is typically around 450°C, this method is difficult to apply to the bonding of preformed ICs. A low temperature bonding technique using sodium silicate glass as an intermediate layer has been reported (16). A strong bonding strength was obtained only at a temperature of 200°C.

4.2 Thermal bonding

Thermal bonding techniques using intermediate layers of glasses that melt at low temperatures have been used for packaging etc.(1). The flit glass of lead-glass systems is a conventional material whose bonding temperature is in the range 415°C to 650°C. Boron glass can also be used as the intermediate layer (17). The necessary temperature for bonding SiO_2 to SiO_2 and SiO_2 to p-type silicon is about 450°C. When n- type silicon is bonded, the temperature must be increased up to 900°C. This is because a BPO_4 phase with a high-melting temperature is formed at the surface (17).

4.3 Anodic bonding

Many kinds of microsensors have been fabricated by using electrostatic force for the bonding.

4.3.1 Pyrex glass to silicon

Anodic bonding of silicon to Pyrex glass is the most popular method of bonding and is useful for fabricating hermetically sealed microstructures. The schematic set-up of the bonding process is shown in Fig.5. The bonding has been achieved at temperatures under 300 - 500°C while applying a negative voltage of 500 - 1000 V to the glass. This low temperature process is acceptable for a wafer that processes a circuit. Since the thermal expansion coefficients of silicon and Pyrex glass are close to each other over the range of bonding temperature, the residual stress after bonding is small. This is the reason why Pyrex glass is used for this purpose.

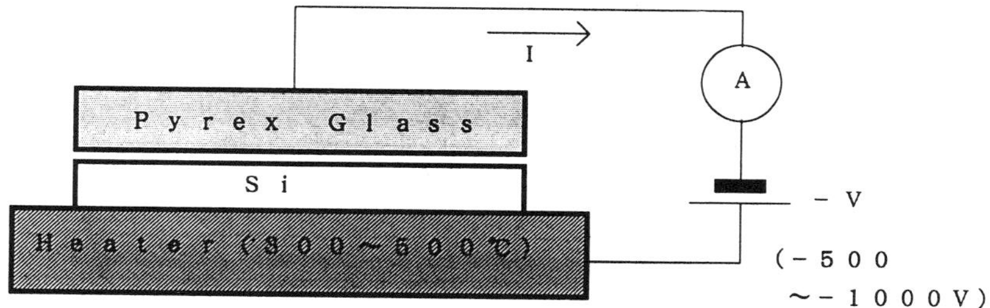

Fig.5 Schematic diagram of the set-up of the anodic bonding method

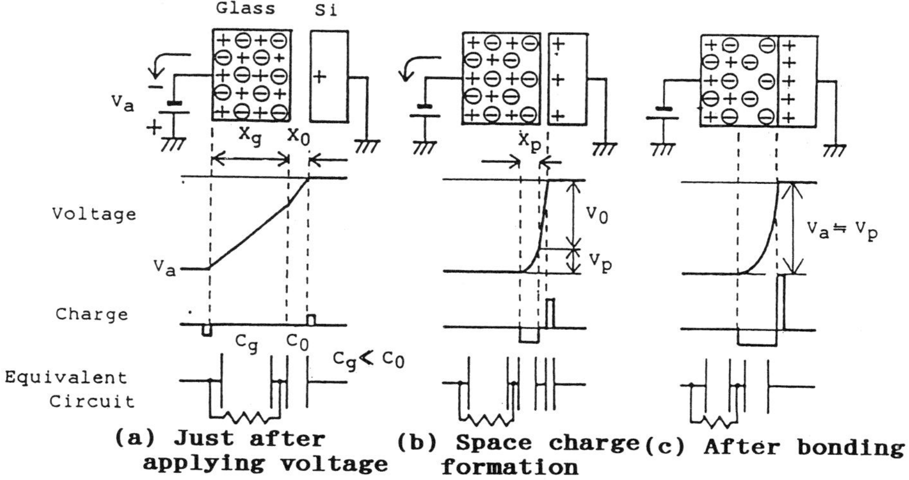

Fig.6 Principle of the anodic bonding of Pyrex glass to silicon

The mechanism of anodic bonding can be explained as shown in Fig.6. Cg, Co, Xg and Xo are the glass capacitance, the gap capacitance, the thickness of the glass and the gap between the glass and the silicon. Just after application of the voltage Va, most of Va is dropped in the glass because the Cg << Co,(Fig.6a). With high voltage and temperature, the mobile sodium ions are attracted by the negative field and move to the negative electrode. This increases the space-charge region at the silicon-glass interface (Fig.6b). The electric field at the gap rises as the space charge layer grows. This increases the attractive force between the substrates, increasing the area of intimate contact. The electrostatic force reaches a maximum value when the induced space -charge voltage is equal to the applied voltage (Fig.6c). The Si-O-Si bonds are considered to be formed at the interface.

4.3.2 Silicon to silicon

Silicon to silicon anodic bonding is also possible using an intermediate layer of sputtered Pyrex glass. A. Hanneborg et. al. have shown that high bonding strength can be obtained by bonding at 400°C with an applied voltage of 50 - 200 V (18).

By using low temperature glass like lead glass as the intermediate layer, two silicon wafers can be bonded at room temperature with a small applied voltage (19). The typical bonding condition is under 60 V with 1.0 atm mechanical pressure at 25°C. This method enables the bonding of two different materials which have very different thermal coefficients. Silicon and stainless steel or silicon and ITO coated glass can be bonded by this method.

5. INTEGRATED SENSORS AND MICROPACKAGING

The integration of sensors and circuits for readout and interface is the first step in the development of intelligent sensors. Fig.7 shows the structure of an integrated absolute capacitive-type pressure sensor (20). It consists of anodically bonded silicon and glass substrate having a hermetically

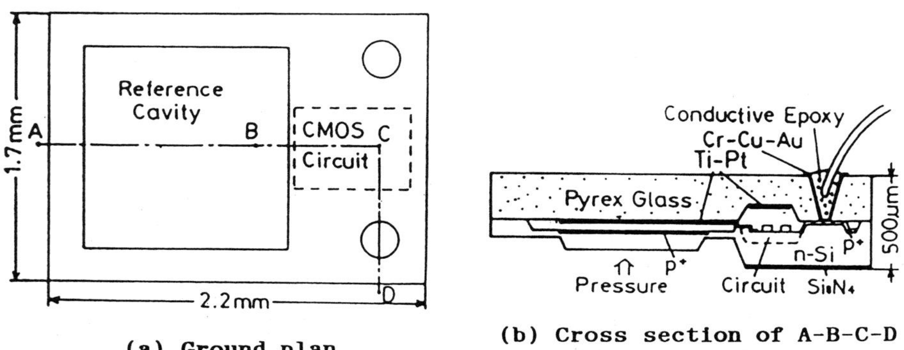

(a) Ground plan

(b) Cross section of A-B-C-D

Fig.7 Integrated capacitive-type pressure sensor

sealed reference cavity. The pressure is detected by the change in capacitance between the silicon diaphragm and aluminum electrode formed on the glass. A capacitance to frequency converter readout CMOS circuit is integrated on to the silicon substrate. In this case the chip itself is a packaged unit because the sensor and circuit are shielded from the environment. This structure enables the packaging to be completed at the wafer level by batch fabrication. Similar hermetically sealed structures are useful for accelerometers and resonant sensors etc. (21,22).

Sealed electrical feedthrough structures are very important for these sensors. The vertical feedthrough structure shown in Fig.7 has been used in another pressure sensor (23). The diffused feedthrough from the cavity is formed on the silicon and a tiny through-hole is formed in the glass. The diffused

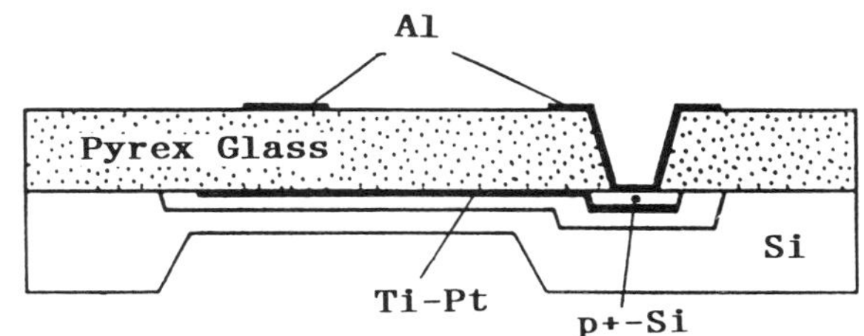

Fig.8 Vertical feedthrough structure.

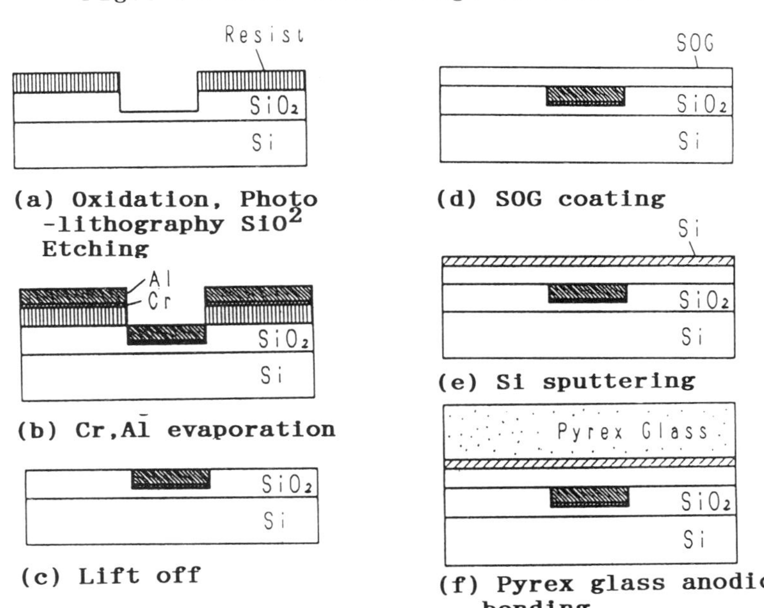

Fig.9 Lateral feed trough structure.

silicon and glass around the hole is anodically bonded and the Cr-Cu-Au layer is deposited through the hole. A lead wire is connected to the pad inside the hole with conductive epoxy. Fig.8 shows another way of achieving of vertical feedthrough using a thin p+ silicon island which is connected with the metal layer on the glass. The silicon island was made by glass -silicon anodic bonding and p+ etching stop in the hydrazine water etchant. This structure enables stray capacitance to be minimized (20).

An example of a lateral feedthrough structure with its fabrication process is shown in Fig.9. The grooves formed in the thick SiO_2 are filled with Al or Cr metal. The same resist mask for SiO_2 etching is used, enabling self alignment and the lift-off process. Then the spin-on-glass is coated to achieve isolation and planerization. The silicon layer is deposited by sputtering and the Pyrex glass is bonded by means of the anodic bonding technique (20).

6.CONCLUSION

Basic techniques of micromachining, etching and bonding have been described. Micromachining is also applicable to the packaging process to realize not only miniaturization but also low-cost encapsulation by batch-process packaging. When micromachining is applied to the fabrication of advanced intelligent sensors, IC compatible processes become more important.

7.REFERENCES

1). R.T.Howe, R.S.Muller, K.J.Gabriel and W.S.N.Trimmer, IEEE Spectrum, July (1990), 29

2). K.Murakami, Y.Wakabayashi, K.Minami and M.Esashi, Proc. of Micro Electro Mechanical Systems, MEMS'93, Florida, (1993), 18

3). H.Seidel, L.Csepregi and A.Heuberger, J. Electrochem. Soc. 137 (1990) 3612

4). H.Seidel, L.Csepregi and A.Heuberger, J. Electrochem. Soc. 137 (1990) 3626

5). L.D.Clark Jr., J.L.Lund and D.J.Edell, IEEE Solid-State Sensor & Actuator Workshop, Hilton Head SC, (1988)

6). M.Mehregany and S.D.Senturia, Sensors & Actuators, 13 (1988) 375

7). U.Schnakenberg, W.Benecke and B.Lochel, Sensors & Actuators, A21-A23 (1990) 1031

8). O.Tabata, R.Asahi and S.Sugiyama, Tech. Dig. 9th Sensor Symposium, (1990) 15

9). O.Tabata, R.Asahi, H.Funabashi and S.Sugiyama, Tech. Dig. Int. Cof. on Solid-State Sensors & Actuators Transducers' 91 (1991) 811

10). U.Schnakenberg, W.Benecke and P.Lange, Tech. Dig. Int. Cof. on Solid-State Sensors & Actuators Transducers' 91 (1991) 815

11). Y.Linden, L.Tenerz, J.Tiren and B.Hok, Sensors & actuators, 16 (1989) 67

12). B.Kloeck, S.D.Collins, N.F.de Rooij and R.L.Smith, IEEE Trans. Elect. Dev., ED-36 (1989) 663

13). S.S.Wang, V.M.McNeil and M.A.Schmidt, Tech. Dig. Int. Cof. on Solid-State Sensors & Actuators Transducers' 91 (1991) 819
14). S.Shoji and M.Esashi, Tech. Dig. 9th Sensor Symposium (1990) 27
15). P.W.Barth, Sensors & Actuators, A21-A23 (1990) 919
16). H.J.Quenzer, W.Benecke and C.Dell, Proc. Micro Electro Mechanical Systems MEMS'92, Travemunde, (1992) 49
17). L.A.Field and R.S.Muller, Sensors & Actuators, A21-23 (1990) 935
18). A.Hanneborg, M.Nese and P.Ohlckers, Tech. Dig. of Micromechanics Europe (1990) 100
19). M.Esashi, A.Nakano, S.Shoji and H.Hebiguchi, Sensors & Actuators A21-A23 (1990)31
20). M.Esashi, N.Ura and Y.Matsumoto, Proc. Micro Electro Mechanical Systems MEMS'92, Travemunde (1992), 43
21). Y.Matsumoto and M.Esashi, Tech. Dig. 11th Sensor Symposium, Tokyo (1992) 47
22). K.Yoshimi, K.Minami and Y.Wakabayashi and M.Esashi, Tech. Dig. 11th Sensor Symposium, Tokyo (1992) 35
23). M.Esashi, T.Matsumoto and S.Shoji, Sensors & Actuators, A21-A23 (1990) 1048

Intelligent Sensors
H. Yamasaki (Editor)

3-Dimensional Integrated Circuits

Takakazu Kurokawa

Department of Computer Science, National Defense Academy
1-10-20 Hashirimizu, Yokosuka 239, Japan

Abstract

A survey of research into the Three-Dimensional (3-D) IC technologies which are considered to be those that will lead the field of integrated circuit technology in the next century is made. As a national project in Japan, several companies have succeeded in fabricating prototypes of 3-D IC chips consisting of four or five active transistor layers. Furthermore, applications of 3-D IC chips are discussed. As a result, many applications of 3-D IC chips such as 3-D DRAM, 3-D SRAM, 3-D Gate Array, 3-D image sensor, 3-D image processor, 3-D logic array, 3-D shared memory, and so on, are proposed. Among them, the 3-D image sensor and 3-D image processor are likely to be the best 3-D IC chips for actual application to control systems. The basic notion of these 3-D IC chips is also discussed.

1. Introduction

Microminiaturization and high density are indispensable factors to consider in increasing the performance of Integrated Circuits (IC)[1-2]. However, various technical problems such as the limitations of chip size and microminiaturization, as well as economic problems come to the fore. In order to alleviate these problems, an attempt has been made to develop Three-Dimensional (3-D) IC technologies[3-5]. This began more than ten years ago[6-7], and now 3-D IC chips consisting of up to four active transistor layers have successfully been manufactured. Furthermore, some 3-D IC systems have been proposed. The research and development of 3-D IC chips is one of the main themes of the national project of Japan named "The Research and Development Project of Basic Technology for Future Industries"- a ten-year program. The aim of this is to develop the most important basic technologies which are indispensable in the establishment of the next generation of electronics and other industries.

A survey of 3-D IC technologies in Japan is given in this paper. Section 2 defines the 3-D IC chips which we will discuss in this paper. In section 3 some representative classification methods for 3-D IC chips are discussed. The fabrication process of a 3-D IC chip is presented in Section 4. In section 5 the benefits of 3-D IC systems as well as many problems in designing 3-D IC systems are enumerated. In section 6 research into 3-D IC technologies in Japan is summarized. In the last section of this paper the application of 3-D IC chips to control systems is discussed.

2. Definition of 3-D IC Chips

The 3-D IC chips that are discussed here are defined as "*vertically integrated IC chips consisting of more than two active device layers*". There are several researchers considering the *one-active-layer (OAL) mode*, which restricts the active devices to the bottom layer and devotes the remaining layers to wire routing, as a kind of 3-D IC chip[3-4][8-9]. There are also several research results which reveal several benefits of this mode. However, this mode is regarded as only an extension of the conventional 2-D IC chips because the technologies needed to fabricate such 3-D IC chips require only multilevel metallization technology. Thus, we shall exclude the OAL mode from the 3-D IC chips treated in this book.

3. Classification Methods for 3-D IC Chips

3-D IC chips may be classified by several methods. Figure 1 shows the two representative classification methods for 3-D IC chips that will be discussed in detail below.

3.1. Functions in Each Layer

According to the usage of each layer, 3-D IC chips can be classified into two classes, *Stacked High-Density 3-D ICs* and *Stacked Large-Capacity, Multi-Functional 3-D ICs*, depending on whether each layer has the same or different functions.

(a) Stacked High-Density 3-D ICs

This kind of 3-D IC chip has the same functions applied to each layer. A 3-D memory, shown in Figure 2, and 3-D systolic arrays, for example, are applications of this class.

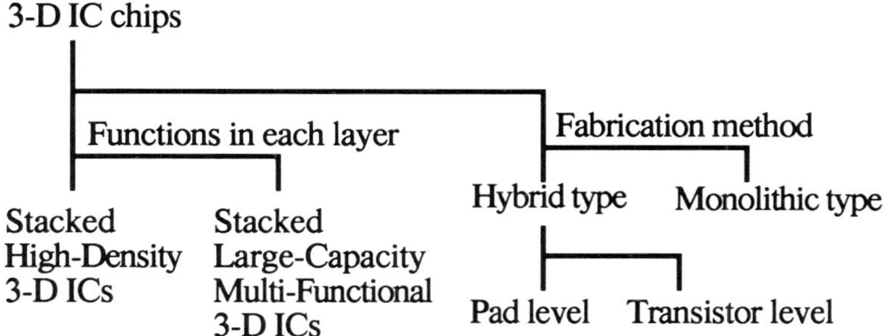

Figure 1. Classification of 3-D IC chips.

(b) Stacked Large-Capacity Multi-Functional 3-D ICs

In this class of 3-D IC chips different functions are applied to each layer. 3-D image processors[10], shown for example in Figure 3, and 3-D I/O terminals are applications of this class. 3-D image processors have been recognized to be one of the most suitable applications as well being the closest thing to a control system actually employing 3-D IC technologies.

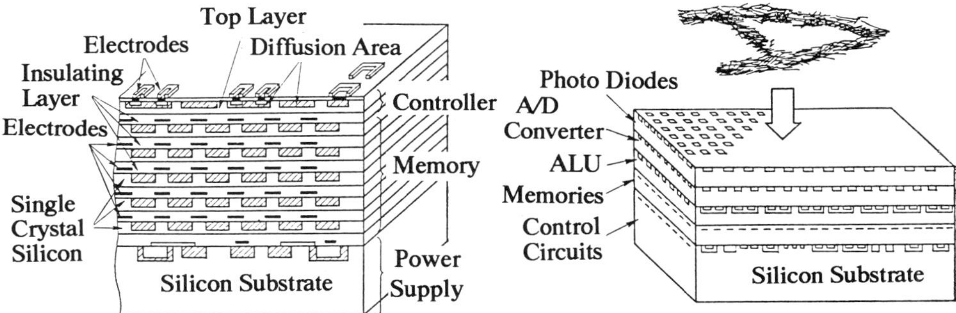

Figure 2. 3-D memory.

Figure 3. 3-D image processor.

3.2. Fabrication Method

There are many methods of fabricating multiple active device layers. Firstly, this classification method can be divided into two classes: *hybrid type* and *monolithic type*.

(a) Hybrid Type

This class of 3-D IC chips are fabricated by attaching independently fabricated IC chips using metal pools and bumps. The benefits of this class as compared to the other classes are:

·easier to fabricate,

·higher yield,

·fewer technical problems, and

·easy addition of many functions.

This class of 3-D IC chips can be subdivided into two classes, the *Pad level* and the *Transistor level,* by virtue of the level of attachment between IC chips. The attachment of IC chips in the former class is at the I/O pad level as shown in Figure 4. The concept of this class may be the same as in the technologies for micro-modules. There are some problems with this class: I/O pads are so large compared to the size of transistors or the wiring that they prevent high density. The small gaps between the IC chips prevent effective cooling.
The attachment of chips in the latter class is at the transistor level. A typical process technology for fabrication of this class of 3-D IC chips is the CUBIC(CUmulatively Bounded IC) technology developed by NEC Corp.[11], shown in Figure 5.

(b) Monolithic type

This class of 3-D IC chips, shown in Figure 6, may be the nucleus of the development of 3-D IC chips in the future. A technology to make the stratum of planar transistors is needed to fabricate monolithic 3-D IC chips. Many problems in developing the fabrication technologies for this class of 3-D IC chips will be discussed in detail in Section 5.

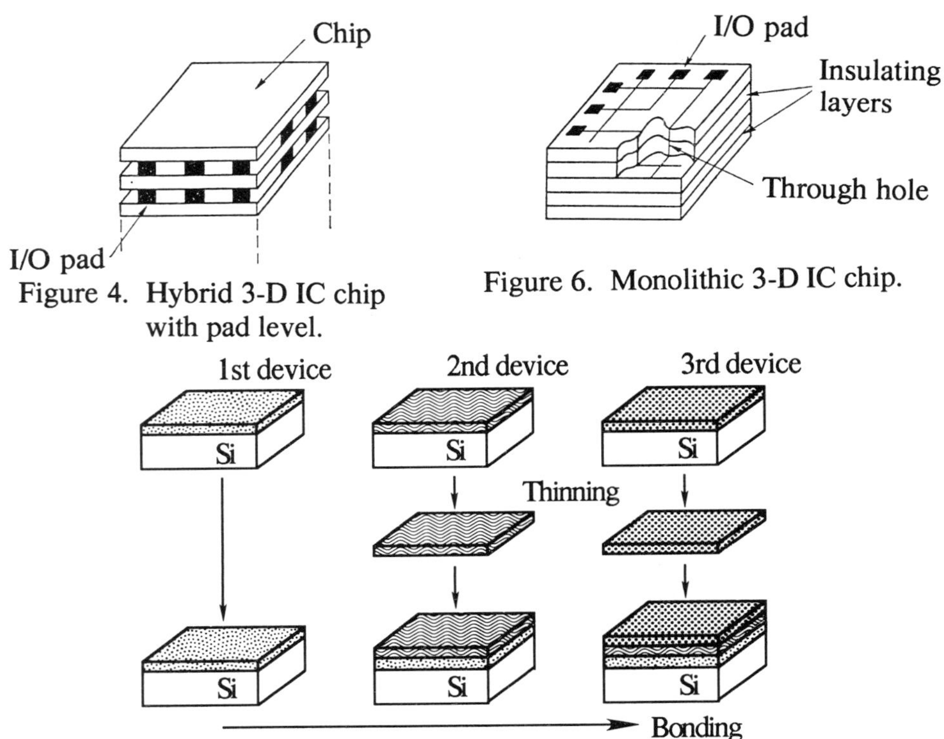

Figure 4. Hybrid 3-D IC chip with pad level.

Figure 6. Monolithic 3-D IC chip.

Figure 5. CUBIC Technology.

4. Fabrication Process for 3-D IC chips

Figure 7 shows a representative wafer fabrication process for forming a second layer on the first or bottom layer for 3-D CMOS IC chips[5] included in the monolithic type classification. In this figure, (1) and (2) show the first step in the process of stacking an insulating layer onto the bottom layer, which is fabricated in just the same way as a conventional 2-D CMOS IC wafer. An important technological process involved in making this insulating layer is the planarization technique. Then, single crystalline Si islands are made for the transistor regions in the 2nd layer by means of recrystallization techniques using a laser or electron beam as shown in (3) through (5). These technologies have several advantages for keeping the wafer temperature low enough so as not to damage the underlying devices. These process technologies for the fabrication of single crystalline Si on an insulating layer are called the *SOI (Silicon On Insulator) technologies*, and are being studied extensively world wide. Transistors in the second layer are fabricated as shown in (6) and (7), which require the same process as the conventional 2-D CMOS IC wafer fabrication. Finally, through holes to accommodate the vertical wirings to connect the 1st and the 2nd layers are made as shown in (7). Thus, the 2nd layer is fabricated on the bottom layer using several important techniques such as planarization, SOI, and vertical wiring. The 3rd layer or the 4th, 5th and so on are fabricated by following the same process from (2) to (7).

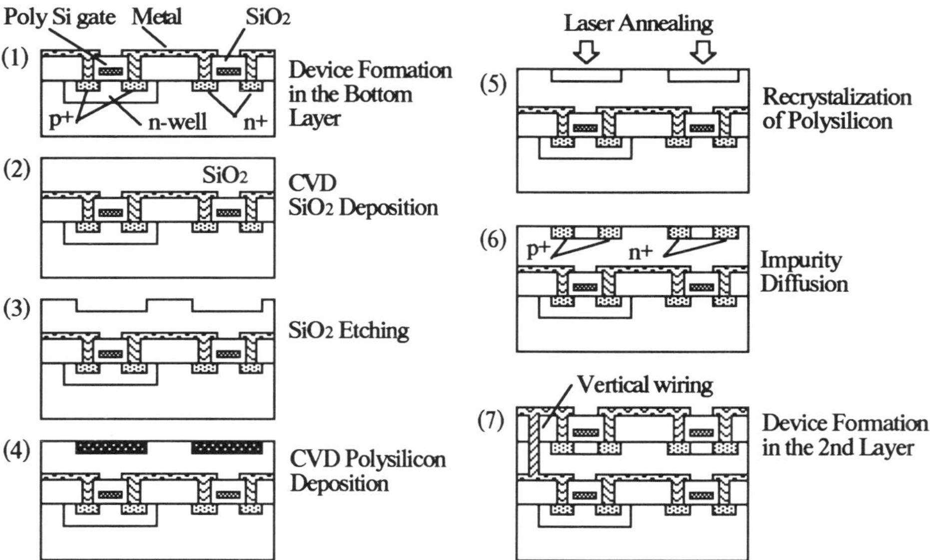

Figure 7. Wafer fabrication process for 3-D monolithic CMOS IC chips.

5. Benefits of and problems with 3-D IC chips

There are many expected benefits of 3-D IC chips, which will be enumerated below. First of all, 3-D IC chips inherit the benefits of 2-D IC chips listed below because they are based on conventional 2-D IC chips:
(1) Small size, (2) Light weight, (3) Mass production, (4) Low price,
(5) High speed, (6) High reliability and (7) Low power dissipation.
Furthermore, compared to the conventional 2-D IC chips, the following benefits can be expected through the effective use of the third dimension:
(1) High density, (2) Material savings and (3) High speed.
There have been several investigations comparing 2-D IC chips with 3-D IC chips to demonstrate the superiority of 3-D IC chips with respect to the three points listed above[3-4][8-9][12-17].

On the other hand, there are enormous problems in making 3-D IC chips for practical use. These problems can be roughly divided into the two classes shown in Figure 8. The first class is a set of problems concerning the development of the fabrication process for 3-D IC chips, and the second class is a set of problems in designing 3-D IC systems. The intensive research being conducted by our government, private companies, and several universities are gradually solving these problems. The research projects together with their research results will be shown in the next section.

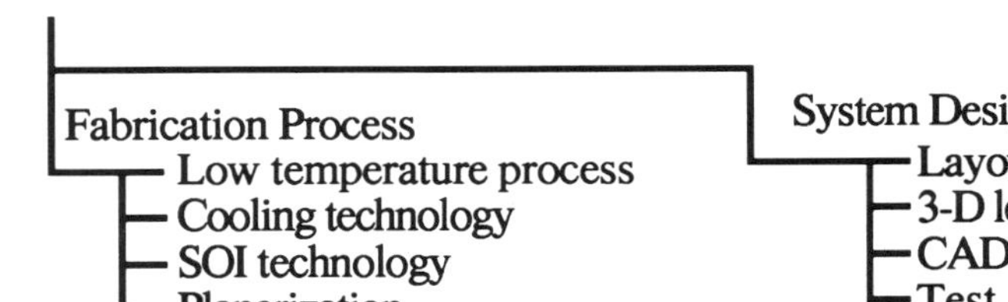

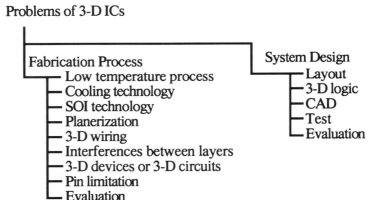

Figure 8. Problems encountered in development of 3-D ICs.

6. Research into 3-D IC chips in Japan

In Japan the Research and Development Project of Basic Technology for Future Industries was started in 1981 by the initiative of the MITI to develop innovative basic technologies for future industries. The research and development association for future electron devices (FED) was also established in 1981 under the license of MITI. They focused on the following five themes: Superlattice Devices, 3-D Devices, Hardened IC's for Extreme Conditions, Bio-electronic Devices, and Superconducting Electron Devices.[18]

Concerning the research and development of 3-D devices, the Electrotechnical Laboratory (ETL) and seven companies (NEC Corporation, Toshiba Corporation, Mitsubishi Electric Corporation, Oki Electric Industry Company Limited, Matsushita Electric Industrial Co., Ltd., Sanyo Electric Co., Ltd., and Sharp Corporation) are participating in this project. Furthermore, other private companies and also some universities are also involved. The schedule of this project was divided into three phases as follows [18]:

(1) Phase 1 (1981-1984)

The SOI technologies with laser beam annealing, electron beam annealing, and lateral solid phase epitaxy were developed.

(2) Phase 2 (1985-1987)

The design of 3-D IC chips, thermally stable wiring technology, and high aspect ratio through hall wiring technology were developed and the prototypes of 3-D IC chips were realized.

(3) Phase 3 (1988-1990)

The final 3-D devices were fabricated. The final phase of this project reassigned the final target to more sophisticated IC systems consisting of four or five layers.

The products involving 3-D IC chips fabricated as the final prototypes of this project by private companies are summarized below.

(1)Stacked High-Density 3-D IC's

(a) NEC Corp. [11]

CUBIC(Cumulatively bonded IC) technology, as shown in Figure 5, for the bonding of a thin film device layer and dual active layer-CMOS technology have been developed. Using these technologies, a 43-stage ring oscillator consisting of six active device layers was fabricated.

(b) Toshiba Corp. [19]

Using an optimized amplitude modulated pseudo-line electron beam method, five-layer stacked SOI films were developed. Furthermore, SOI MOS ring oscillators whose channel length is less than 0.2μm were also fabricated on a recrystallized layer. The delay-time at room temperature was 37ps per stage where the supply voltage was 3.0v.

(c) Oki Electric Ind. Co.,Ltd. [20]

Bridging epitaxial growth method by low pressure MOCVD was developed for the fabrication of a stacked structure. As an application of this method, a 3-D IC chip of an InGaP visible LED and its driver using GaAs MESFET was fabricated on Si substrate. A red emission with sufficient intensity was confirmed under room lighting.

(2)Stacked Large-Capacity Multi-Functional 3-D ICs

(a) Mitsubishi Electric Corp. [21]

A 3-D image processor with four active device layers as shown in Figure 9 was

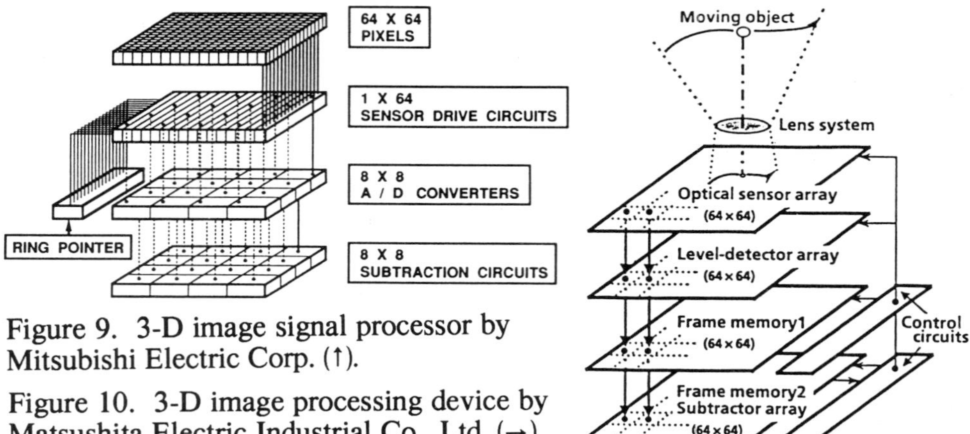

Figure 9. 3-D image signal processor by Mitsubishi Electric Corp. (↑).

Figure 10. 3-D image processing device by Matsushita Electric Industrial Co., Ltd. (→).

fabricated and confirmed its good operations as an intelligent image sensor for range sensing applications. This chip consists of 64×64 pixels in the top layer, 1×64 sensor drive circuits in the second layer, 8×8 A/D converters in the third layer, and 8×8 subtraction circuits in the bottom layer. The real time application of range sensoring, which cannot be performed by a conventional CCD camera using a 2-D IC chip, was first realized by this 3-D IC chip.

(b) Matsushita Electric Industrial Co.,Ltd. [22]

As a sensor for moving objects, a 3-D IC chip with four active device layers as shown in Figure 10 was developed. This chip consists of 64×64 pixels in the top layer forming an optical sensor array, a 64×64 level detector array in the second layer, a 64×64 frame memory with control circuits in the third layer, and a 64×64 frame memory and subtractor array with control circuits in the bottom layer.

(c) Sharp Corp. [23]

A 3-D IC chip with four active device layers for character recognition as shown in Figure 11 was fabricated. This chip consists of 55,000 diodes and 220,000 transistors on a 14.3mm×14.3mm single die. It can sense 12 characters at the same time, and recognize 64 kinds of characters containing both upper case and lower case letters, Arabic numerals , and several other symbols. Each character used for recognition by this 3-D IC chip consists of 10×14 bit matrix.

(d) Sanyo Electric Co.,Ltd. [24]

Using Lateral Solid Phase Epitaxy (L-SPE) technology, a multi-layer stacked structure was developed. A high efficiency photosensor array for long wavelengths employing a Si-Ge hetero-junction device was fabricated in the top layer of a stacked SOI structure.

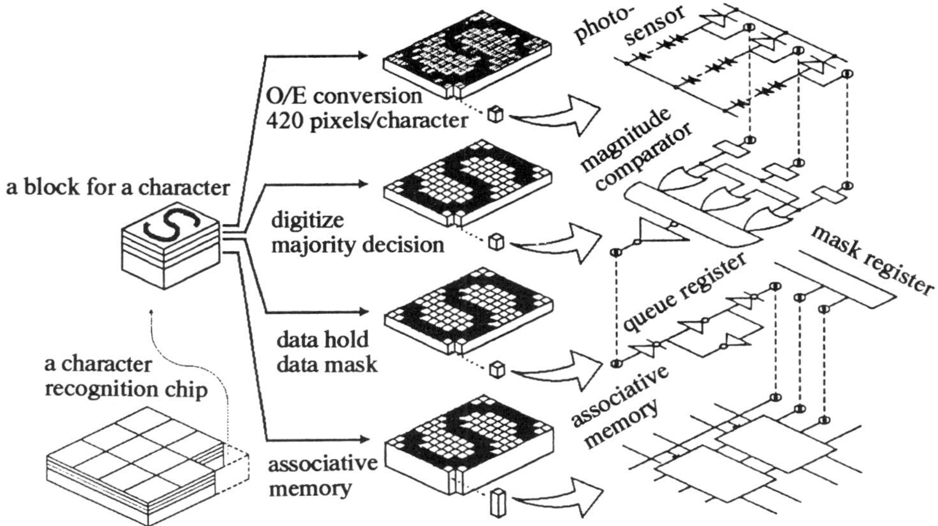

Figure 11. Character recognition image sensor by Sharp Corp..

7. Applications of 3-D IC chips to control systems

Several 3-D IC systems, such as smart image sensors, 3-D common memory, A/D converters, gate arrays, interconnection circuits, neural networks, and so on, have already been proposed as applications of 3-D IC technologies. Of these, smart image sensors are regarded as the closest approach to the actual application of 3-D IC systems. They are also the best example of the application of 3-D IC chips to control systems. Here, the overall prospect of smart image sensors using 3-D IC chips will be summarized.

The functions needed for an image sensor are photodiodes to change information in light form into analog electrical data, an A/D converter to convert analog data to digital data, ALU and memory if needed for the calculation, and control circuits, as shown in Figure 3. These functions can work in a pipeline fashion. Furthermore, 3-D image sensors have more benefits such as data parallelism in image computation, and high speed recognition in data matching. Thus, image sensors are very suitable applications of stacked large-capacity multi-functional 3-D IC systems. With such a background, several private companies in Japan such as Mitsubishi Electric Corp., Matsushita Electric Industrial Co.,Ltd., and Sharp Corp. have been making efforts to fabricate 3-D smart image sensors. Their individual research results including their basic structures have already been discussed in the previous section.

8. Conclusion

A survey of 3-D IC technologies, especially in Japan, has been made. In Japan, there has been intensive progress in 3-D IC technologies as a result of the national project. Since research and development in 3-D IC technologies began about ten years ago, the only 3-D IC chips which have been fabricated as prototypes are the MSI or LSI forms. However, already many possibilities and benefits of 3-D IC chips have been clarified. Furthermore, several realistic 3-D IC systems have been proposed. Among these, the 3-D smart image sensors are the 3-D IC systems which are most likely to be in actual use within several years. There are also several problems which are inherent in 3-D IC technologies, and we hope that there will be continuous progress in the research and development of 3-D IC systems in order to overcome these problems.

9. References

1 J.Smith, Electronics: Circuits and Devices, Wiley, New York, 2nd ed., (1980).
2 S.Muroga, VLSI System Design, Wiley, New York, (1982).
3 A.L.Rosenberg, Three-dimensional integrated circuitry, VLSI Systems and Computers, Computer Science Press, Rockville, MD, pp.69-80 (1981).
4 A.L.Rosenberg, Three-dimensional VLSI: a case study, J.ACM, **30**, 3, pp.397-416 (1983).
5 A.Kokubu, Three-dimensional device project, Tutorial Textbook of the 13th Ann. Int. Symp. Computer Architecture, pp.75-101 (1986).
6 A.Gat, L.Gerzberg, J.F.Gibbons, T.J.Magee, J.Peng, and J.D.Hong, CW laser anneal of polycrystalline silicon, crystalline structure, electrical properties, Appl. Phys. Lett., **41**, pp.775-778 (1978).
7 J.F.Gibbons, SOI-a candidate for VLSI?, VLSI Design III, pp.54-55 (1982).
8 F.P.Preparata, Optimal three-dimensional VLSI layouts, Math. Systems Theory, **16**, pp.1-8 (1983).
9 F.T.Leightoon and A.L.Rosenberg, Automatic generation of three-dimensional circuit layouts, Proc. ICCD, pp.633-636 (1983).
10 K.Taniguchi, Three dimensional IC's and application to high speed image processor, Proc. IEEE 7th Symposium on Computer Architecture, pp.216-222 (June 1985).
11 Y.Hayashi, Evaluation of CUBIC (CUmulatively Bonded IC) Devices, Extended Abstracts of 9th Symposium on Future Electron Devices, pp.267-272 (Nov. 1990).
12 T.Kurokawa and H.Aiso, Polynomial transformer, Proc. IEEE 7th Symposium on Computer Architecture, pp.153-158 (June 1985).

13 M.Hasegawa and Y.Shigei, $AT^2=O(N\log^4N)$, $T=O(\log N)$ fast Fourier transform in a light connected 3-dimensional VLSI, Proc. 13th. Ann. Symp. Computer Architecture, pp.252-260 (1986).

14 T.Kurokawa, K.Ogura, and H.Aiso, The Evaluation of Three-Dimensional VLSI by Studying the Layout of Interconnection Networks, Proc. of JTC-CSCC'88, pp.342-347 (1988).

15 T.Kurokawa, Y.Ajioka, and H.Aiso, Matrix Solver with Three-Dimensional VLSI Technology, Electronics and Communications in Japan, Part 2, **72**, 8, pp.34-44 (1989).

16 T.Kurokawa and H.Aiso, A Proposal for Polynomial Transformer and Its Design using 3-D VLSI Technology, Electronics and Communications in Japan, Part 2, **72**, 9, pp.56-64 (1989).

17 Y.Ichijo, T.Kurokawa, and J.Inada, Evaluation of Suitability of Sorting Circuits as 3-D Integrated Circuits, Electronics and Communications in Japan, Part 2, **75**, 2, pp.101-113 (1992).

18 T.Sakamoto, Introduction of Future Electron Device Projects, Extended Abstracts of 9th Symp. on Future Electron Devices, pp.7-11 (Nov. 1990).

19 S.Kambayashi, M.Yoshimi, M.Kemmochi, H.Itoh, M.Takahashi, Y.Takahashi, H.Niiyama, S.Onga, H.Okano, and K.Natori, Large area recrystallization of a multi-layered SOI and a study of 1/4 μM SOI MOSFETs, Extended Abstracts of 9th Symp. on Future Electron Devices, pp.245-250 (Nov. 1990).

20 T.Ueda, LED arrays stacked on GaAs MESFETs on Si substrates, Extended Abstracts of 9th Symp. on Future Electron Devices, pp.251-254 (Nov. 1990).

21 Y.Inoue, K.Sugahara, M.Nakaya, and T.Nishimura, Evaluation of SOI devices and the 4-layer 3-D IC with a range sensor capability, Extended Abstracts of 9th Symp. on Future Electron Devices, pp.285-289 (Nov. 1990).

22 K.Yamazaki, Y.Itoh, A.Wada, K.Morimoto, and Y.Tomita, 4-layer 3-D image processing IC, Extended Abstracts of 9th Symp. on Future Electron Devices, pp.279-284 (Nov. 1990).

23 K.Kioi, S.Shinozaki, S.Toyoyama, K.Shirakawa, K.Ohtake, and S.Tsuchimoto, Design and implementation of a four-story structured image processing sensor, Extended Abstracts of 9th Symp. on Future Electron Devices, pp.273-278 (Nov. 1990).

24 H.Ogata, S.Nakanishi, and K.Kawai, Transmission electron microscopic observation of defect structure in a multi-layer-stacked structure fabricated using solid phase epitaxy, Extended Abstracts of 9th Symp. on Future Electron Devices, pp.239-244 (Nov. 1990).

Intelligent Sensors
H. Yamasaki (Editor)

Image Processor and DSP

Teruo Yamaguchi [a]

[a] Department of Mechanical Engineering, Faculty of Engineering, Kumamoto University, 2-39-1, Kurokami, Kumamoto 860, Japan

Image processors are required in intelligent sensor technology to deal with array sensor signals. Recent improvement in microelectronics has made it possible to realize the elaborate methods of visual sensing which were once too complicated to be carried out by analog signal processing.

In this paper, typical techniques of signal processing (FFT and digital filtering) are explained, which most DSPs are designed to execute effectively. Image processors are discussed associated with signal input/output interface which is one of the most difficult obstacles to introduce to the processor for the smart image sensing. Our trial to overcome such a difficulty is also described.

1. INTRODUCTION

The biological vision system exhibits marvelous performance and its elaborate mechanism and compact structure are far beyond our understanding. Even a small insect like a fly can compute its own velocity from the visual scene and turn its flight direction to avoid collision with a wall or the ground. Human vision easily marks a better score in pattern recognition than any artificial systems have ever attained. Although the entire mechanism of the biological vision system has not yet been elucidated, many hypothetical algorithms and mechanisms for some functions of biological vision have been proposed in various fields including engineering. Most of them are quite computational and even feasible for artificial vision systems. But so far, few artificial vision systems can adopt such a visual processing scheme. This is mainly because of the difference between biological and artificial information processing paradigms. In Table 1, some of these fundamental differences are listed.

It is not always good strategy to completely imitate the biological vision mechanism to realize artificial vision without considering the differences between them. To achieve high-performance visual sensing, we must learn not only the nervous mechanism of the biological vision system, but also the information processing principle and algorithm that the system uses. They should be described in physical or mathematical words, e.g. the theory of information, signal processing or computational mathematics. From the viewpoint of engineering, it is necessary to implement the vision system by machine-oriented techniques, and moreover by combining newly developed technology. To realize an elaborate vision system like the biological one, elementary techniques for image sensing are now being developed. Fortunately, recent development in microelectronics is helpful for

Table 1
Differences between biological and artificial vision

Biological vision	Artificial vision
Hierarchical from early vision to cortex	Overemphasis to signal processors
Distributed processing	Concentrated processing
Parallel processing	Serial processing
Parallel interfacing	Serial interfacing (yet)
Continuous system in temporal domain and discrete in spatial domain	Discrete† in temporal domain and discrete in spatial domain
Space-variant sampling	Space-invariant sampling

† Frame period is 1/30 s in NTSC

such intelligent sensing, and in the not distant feature digital image processing will be realized as easily as current audio signal processing.

In this paper, we will discuss image processing and its implementation by the image processor or the digital signal processor (DSP). In Table 1, we point out that a serial and discrete signal processing scheme is inevitable for the artificial vision system. This is chiefly due to the restriction of present microelectronics. It is preferable to process an image in parallel, but here we will confine our study to the current signal processing devices and their related technology, with a longing for a better choice.

If the image processing is to be carried out by the digital and serial signal processing method, it is logical to use image processors for the following reasons:

- Microelectronics is one of the most extensive fields of scientific technology and it is profitable for intelligent sensors to adopt the outcome of it, especially from the viewpoint of miniaturization and integration of the sensor.
- According to the sampling theorem, it has been proved that the original continuous signal can be reconstructed from a sampled signal sequence under a given sampling condition. This basic theory guarantees the validity of discrete signal processing on microprocessors.
- It is possible to implement a visual computational algorithm by coding a signal processing program. It is also expected to promote sensor development efficiency.

It should be mentioned that even though many image processors and DSPs have already been developed and sold commercially, the requirement for the signal processor is often beyond their capabilities, especially with regard to image processing. Moreover, from the viewpoint of the intelligent sensor, the data-passing method from the primary sensor to the processor must still be developed. This is because the requirements of sensing and computing are different and current processors are developed solely for computing.

2. FUNDAMENTAL THEORY OF DIGITAL SIGNAL PROCESSING

2.1. Sampling theorem

Before we discuss image processing equipment, we must brush up the basic signal processing theory to support the validity of digital signal processing.

For simplicity, we will suppose that the signal is one-dimensional. A continuous signal must be sampled and quantized into a digitally discrete signal sequence so that it can be processed by digital microprocessors. The proposition in question is whether the sequence can reproduce the original signal. The sampling theorem described below guarantees that the original signal can be reconstructed from a digital signal sequence under the appropriate condition.

Sampling theorem Let the continuous signal be $x(t)$ and suppose its bandwidth is limited within 0 to W, then it is possible to reconstruct the original signal from a sequence sampled in period T where $T < \frac{1}{2W}$ (the proof is omitted here).

The original signal is reconstructed from the sampled sequence $\{x^*(nT)\}_{n=-\infty}^{\infty}$ by the following interpolation formula:

$$x(t) = \sum_{n=-\infty}^{\infty} x^*(nT) \frac{\sin \pi(n - \frac{t}{T})}{\pi(n - \frac{t}{T})}. \tag{1}$$

This theorem ensures that the digital signal can represent the continuous signal without missing any information. Nevertheless it is true only if the prerequisite of the theorem is precisely followed. For example, we can easily introduce the corollary that if the sampled sequence is finite, the reconstructed signal will contain a truncation error. Another typical erroneous case is that the upper boundary of the original signal W is greater than the Nyquist frequency $f_N(= \frac{1}{2T})$, where an unwanted artifact, called the aliasing error, is mixed into the estimated signal. Sampling frequency must be over $2W$, and in practice, it should be several times larger.

2.2. FFT and filtering

FFT Fast Fourier transform (FFT) and its associates like fast cosine transform (FCT) are among the most frequent uses of digital signal processors. FFT is considered to be an effective implementation method of the discrete Fourier transform (DFT), and due to the reduction of calculation time, it has become a popular method for estimating spectra in various applications.

DFT for the N point sampled sequence is described as follows:

$$X^*(kF) = \sum_{n=0}^{N-1} x^*(nT) W_N^{nk}, \tag{2}$$

where $W_N = e^{-\frac{2\pi j}{N}}$ and $F = \frac{1}{NT}$. The inverse transform of DFT is easily deduced from eq. (2).

$$x^*(nT) = \frac{1}{N} \sum_{k=0}^{N-1} X^*(kF) W_N^{-nk} \tag{3}$$

If eq. (2) is directly calculated, it requires N^2 multiplications. FFT enables us to obtain the spectra by about $N \log_2 N$ multiplications. This effect is particularly effective in cases of large N. Part of the basic concept of FFT is the division of the sequence $x(nT)$ shown in eq. (2) into two sequences, as indicated in eq. (4). Let the sequences be $x_0^*(nT)\,(=x^*(2nT))$ and $x_1^*(nT)\,(=x^*((2n+1)T))$, respectively. Suppose N satisfies $N=2^p$, eq. (2) can then be described as

$$\begin{aligned} X^*(kF) &= \sum_{n=0}^{N-1} x^*(nT)W_N^{nk} \\ &= \sum_{n=0}^{N/2-1} x_0^*(nT)W_N^{2nk} + W_N^k \sum_{n=0}^{N/2-1} x_1^*(nT)W_N^{2nk} \\ &= \sum_{n=0}^{N/2-1} x_0^*(nT)W_{N/2}^{nk} + W_N^k \sum_{n=0}^{N/2-1} x_1^*(nT)W_{N/2}^{nk} \\ &= X_0^*(kF) + W_N^k X_1^*(kF). \end{aligned} \tag{4}$$

By repeating this operation p times, the total number of multiplications is reduced to $N\log_2 N$, as it is shown in Figure 1. This figure illustrates an example of an FFT operation when $N=8$.

FFT is one of the most commonly used spectrum estimation methods. It should, however, be mentioned that the transformation pair of the two series are not equivalent to the original continuous Fourier transform pair. That is, discrete Fourier spectrum is not the true spectrum, but a good approximation of it. Therefore, to get a meaningful result, the sampling frequency should be several times larger than the bandwidth of the original signal.

Digital filter Another requirement for digital signal processing is filtering, including a simple finite impulse response (FIR) or an infinite impulse response (IIR) filtering and convolution. These filters are now required in two-dimensional image processing, but current image processors or DSPs are not powerful enough to execute them in real time. Some commercial video sets are equipped with digital processing devices to reproduce a high-precision video signal. Most of those signal processing chips, however, are application-specific and attain real-time speed in exchange for programmability. General purpose signal processors powerful enough to carry out arbitrary filtering are still expected. Here, we will discuss the implementation methods of filtering to realize such an effective performance.

In general, one-dimensional digital filters are described as

$$g(m) = \sum_{i=0}^{L} a(i)f(m-i) - \sum_{i=1}^{L} b(i)g(m-i), \tag{5}$$

where L is the order of the filter and $a(m)$ and $b(m)$ are coefficients. A block diagram of this filter is illustrated in Figure 2.

It is possible to implement the digital filter by directly calculating eq. (5) if the target signal is a one-dimensional speech or sound signal with a limited bandwidth of 20 kHz (i.e., sample period is about 25 μs). This is entirely due to the development of current DSPs which can execute multiplication-accumulation within a cycle time less than 100 ns. It is, however, difficult to extend the method directly to an image signal, because the cycle

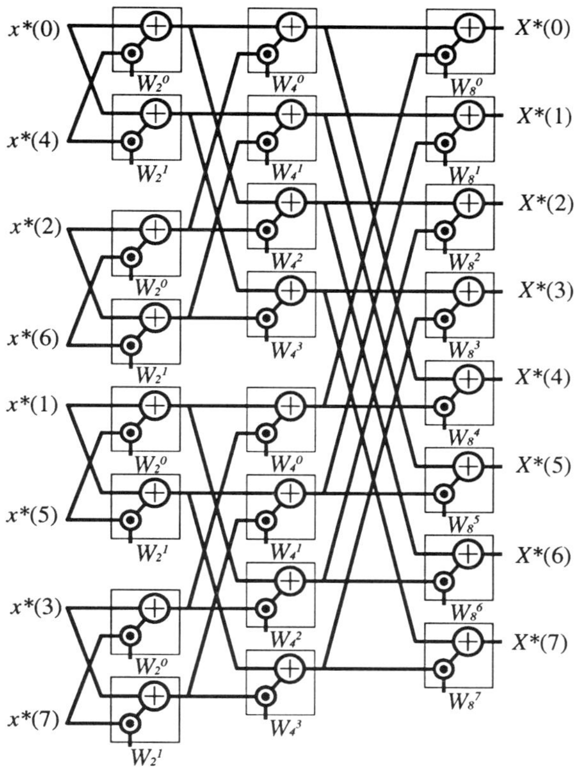

Figure 1. A diagram of FFT operation for $N = 8$ [1]. Each square block indicates a butterfly operation. Note that some coefficients have identical or opposite signs (e.g. $W_4^0 = W_8^0$ and $W_8^5 = -W_8^1$). In practice, some modifications enable more reduction of multiplicative operations.

time of the image data is as short as that of the processor. Moreover, the digital filter for the image is inevitably two-dimensional. Therefore, it is necessary to introduce some improvements to image filtering, which will include more elaborate parallel algorithms based on multiprocessor architecture. Here, we will exemplify one of these techniques [2].

If the class of two-dimensional filters in question is limited to causal IIR (or FIR) filters, it can be described as follows:

$$g(m,n) = \sum_{i=0}^{L}\sum_{k=0}^{L} a(i,k)f(m-i,n-k) - \sum_{\substack{i=0 \\ i+k>0}}^{L}\sum_{k=0}^{L} b(i,k)g(m-i,n-k) \tag{6}$$

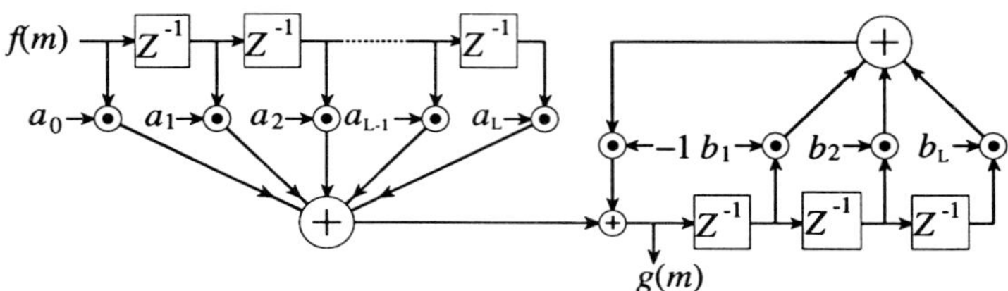

Figure 2. Block diagram of directly implemented digital filter. The two accumulators shown in the figure are most effectively implemented by $\log_2 L$ adder stages and it takes so many clock cycles to get an output. This restriction is reduced in the state-space method.

The number of multiplication-accumulation operations increases to the order of $O(L^2)$ and therefore its cycle time should be much shorter. It is impossible to carry out the operation with a single microprocessor: it is necessary to use multiprocessor-based filter implementation. One of the most promising techniques to solve this problem is the state-space description of the digital filter.

According to the state-space method, eq. (6) can be rewritten as follows:

$$\begin{aligned}
y(m,n) &= r_{L-1}(m-1,n) + q_{L-1}(m,n-1) \\
g(m,n) &= a_0 f(m,n) + y(m,n) \\
r_k(m,n) &= c(k)f(m,n) + d(k)y(m,n) + r_{k-1}(m-1,n) \quad \text{for} \quad 0 \le k \le L^2+L-1 \\
q_i(m,n) &= c_q(i)f(m,n) + d_q(i)y(m,n) + r_l(m-1,n) + q_{i-1}(m,n-1) \\
&\qquad \text{for} \quad 0 \le i \le L-1, l = L^2+L-1-iL,
\end{aligned} \tag{7}$$

where $r_k(m,n)$ and $q_i(m,n)$ are the kth state variable-related horizontal delays and ith state variable-related vertical delays, respectively ($q_{-1}(m,n-1) = 0$ and $r_l(m-1,n) = 0$, where $l = pL-1$ for $0 \le p \le L$). Parameters $a_0, c(k), c_q(i), d(k)$, and $d_q(i)$ are modified filter coefficients.

In eqs (5) and (6), the multiplier-accumulator is calculation-primitive, while in eq. (7) the calculation primitive is composed of two multipliers and three adders as shown in Figure 3. All variables that appear in eq. (7) can be calculated by this primitive. The advantages of this implementation are:

- It is possible to calculate the state variables in parallel. This means that if parallel processing based on microprocessors is available, one input data is completely processed in a few constant cycles, while direct implementation must take at least the same number of cycles as the logarithm of the filter order.
- It is possible to realize any filter order by using the same primitive composed of a few processing elements (in one-dimensional cases, the necessary of elements is 4, and in two-dimensional cases 5).

These features are favorable for image filtering.

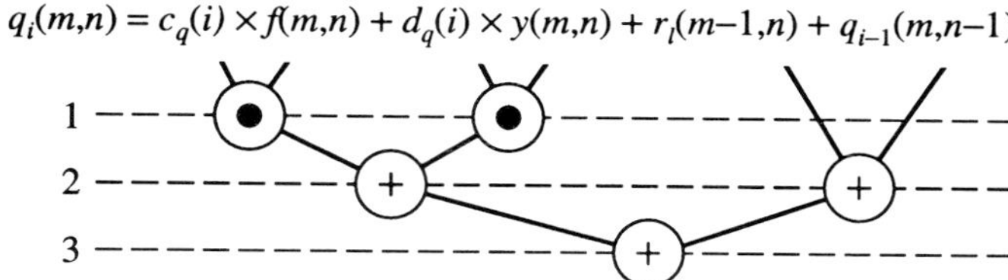

Figure 3. Tree structure for computation of a state variable in case of two-dimensional digital filters [2].

3. KEY POINTS OF SIGNAL PROCESSING BY DIGITAL PROCESSOR

3.1. Input/Output

The essential point of the design of image processors or DSPs lies in the input/output configurations, because their capacity is one of the most crucial factors directly affecting system performance. This is more crucial for intelligent sensing, because it is necessary to deal with measured data as much as possible to maintain good measurement. The input/output is, however, one of the weakest points for most processors, especially for the faster chips, including RISC microprocessors. They suffer from input/output even between memories, because the machine cycle is several times shorter than the memory cycle. Interfacing to the sensor is more difficult because of the variety of sensors, the vast amount of data to be dealt with, unpredictable and asynchronous output timing, and so on. Nevertheless, there have been few studies that have tried to overcome these difficulties, and input/output will remain the hardest obstacle for image processors used for visual intelligent sensing.

In this section, we will discuss the present state of standard interfacing techniques between sensors and microprocessors. Input/output data interfaces currently available can be classified into two categories from a viewpoint of data line wiring: serial interface and parallel interface.

Serial interface This is commonly used in many DSP chips or microprocessors for signal handling from sensors via analog to digital converters. These processors have a synchronous clock line and a pair of transmission and reception lines to control the peripheral and to receive the result status. This interface enables multiple sensor selection by using auxiliary addressing lines to select a single peripheral. In general, this is not profitable for large data transactions because the bandwidth of the line is relatively small, and there are no systems adopting this interface to transmit image data.

Parallel interface There are no standards for the intelligent sensor bus. In most cases, a dumb parallel interface is used, and general purpose standardized parallel buses such as GP-IB or SCSI are seldom used. So far, it is unusual to extend the internal computer bus to the location of the sensor. For the intelligent sensor approach, however, it is possible to extend a fast and wide bus to that location because sensors and processors are likely to be arranged closely. This method will potentially realize fast data passing from multi point devices.

It is possible to classify data interfaces from another viewpoint of data transfer operation using, for example, batch processing or pipeline processing.

Batch processing This method is characterized by two alternate processor operation modes: data acquisition mode and data processing mode. Most processors must have large overheads to switch the two modes. Moreover, if the turnaround time is longer than the video frame period, image data will be missed. This problem can be solved, for example, by using the two processors alternately, however, they will need additional control mechanisms to distribute input data.

Pipeline processing Using this method, input data are passed into a processor without switching the context. Only a processor composed of pipeline stages can get and process the data without stopping. This method usually requires an asynchronous handshake between input/output devices and a short-time data buffering via FIFO because the data rate of the sensor and processor may be different and time-variant. This method is suitable for the intelligent sensor data-passing bus for the following reasons.

- In a sensing system, measured data is transferred in one direction from sensor to processor.
- It is likely that input data from a given sensor are all processed in the same way.
- Particularly for image data, the order or vicinity of sequential data is significant for signal processing.

Moreover, this method has the following advantages.

- Bus transaction is relatively simple and it is possible for the sensor to initiate a bus transaction when it needs to pass data to the processor.
- This method is compatible with traditional digital signal calculators like multiplier-accumulators (they were formerly called "digital signal processors") by adding some hardware.

However, this method has the following disadvantages.

- Delay due to buffering may be serious in some cases.
- If the lower side (data accepter) takes more processing time, the upper side may cause data overflow.

Most of the general image processors and signal processors available are still categorized in batch processing, while some custom image processors and a few general purpose image processors are classified in pipeline processing.

Compared with the processor-to-memory connection which has been extensively developed in recent computers, the sensor-to-processor connection is seldom studied. Unless dramatic improvement is introduced, digital image processing in intelligent sensors will not be able to activate its own power. Thus, it will be more important to process the primary signal utilizing the physical or chemical characteristic of the sensor before it is converted to a digital signal. In fact, these two signal processing schemes will complement each other.

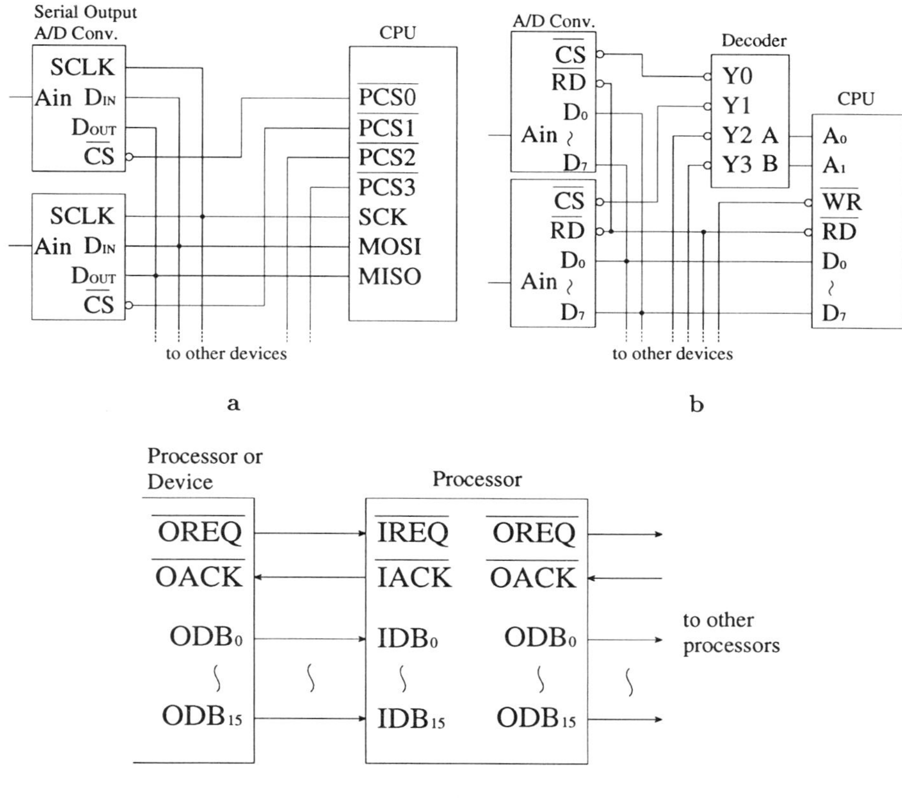

Figure 4. Interface between sensor and processor. **a** Serial bus interface. **b** Parallel bus interface. **c** Data flow interface.

3.2. Data storage and processing unit

Most of key points for high-speed signal processors are now the same as those for general purpose microprocessors. For example, a short machine cycle, extension of data width, independent instruction/data bus, and hardware floating point calculation are also effective for signal processors. In addition to these techniques, there are various techniques specific to signal processors.

- Extra multiplier-accumulators other than ALU. Multiplier-accumulators can execute primitive operations used in digital signal processing like FFT and filtering described in section 2. A set of multiplier-accumulators or a pipelined multiplier-accumulator is also preferable, especially for sound or image processing.

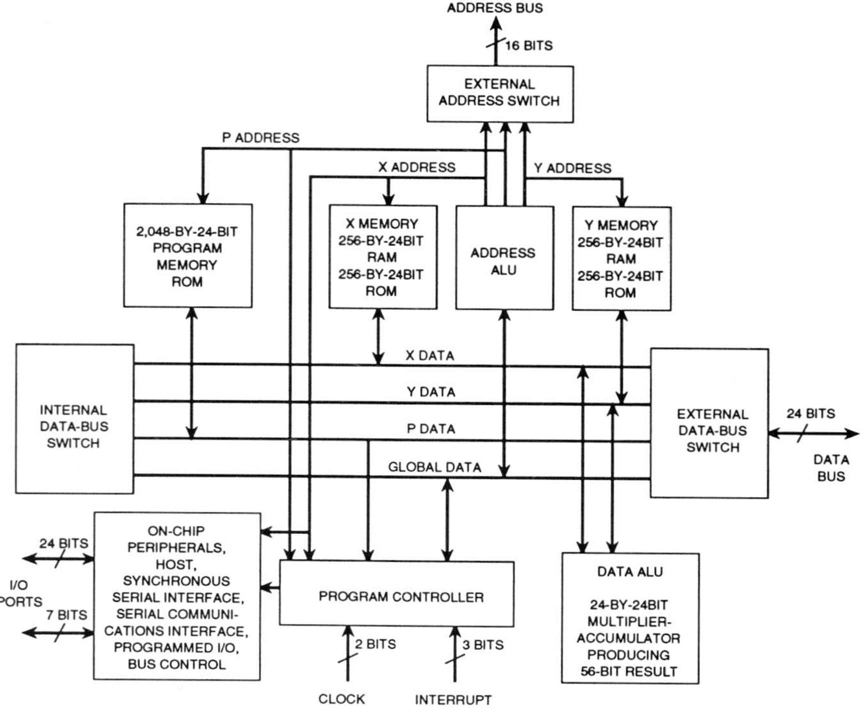

Figure 5. Internal structure of Motorola DSP56000 [3].

- Large on-chip memory. Many DSPs are equipped with several kilobytes of instruction and data memory. To process data at the highest speed, it is necessary to eliminate the overhead of external memory access. This is also profitable to miniaturize the sensing system.
- Special addressing-mode for signal processing. Most signal processors include an instruction set for the cyclic or parallel access to the memory. The cyclic access mode is convenient for reference to the coefficient table. Multiple bus set is profitable for simultaneous data input/output, e.g. to filter the data coming into the processor consecutively.

In Figure 5, the internal block of a typical digital signal processor (Motorola DSP56000) is shown. This processor has many of the features mentioned above and can complete 1024-point complex FFT in 3.7 ms. Most recent DSPs also exhibit much the same performance. It is notable that the advantage of DSP does not consist only in its processing power, but also in its software's ability to implement various processing algorithms. When a signal processor is used as a tool for sensing, this is one of the most attractive features because of its convenience to the system designer.

4. APPLICATION OF DIGITAL IMAGE PROCESSOR

4.1. Recent application of image processor and DSP

In the past few years, many new applications of DSP have appeared in consumer products, e.g. digital portable telephones and multimedia computers [5]. In these applications, however, DSPs are used for speech data encoding/decoding or moving picture compression/uncompression based on filtering or FFT. This has resulted in the development of many special chips for these purposes. Nevertheless these specific processors do not always fill the requirements of an intelligent sensor system. Such systems often require more versatile ability to deal with unexpected events and to integrate many sensors.

Future image processors or DSPs for intelligent sensing will have to have the following abilities:

(1) interfacing to sensor;
(2) real time data processing; and
(3) adaptability to various applications.

Since current generic image processors are rarely intended to cope with real-time data processing, we must design an effective input/output mechanism. Our image sensing systems intended to realize real-time image sensing will be explained below.

4.2. Intelligent visual camera

This system can directly handle a visual signal from a monolithic image sensor in its image processing unit, which consists of four image pipelined processors (ImPP). The entire setup is shown in Figure 6a. This system can calculate the location of a target in motion within the camera's field of view.

The camera head is equipped with a MOS area image sensor. The area sensor can be scanned in its rectangular subregion with regard to arbitrary position, resolution and size. This mechanism realizes the gaze without actual camera movement. It is possible to watch the location of interest without putting an extra load on the processor and data transmission channel.

The image sensor can capture two image regions at the same time. One of them is assigned to the "gazing region" mentioned above and the other to the "browsing region" to detect a moving object by scanning the whole field of view with a relatively coarse resolution. These two images are then put together with a tag field and made up into data tokens for ImPP. Data from the two channels are sent to the first ImPP processor via the FIFO. The FIFO works as a buffer between the camera and the processor.

The microcontroller in the camera head controls the complicated scanning sequence described above. The processing unit consists of four ImPPs, one MAGIC (ImPP supporting LSI), and a local image memory which is used as the frame memory to display the results on CRT. This system can detect the location of moving objects in the camera field[6] and can display two gazing and browsing fields simultaneously. By loading another program on the processor, this system can apply a smoothing filter or 2nd-order image derivative filter to the input image, for example. This flexibility is obtained by conjugating signal

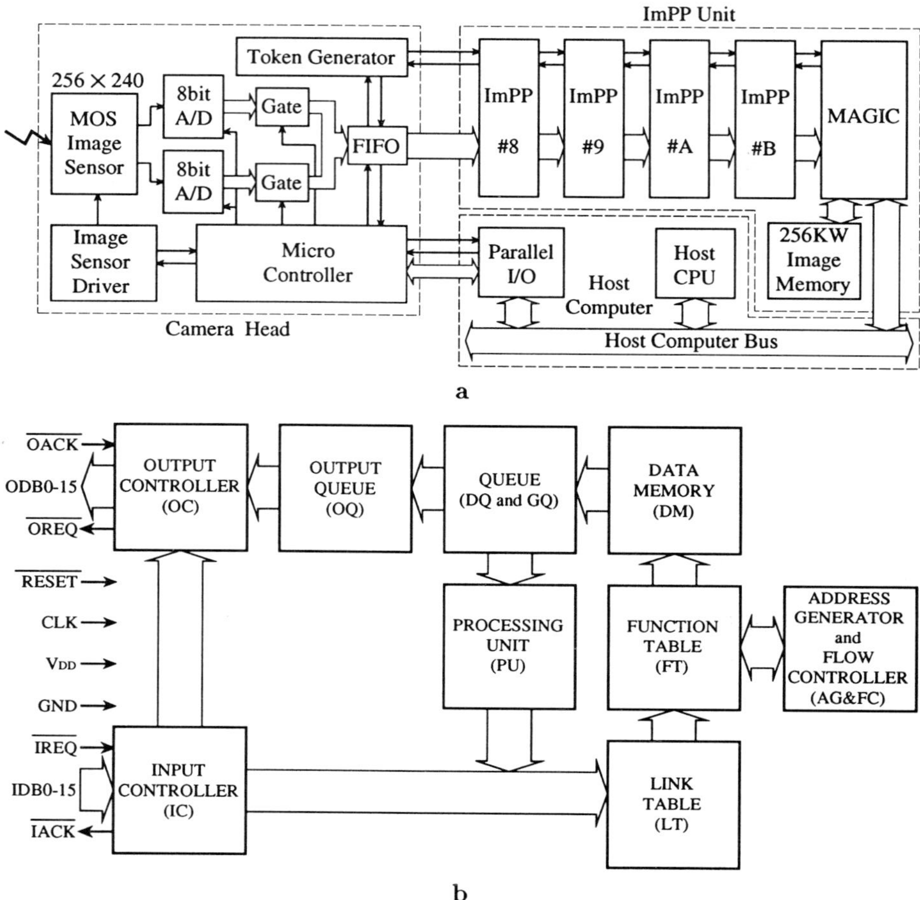

Figure 6. **a** Intelligent camera with data flow processors. **b** Internal structure of ImPP processor [4].

processors to sensing system.

The internal structure of ImPP is shown in Figure 6b. This chip is an ImPP based on data flow architecture. The main processing loop (pipeline ring) consists of a link table, a function table, AG&FC, data memory, a queue, and a processing unit (see right hand side of Figure 6b). It can acquire tagged data (token) in every pipeline cycle via the input controller and the data is then passed to the link table in the next cycle. The input data are handled by the processing unit while traveling around the pipeline ring several times.

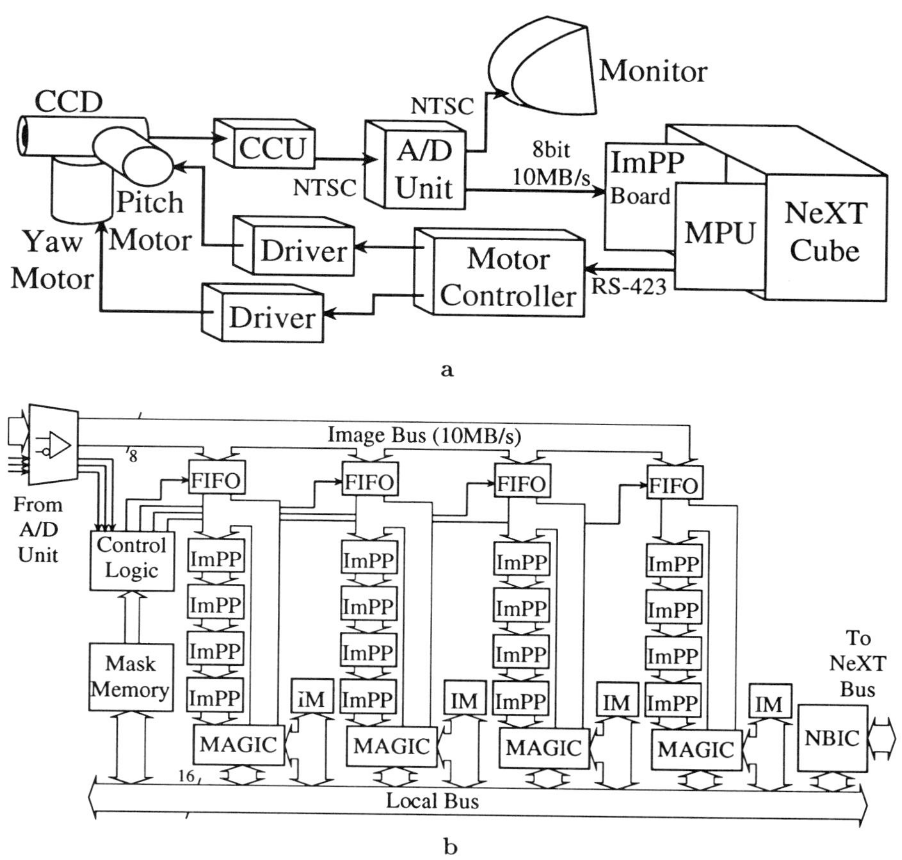

Figure 7. **a** Intelligent visual system with gaze control. **b** Internal structure of ImPP board. 16 ImPPs are built in and enables 160 MOPS image processing.

Computed results are sent via the output controller to the output bus. The output data is also a token and can therefore be directly accepted by another ImPP cascaded to the output bus (see also Figure 4c).

4.3. Image processing camera with gaze control mechanism

This system can control the gaze direction according to the result of image processing in real time. The entire system is shown in Figure 7a and the configuration of its own ImPP board is shown in Figure 7b.

The gaze direction of the camera head can be rotated both horizontally and vertically by two attached motors. The camera is a head-separated CCD black-and-white camera. Its video output is converted to 8-bit digital data and fed to the ImPP board at a rate of more than a pixel per 100 ns. The ImPP board has four processing channels, and each of them consists of four ImPPs. The combination of the destination channel of the input data is selected by a 4-bit word data stored in the mask memory. Each bit of the data corresponds with a channel and each mask data to a pixel in image. This enables arbitrary regional division and selection to the four processing channels.

For example, to realize a spatiotemporal filtering to discriminate whether the viewed object is in motion or not, temporal domain filtering is assigned to two channels and spatial domain filtering is assigned to another channel. Another example is that to calculate the velocity field from the gradient of image pattern (differentiated values of image brightness), the input image is divided into sixteen stripes (each stripe is 15 pixels wide), arranged in groups of four stripes, and sent to the four channels[7]. The filtered image or velocity distribution are then transferred to each image memory (IM), which is always accessible from both MAGIC and local bus asynchronously. The host computer can then take the filtered results via the shared image memory. The computer can instruct the motor controller to move the camera head so that it can look at the target.

5. CONCLUSIONS

Digital signal processing based on microprocessor technology is now extending its applicable range to the same image processing that biological vision systems can perform. Various digital signal processors have been developed and digital-based image processing is becoming more easy to realize. These processors are, however, less powerful than the intelligent visual sensor because they cannot realize even a low-order FIR filter in real time, while special hardware or an analog signal processing circuit would complete the task more easily. The breakthrough on this problem is quite uncertain, but real-time image processing is definitely required for visual sensing.

To use the digital signal processor on an integrated intelligent sensor, the processor must be available as a core of custom LSI chip with fast input/output peripherals. Then, the digital processor will take the place of traditional analog signal processing hardware because of its convenience.

REFERENCES

1. T. Motooka (ed.), VLSI Computer II, Iwanami, Tokyo, 1985.
2. J. H. Kim and W. E. Alexander, IEEE Trans. on Computers, **C–36**, No. 7 (1987) 876.
3. Motorola's sizzling new signal processor, Electronics, McGraw-Hill, March 10, 1986.
4. NEC Corp., μPD7281 users manual, NEC Corp., Tokyo, 1984.
5. M. Kato and S. Nakayama, Nikkei Electronics, No. 553 (1987) 111.
6. T. Yamaguchi and H. Yamasaki, Proc. 2nd Int. Sympo. Measurement and Control in Robotics, Tsukuba, Japan (1992) 257.
7. T. Yamaguchi and H. Yamasaki, Proc. IEEE Int. Conf. Multisensor Fusion and Integration for Intelligent Systems, Las Vegas, USA (1994) 639.

Intelligent Sensors
H. Yamasaki (Editor)

Biosensors for molecular identification

Masuo Aizawa

Department of Bioengineering, Tokyo Institute of Technology
Nagatsuta, Midori-ku, Yokohama, 227 Japan

Abstract

The intercellular communication depends on the receptor protein embedded on the cellular membrane surface for its selective identification of information molecules. In contrast, the molecular identification in olfaction and taste is primarily determined by the nonselective reception of molecules followed by pattern recognition in the neuron networks. Biosensors are designed to mimic these biological mechanisms; (1) biocatalytic sensors, (2) bioaffinity sensors, and (3) taste and olfactory sensors. Research and development of these biosensors for molecular identification are described.

1. MOLECULAR IDENTIFICATION MECHANISMS IN BIOLOGICAL SYSTEMS

Hormone molecules, such as insulin, are excreted from specific cell circulating throughout our body. The role of these molecules is to transfer the corresponding message from the excreting cell to a specific targetting cell. The intercellular communication is accomplished in biological systems by exchanging these information molecules. The bloodstream works as the information network for intercellular communication.

Another intercellular communication may be found in a synapse, which is a junction between neurons. The terminal of a neuron excretes a neurotransmitter molecule. Since the adjacent neuron is linked by a gap of about 3 nm, the excreted neurotransmitter diffuses to the adjacent neuron thus transferring the information.

As is clearly indicated by these examples, intercellular communication requires, in general, excretion, transportation, and reception mechanisms for the information molecule. Of these mechanisms, reception mechanisms are essentially the most important. In reception mechanisms, each specific information molecule can be selectively identified. It is a receptor protein that can identify the corresponding information molecule at a targetting cell.

Rapid progress in bioscience has given us a distinctive insight in the receptor-molecule assemblies. Several categories of receptor-molecule assemblies have been described and clarified by their structure. The first is a category of ion channel-associated receptors. The corresponding ligand molecule is recognized and bound by a specific site of on receptor, which is followed by the induction of either the opening or the closing of the ion gate. Receptors for neurotransmitters may fall in this category. The second category of receptors is comprised of G-protein-associated receptors, while the third category includes phosphokinase-associated receptors.

Although some other categories of receptors can be found, it should be kept in mind that an information molecule is recognized at a specific site on the receptor protein, and the induced conformational change of the receptor protein results in the amplified transduction of the input information by the adjacent protein-molecule assembly. These molecular identification mechanisms provide us with a design principle of biosensors for molecular identification, which will be described in the following section.

It must be noted that biological systems have quite different mechanisms for molecular identification in olfaction and taste sensing. Some oder and taste molecules are recognized by receptor cells in the nose and tongue. The molecular identification mechanism of these molecules should correspond to that of hormones and neurotransmitters. However, many receptor cells respond to a variety of molecules. These receptor cells transmit signals to the connected neuron networks. This input information may then be processed in several steps through the neuron networks and finally projected in a specific area of the brain. The brain can identify each oder and taste by pattern recognition, probably due to a sufficient learning process of the neuron network.

Current research on artificial noses and tongues has been oriented to imitate the molecular identification mechanism connected with neuron network.

2. DESIGN PRINCIPLE OF BIOSENSORS FOR MOLECULAR IDENTIFICATION

Biosensors for molecular identification are designed to mimic the molecular identification mechanism in biological systems. Therefore, as indicated before, there are two ways of designing biosensors as schematically illustrated in Fig.1. The one is to implement a receptor protein selective for a specific determinant, and the other is to integrate a nonspecific receptor and a neuron network.

2.1 Design of biosensors with specific receptors

This type of biosensor has been intensively investigated since the early 1970s. The most important part of these biosensors is the receptor for molecular recognition in which biosubstances are implemented. There are two groups of biosubstances that can be used as receptors: (1) biocatalytic substances exemplified by enzymes and (2) bioaffinity substances

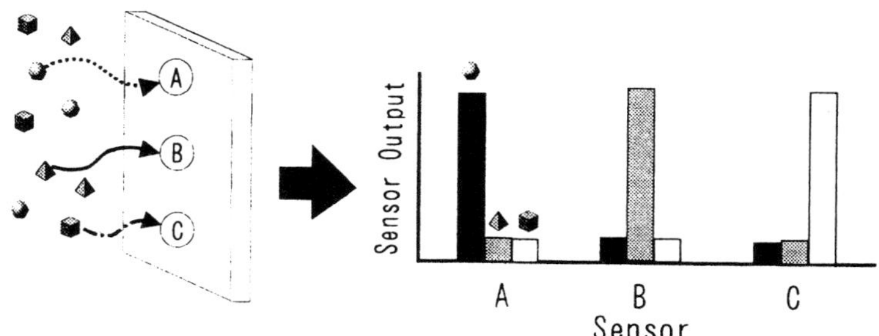

A. Sensor array specific for individual molecule

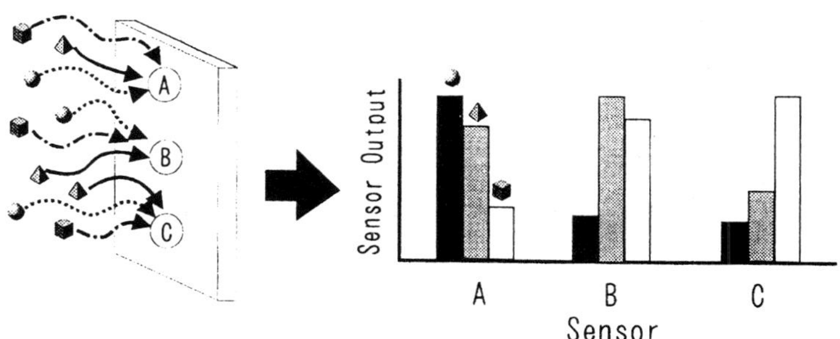

B. Sensor array specific for a group of molecules

Fig.1 Designing Principles of Biosensors for Molecular Identification

including antibody and binding proteins [1].

The biocatalytic substances can recognize the corresponding substrate molecule and form a transit complex, which is followed by an immediate dissociation to products. Therefore, the signal transduction part of biosensors should be designed in such a manner that either a transit state of complex or a product is transduced to the output signal. The signal transduction may be conducted with amperometric, potentiometric, optical, thermal, or piezo acoustic devices. This type of biosensor is most commonly constructed by immobilizing the biocatalytic substances on the surface of the signal transducing device. In the biocatalytic biosensors, the molecular recognition of a specific determinant is completed by the receptor.

The bioaffinity biosensor is designed in a similar manner as the biocatalytic biosensor. Due to their strong affinity, bioaffinity substances form a very stable complex with the corresponding determinant. The key issue is, thus, how to prohibit the nonspecific interaction of contaminants on molecular recognition.

2.2 Design of biosensors with nonspecific receptors

Nonspecific receptors can respond to a certain range of molecules. Many synthetic substances as well as biosubstances may work as nonspecific receptors. These substances, however, are required to respond in a slightly different manner by reflecting the characteristics of each species of determinant molecule. Several nonspecific receptors are usually employed to identify a specific oder or taste.

Each nonspecific receptor is conjugated with a signal transducing device. The total sensing system consists of several segments incorporating different characteristic nonspecific receptors. The output signals of these segments may not be equal to a specific molecule.

The output signals of the sensing segments are transmitted to a neuron network. After learning, the total sensing system can recognize and identify a specific molecule with the help of nonspecific recognition of the receptors and pattern recognition of the neuron networks.

3. BIOCATALYTIC BIOSENSORS FOR MOLECULAR IDENTIFICATION

A biocatalytic biosensor consists of two functional parts whose roles are molecular recognition and signal transduction. Enzymes are most commonly used for the molecular recognition part of biocatalytic biosensors. The catalytic site of an enzyme is precisely designed to recognize the corresponding substrate by specific interaction at several points, followed by a specific catalytic reaction. In biocatalytic biosensors, a determinant molecule is identified by the biocatalytic substance. The determinant molecule forms a transit state of complex with the biocatalytic substance at the active site, which results immediately in dissociation of the complex.

Molecular identification is followed by signal transduction using electrochemical, optical, thermal, or acoustic devices, because biocatalytic reactions are accompanied by either electrochemical, optical, thermal or acoustic changes. The biocatalytic biosensors are classified into (1) electrochemical, (2) optical, (3) thermal, and (4) acoustic sensors depending on their signal transduction principles.

Due to molecular identification by biocatalytic substances like enzymes, determinants of the biocatalytic biosensors are limited to enzyme substrates and some others. Although there are thousands of enzymes available, oxidases and dehydrogenases have been implemented successfully in biocatalytic biosensors. The determinants are exemplified by (1) glucose, fructose, sucrose, (2) glutamate, leucine, alanine, (3) cholesterol, neutral lipid, (4) urea, and (5) ATP.

3.1 Biocatalytic biosensors for glucose

Glucose is identified by glucose oxidase that catalyzes the oxidation of β-D-glucose and oxygen to gluconolactone and hydrogen peroxide. Glucose oxidase strictly identifies β-D-glucose from other molecules.

There are several signal transduction mechanisms for the selective detection of glucose: amperometric signal transduction from hydrogen peroxide or oxygen; amperometric signal transduction from electron transfer of glucose oxidase; and potentiometric signal transduction by a field effect transistor (FET).

(1) Amperometric signal transduction from hydrogen peroxide or oxygen. Either hydrogen peroxide or oxygen can be quantitatively determined by electrochemical oxidation or reduction, respectively, at the surface of a potential-controlled electrode. The electrode surface is commonly covered by a permselective polymer membrane to prohibit interference. Membrane-bound glucose oxidase is fixed on the permselective polymer membrane on the electrode surface. A test solution permeates through the enzyme membrane of the sensor, but not through the permselective polymer membrane. Only β-D-glucose is selectively converted into gluconolactone within the enzyme membrane, which results in an increase in hydrogen peroxide and a decrease in oxygen in the vicinity of the enzyme membrane. The changes in hydrogen peroxide and oxygen reflect an increase in the output of a hydrogen peroxide electrode and a decrease in the output of an oxygen electrode. As a result, glucose is selectively determined by measuring the current of the biocatalytic biosensor.

The amperometric type of glucose sensor was commercialized in the mid1970s. The commercialized product is, in general, installed in a bench-top glucose analyzer. The most popular application of these analyzers may be found in clinical analysis for blood glucose diagnosis and food analysis.

Although glucose oxidase has an extremely excellent selectivity for glucose, the selective determination by the glucose sensor is sometimes disturbed by blood components when it is exposed to whole blood. Several improvements have been made to protect the biosensor from contamination by whole blood. It is, however, a serious problem when using a glucose sensor for in-vivo continuous monitoring of blood glucose, primarily due to adsorption of body fluid components on the sensor surface.

(2) Amperometric signal transduction from electron transfer of glucose oxidase. In the biocatalytic reaction, glucose oxidase is reduced by the oxidation of glucose. The reduced form of glucose oxidase is regenerated to the oxidized form by transferring an electron to dissolved oxygen, which results in the generation of hydrogen peroxide. Instead of oxygen, an electrode can work as an electron acceptor. If the reduced form of glucose oxidase can transfer an electron to the electrode, glucose is determined by measuring the oxidative current from the glucose oxidase.

Due to steric hindrance of glucose oxidase, however, the direct electron transfer from glucose oxidase to an electrode is not so easy. "Molecular interfaces" have been proposed to promote the electron transfer between the enzyme and the electrode surface [2-6]. The molecular interfaces are an electron mediator, a conducting polymer, and a conducting organic salt. The molecular-interfaced glucose oxidase makes a smooth electron transfer when it is reduced by the oxidation of glucose.

The electron transfer type of glucose sensor has several advantages in its fabrication and characteristics over the hydrogen peroxide and oxygen electrode-based sensor. The advantages include the simplicity of the sensor structure and its processability in planar technology. Mass production of disposable glucose sensors has been commercially available because of these advantages.

A disposable glucose sensor can be produced by screen-printing a carbon paste on a plastic plate, followed by another screen-printing of an enzyme-containing polymer on the carbon-paste electrode. The required electron mediator may be mixed in the carbon paste. Ohmic contact formation and packaging, of course, should follow these screen-printing processes.

A credit card-sized glucose analyzer has been widely used for personal monitoring of individual blood glucose. A drop of blood is enough to determine blood glucose with this sensor chip.

(3) Potentiometric signal transduction by FET. A pH-sensitive FET can be used as a signal-transducing device for a glucose sensor because the biocatalytic reaction results in the increase in proton concentration. A glucose oxidase layer is formed on the surface of the pH-sensitive gate of the FET. Glucose identification therefore is followed by a change in the output signal of the sensor.

The buffer action of the test solution causes severe retardation of the sensor response. Careful attention should be paid to avoid such an interface.

3.2 Biocatalytic biosensors for ethanol

Ethanol is selectively identified by alcohol dehydrogenase from some specific sources. These alcohol dehydrogenases are associated with the cofactor nicotinamide adenine dinucleotide (NAD) which is not firmly fixed to the active site of the enzyme. Because of this, it has been difficult to implement alcohol dehydrogenase and NAD into a sensing device. Therefore, no biocatalytic biosensor for ethanol has been available.

Aizawa et al. has succeeded in the fabrication of an ethanol sensor using an electrochemical process [7]. A platinized platinum electrode is soaked in a solution-containing alcohol dehydrogenase, which is complexed with meldola's blue and then transferred to a solution containing pyrrole. The electrochemical polymerization of pyrrole starts with a constant potential to the electrode and stops when the thickness of the polypyrrole membrane reaches the required value. The polypyrrole membrane-bound enzyme is then covered by a gas-permeable polymer membrane.

Ethanol in either the gas or the aqueous phase permeates through the membrane and is identified by membrane-bound alcohol dehydrogenase. An electron transfers from ethanol to NAD via alcohol dehydrogenase at the active site of the enzyme. NAD becomes reduced due to the electron transfer from ethanol. Meldola's blue and polypyrrole make an electron-transfer network between the reduced form of NAD and the electrode.

Ethanol is selectively determined by measuring the current at a fixed potential. The ethanol sensor can work in both aqueous solutions and gas phases.

This biosensing principle has been applied to construct biosensors for alanine, leucine, and some others, because these determinant molecules are encountered by the NAD-assocaited enzymes.

3.3 Biocatalytic biosensors for ATP

Adenosine triphosphate (ATP) is selectively identified by ATPase and firefly luciferase. ATPase catalyzes the decomposition of ATP into adenosine diphosphate (ADP), while firefly luciferase generates luminescence in the presence of lucipherine and ATP. Because of its extremely high affinity to ATP, firefly luciferase is appropriate for the highly sensitive determination of ATP.

An ATPase-based biosensor for ATP has been constructed by coupling immobilized ATPase with a thermistor. Since heat is generated quantitatively by the biocatalytic reaction, ATP identification by ATPase is transduced by the thermistor.

Firefly luciferase, on the other hand, is coupled with a photon counter or a photodiode through an optical fiber or an optical guide.

4. BIOAFFINITY BIOSENSORS FOR MOLECULAR IDENTIFICATION

4.1 Optical immunosensors for homogeneous immunoassay

The selectivity of immunosensors may be based on the selective molecular recognition of an antibody [8]. Owing to the extremely high affinity of an antibody to the corresponding antigen, immunoassay techniques in general provide high selectivity. However, it is technically difficult to attain ultimately high sensitivity without an appropriate label. A label such as an enzyme enhances the selectivity on the basis of chemical amplification. To attain the ultimate sensitivity, various labels have been incorporated in immunosensors.

Although a label may provide an immunosensor with enhanced sensitivity, a tedious process including bound-free (B-F) separation is commonly required. Homogeneous immunoassay, in contrast, requires no B-F separation process, because the free form of the label differs in its physical characteristics from the bound form. An immunosensor for homogeneous immunoassay has thus long been expected.

Of the various types of labels, an optical label seems advantageous for realizing an immunosensor for homogeneous immunoassays. Several principles have been proposed, which can be exemplified by two prominent types of immunosensor. One of which is based on a fluorescence label. The core of an optical fiber is modified by a layer of antibody. Labelled antigen is present with free antigen to be determined in a solution. Both the labelled and free antigens react competitively with the immobilized antibody. If an excitation beam passes through an optical fiber, an evanescent wave penetrates in the vicinity of the core surface. The penetration depth of the evanescent wave is limited to the length of two or three molecules of protein. It is probable that the labels attached to the surface of the core-immobilized antibody will be excited by the evanescent wave. The bound form of the label may thus be discriminated from the free form. The development of such a type of immunosensor is expected to be completed in the not-too-distant future.

Aizawa and co-workers [9, 10] proposed an optical immunosensor for homogeneous immunoassay based on a different principle. The immunosensor consists of an optical fiber electrode in which the end of the optical fiber is coated with an optically transparent thin film of platinum. The label used for this immunoassay generates electrochemical luminescence when it reacts electrochemically on the electrode surface. Both labeled and free antigens are added to a solution containing the corresponding antibody to competitively form the immunocomplex. Only the free form of the label can generate electrochemical luminescence when an appropriate potential is applied to the electrode.

An electrochemical luminescence immunosensor has also been fabricated using an optical fiber, one end of which consisted of an optically transparent platinum electrode formed on the surface of the optical fiber by sputtering. The optical fiber electrode was connected to a photon counter and a potentiostat with a function generator.

Possible candidates for the label include luminescent substances such as pyrene, luminol and luciferin. Of these, luminol gave excellent characteristics in constructing a homogeneous immunoassay system. Electrogenerated chemiluminescence of luminol has been intensively investigated specifically around neutral pH.

Differential pulse voltammetry of luminol in a neutral solution gave a sharp anodic peak which could be ascribed to the oxidation of luminol. If the electrode potential was kept at a positive value above the peak potential in differential cyclic voltammetry and then shifted to a very negative potential, luminescence was detected when dissolved oxygen was purged. It was therefore concluded that electrochemically generated hydrogen peroxide from dissolved oxygen causes the electrochemically generated active species of luminol to emit photons.

The electrogenerated chemiluminescence of luminol was efficiently generated at a positive potential in the presence of hydrogen peroxide. Hydrogen peroxide showed excellent enhancement of luminol luminescence. On the basis of these results, electrogenerated chemiluminescence of luminol was obtained with the electrode potential maintained at +0.75 V vs. Ag/AgCl and hydrogen peroxide was added at a concentration of 2 mM. The luminescence intensity was correlated with the luminol concentration. The detection limit of luminol was 10^{-12} M.

Homogeneous immunoassay is performed with an optical immunosensor. A fixed amount of luminol-labelled antigen is added to a sample solution containing the antigen to be determined. After addition of the antibody, the solution is incubated and assayed by measuring the electrogenerated chemiluminescence of the label with the optical fiber electrode. The immunocomplexed label emits negligible luminescence, so that the total luminescence can be attributed to the free label. In other words, the immunocomplexed label is discriminated in electrochemical luminescence from the free label. For example, anti-IgG antibody can be determined in the concentration range 10^{-12}–10^{-8} M by homogeneous immunoassay based on electrogenerated chemiluminescence.

4.2 Electrochemiluminescent sensing for the characterization of DNA-interacting antitumor and antiviral agents

Tris(1,10-phenanthroline)ruthenium(II) generates electrochemiluminescence (ECL) upon electrochemical oxidation. The luminescence is remarkably enhanced when oxalate is present. Recently, it has been shown that one of the phenanthroline ligands intercalates between the base pairs of double helical DNA. The bound Ru complex stays in the major groove of DNA and emits no luminescence, because the Ru complex suffers from steric hindrance of the DNA molecule.

The authors have exploited the ECL of the Ru complex since it can work as a sensing probe for investigating the binding mode of DNA-binding antiviral or antitumor agents. The Ru complex (1 μM) and various concentrations of DNA were mixed, and the luminesecence of the mixture was measured. The luminescence of the Ru complex decreased with an increase in the amount of DNA up to 20 μM. When the Ru complex bound to DNA, electron transfer between the Ru complex and the electrode was inhibited by steric hindrance of the DNA molecule. On the other hand, a decrease in luminescence was not observed when the DNA concentration exceeded 20 μM. The saturated binding ratio between DNA base pairs and the Ru complex is approximately 20:1.

To evaluate the possibility of using the Ru complex as a sensing probe for investigating the binding mode of DNA-interacting agents, ECL of the Ru complex in the presence of DNA and these agents was studied. Cisplatin is known to bind to DNA tightly; however, the binding mode to DNA is unknown. When the cisplatin-DNA complex was used, ECL of the Ru complex increased with an increase in the amount of antitumor agent. The binding of the Ru complex to DNA was inhibited due to the tight binding of cisplatin to the major groove of DNA.

A similar increase in luminescence was observed when daunomycin was incubated with λDNA. Daumonycin may also be bound to DNA in the major groove, leaving few vacant sites for the Ru complex. On the other hand, no increase in luminescence of the Ru complex was observed when actinomycin D was incubated. This may suggest that the drug is bound to the minor groove of DNA, leaving the major groove of λDNA with vacant sites for drug binding. This results in a decrease in the free Ru complex.

The luminescence of chromomycin A3 was similar to that for actinomycin D. Since chromomycin A3 also binds to the minor groove of DNA, inhibition of the Ru complex binding to DNA did not occur.

In summary, the DNA-binding modes of the agents are classified by their intercalation to double-stranded DNA, binding to either the major or the minor groove of the double helix. The present ECL-based sensing was also shown to be feasible for the estimation of the detailed mode of binding of antiviral or antitumor agents.

4.3 Luminescent biomonitoring of benzene derivatives in the environment using recombinant *Escherichia coli*

The insertion of marker genes such as the genes for enzymes and binding proteins allows the tracking of various substances of biological importance. Although these fusion proteins have enormous implications for the study of protein engineering, the main impetus has been the potential technological benefits of genetically engineered microorganisms capable of demonstrating novel functions, such as monitoring of industrial wastes in the environment.

The introduction of luminescent enzymes enables specific microorganisms to detect industrial pollutants *in situ*, without extracting the marker enzymes. Bioluminescence-based techniques offer several advantages, such as the nondestructive detection of marked substances in sewage water and soil samples. These techniques involve the introduction of genes for the recognition of the marked substances and for luminescence generation, originally cloned from *Pseudomonas putida* and the firefly, respectively [11]. TOL plasmid carries a series of genes which is required in the digestion of benzene derivatives. In the plasmid, the gene product, xylR protein, which stimulates the transcription of subsequent genes was found by activating the promotor. We have followed the strategy of fusing the controlling genes of the TOL plasmid to the firefly luciferase gene, because we do not need the induction of all the enzymes required for the digestion of benzene derivative; only the induction of the reporter enzyme.

Luminescent monitoring of environmental pollutants by photon-emitting microorganisms may be performed with a photon-counting device, since the techniques for measuring photons have a high sensitivity and provide a linear response over several orders of magnitude. In addition, a new sensing system for the protection of the environment will be developed by immobilizing the luminescent microorganisms on the surface of a fiber optic.

Luminescence related to a series of aromatic compounds was studied by incubating *E. coli* in a broth containing the corresponding chemical. A bacterial sample with A_{660} = 1 was collected and the relative luminescence intensity of each sample was compared. The *E. coli* HB101-bearing pTSN316 exhibited greater luminescence with benzene and toluene; however, it produced less light with ethyl derivatives of toluene. The ppm level of detection was carried out by employing the plasmid-bearing *E. coli*.

The introduction of foreign genes into *E. coli* for sensing a benzene derivative with bioluminescence provides a means of environmental monitoring. In the case of methylbenzyl alcohol, a lower detection limit of several ppm of the compound could be achieved. Further improvement in sensitivity may be performed by a reduction in the levels of background light by a more sensitive photomultiplier tube. A stronger and more selective promotor may activate the genes for the recognition of a specific derivative with much higher sensitivity.

In addition, the method described here possesses several advantages over other genetically engineered methods. The firefly luciferase gene and the TOL plasmid have been fully sequenced, allowing the construction of a strongly inducible promotor. In principle, benzene, toluene, xylene, and their derivatives can be monitored by the present strategy; other pollutants such as aromatic hydrocarbons and halogenated aromatics should also be detected. Therefore, several microorganisms responsible for a variety of industrial wastes should be prepared for total monitoring of the environment.

5. BIOSENSING FOR ODORANTS AND TASTE

Perfumes are prepared by mixing several ingredients to create an enchanting oder. The mixing of these ingredients is carried out in a similar manner as a harmonized tone is synthesized from elemental tones. Smell and taste may be recognized as good and awful not only by the identification of each ingredient molecule, but also by the combination of these ingredients. It is, therefore, generally accepted that smell and taste are recognized by the molecular identification of receptor cells with the help of neuron networks.

Several attempts have been made to realize biosensors for

odorants and taste. There are two approaches in realizing smell and taste sensors. One approach is to integrate many selective sensor elements for multimolecular information. The other approach is to integrate nonselective sensor elements. Both types of sensors require information processing for simultaneous recognition of multicomponents.

We have attempted to assemble an optical taste sensor for "umami" by integrating nonselective sensing elements [12]. These sensing elements consisted of an optically transparent substrate and several layers of Langmuir Blodgett (LB) layers, each of which contains different fluorophores. Fluorescence of an LB film may be quenched or enhanced when the film comes into contact with the corresponding quencher or enhancer. Since each fluorophore fluoresces at a different characteristic wavelength, the responses of the films can be differentiated.

A solution containing 9,12-anthroyloxy stearate and stearate (molar ratio, 0.2) was spread on the water surface of a 0.25 mM $CdCL_2$ aqueous solution in a trough. Two layers of anthroyloxy stearate and stearate composite film were deposited on a nonfluorescent quartz glass plate at a constant surface pressure. Two layers of stearate film were then deposited. In addition, four layers of perylene-arachic acid film, four layers of pyrene butyrate-arachidic acid (molar ratio 0.2), and two layers of arachidic acid film were sequentially deposited at a constant surface pressure. The fluorescence of each fluorophore embedded in the LB films had a different characteristic wavelength, as was expected.

The sensor was placed in contact with a solution containing umami substances, such as ATP, adenosine monophophate (AMP), guanosine monophosphate (GMP), inosine monophosphate (IMP), and monosodium glutamate (MSG). These umami substances gave intensive quenching effects on the fluorescence of the LB films. The characteristic fluorescence quenching effects were induced by each umami substance. Therefore, umami substances can be recognized by the characteristic pattern of the fluorescence response of the sensor.

Umami was extremely enhanced when nucleotides, such as GMP and IMP, are in the presence of MSG. Such a multiplied effect is an important factor in causing foods to taste delicious. A unique characteristic was derived in response to the coexistence of GMP and MSG. These two umami substances enhanced fluorescence in the wavelength range 360-410 nm, although each of them had a quenching effect when alone. The multiplied umami effect was successfully detected on the basis of this fluorescence enhancement. The components of the umami substances were identified by the response pattern in the other wavelength region.

REFERENCES

1. M.Aizawa, Anal.Chim.Acta, 250 (1991) 349.
2. M.Aizawa, IEEE Eng.Med.Biol., Feb/March (1994) 94.
3. A.E.Cass, D.C.Francis, H.A.O.Hill, W.O.Aston, and J.Hig-

gins, Anal.Chem., 56 (1984) 667.
4. Y.Degani and A.Heller, J.Phys.Chem., 91 (1987) 6.
5. N.C.Foulds and C.R.Lowe, J.Chem.Soc., Faraday Trans.1, 82 (1986) 1259.
6. G.F.Khan, E.Kobatake, H.Shinohara, Y.Ikariyama, and M.Aizawa, Anal.Chem., 64 (1992) 1254.
7. T.Ishizuka, E.Kobatake, Y.Ikariyama, and M.Aizawa, Tech. Digest 10th Sensor Symp. (1991) 73.
8. M.Aizawa, Philos.Trans.Royal Soc.London, Ser.B. 3316 (1987) 121.
9. Y.Ikariyama, H.Kunoh, and M.Aizawa, Biochem.Biophys.Res. Commun., 128 (1985) 987.
10. M.Aizawa, M.Tanaka, Y.Ikariyama, and H.Shinohara, J.Biolu-minesce. Chemiluminesce., 4 (1989) 535.
11. Y.Ikariyama, S.Nishiguchi, E.Kobatake, M.Aizawa, M.Tsuda, and T.Nakazawa, Sensors and Actuators B, 13 (1993) 169.
12. M.Aizawa, K.Owaku, M.Matsuzawa, H.Shinohara, and Y.Ikariya ma, Thin Solid Films, 180 (1989) 227.

Intelligent Sensors
H. Yamasaki (Editor)

Adaptive Sensor System

Kazuhiko Oka[a]

[a]Faculty of Engineering, Hokkaido University, Sapporo 060, Japan

The concept and the features of the adaptive sensor system are described. Emphasis is given to the fact that sensing performance is automatically optimized in the adaptive sensor system. Some examples, including the adaptive velocity sensing system based on the spatial filtering technique, are also discussed.

1. INTRODUCTION

In recent years, various kinds of intelligent instruments have been developed in response to the rapid progress in the field of microprocessors. A lot of sensors are used in these instruments to provide information essential to control the instrument. It follows that the performance of the instrument is heavily dependent on the characteristics of the sensors.

For the stable operation of the instrument, the sensors are required to maintain their performance unaffected by the environmental conditions. One of the approaches dealing with this problem is to fabricate the sensor in a 'rigid' body. Although this approach is essential to reduce the effect of perturbations, it sometimes results in a deterioration of the sensitivity. Another approach is to use a sensor in a 'flexible' structure; the parameters of the sensor are adaptively optimized according to the changes in the environmental conditions. Such a flexible sensor is called an 'adaptive sensor.' The objective of this report is to describe the concept and to give examples of adaptive sensor systems.

2. CONCEPT OF THE ADAPTIVE SENSOR SYSTEM

2.1. One simple example – adaptive voltage sensing

In order to understand the concept of the adaptive sensing system, we first consider the simple example of a voltmeter. Recently, many types of analogue-to-digital (A/D) converters have become commercially available for use in industrial applications. A simple digital voltmeter can be fabricated using such a converter followed by a liquid crystal display (LCD), as shown in Figure 1a. The performance of this type of voltmeter is mainly determined by the A/D converter used. For example, use of a 12-bit A/D converter reduces the dynamic range to about 4000. Such a narrow dynamic range is, however, not sufficient in many applications. To enhance the dynamic range, an adjustable voltage divider is frequently employed in front of the A/D converter, as shown in Figure 1b. The sensitivity of the input voltage can be adjusted by the dividing ratio, which allows us to widen the dynamic range. Care should be taken since this configuration needs manual adjustment of the dividing ratio according to the input voltage; the ratio needs to be readjusted when the input voltage exceeds the dynamic range of the A/D converter.

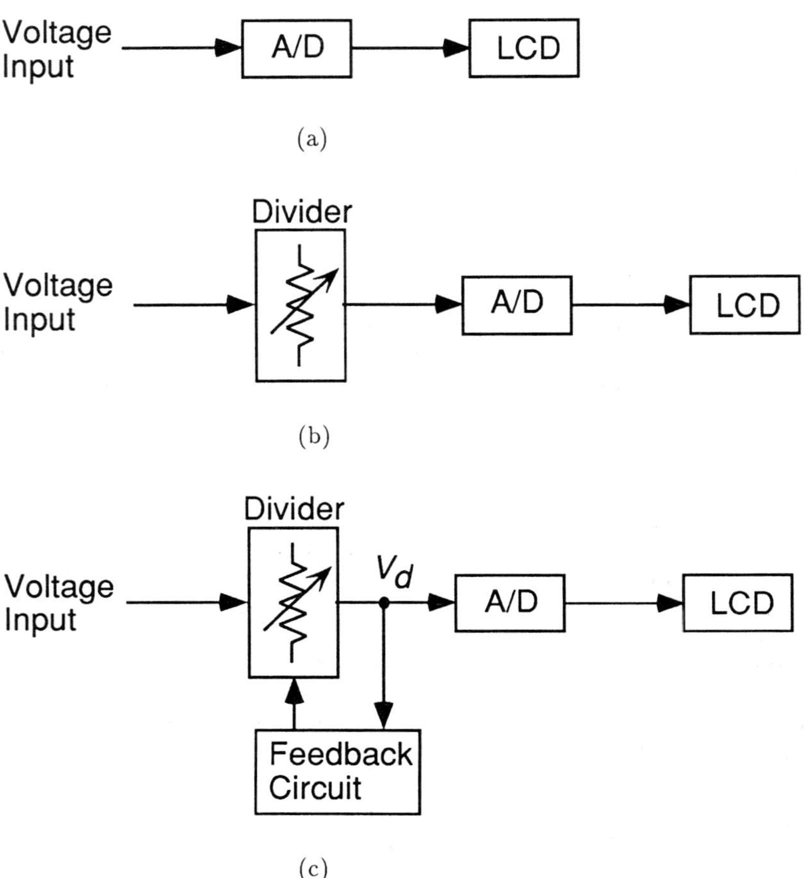

Figure 1. Voltage sensing system.

Automatic adjustment of the voltage divider can be achieved using the configuration shown in Figure 1c. The voltage output V_d from the divider is also fed into the feedback circuit. As soon as V_d exceeds the dynamic range of the A/D converter, the feedback circuit controls the dividing ratio so that V_d returns to the range of the converter. In this configuration, the dividing ratio is always kept to the optimum value. Thus, we can use this voltmeter without worrying about the dividing ratio.

The first configuration (Figure 1a) has a 'rigid' structure for the variation of the input voltage. Although this configuration is simple, adjustment of the sensitivity is impossible.

The second configuration (Figure 1b), on the other hand, is 'flexible' against the input voltage change. The incorporation of the voltage divider introduces flexibility in the sensitivity. However, the adjustment of the sensitivity must be made manually. In contrast to this, the sensitivity of the third configuration (Figure 1c) is automatically adjusted in response to the variation of the input voltage. In other words, the sensitivity is adaptively optimized to match the input voltage. Such a system is called an 'adaptive sensor system.'

2.2. Features of the adaptive sensor system

In general, the characteristics of a sensor depend on various kinds of environmental parameters including the measurands as well as the disturbances. In many cases, the performance of a sensor is restricted by problems due to the variation of these parameters. Accordingly, the adaptive adjustment of the sensor in response to changes in environmental parameters will yield many merits, such as

(1) automatic optimization of the sensitivity,
(2) enhancement of the dynamic range,
(3) enhancement of the signal selectivity,
(4) noise reduction,
and
(5) automatic tracking of the object motion.

In order to develop an adaptive sensor system, it is essential to select transducers with a flexibility for the adjustment of their parameters. Care should be taken since most of the transducers require mechanically movable elements to adjust their parameters. The incorporation of movable elements may cause many problems including vibration, misalignment, lack of stability. It should be noted that extensive study has recently been made on the development of various kinds of functional devices. Use of these devices should solve the problems.

In the adaptive sensor system, automatic adjustment of the sensor parameters is usually performed by means of a feedback loop made of analogue or digital circuits, which has the following functions:

(1) finding the mismatching of the sensor parameter,
(2) determining the optimum parameter,
and
(3) adjusting the parameter.

Auxiliary sensors are sometimes used to offer significant information about the environmental conditions. In many cases, adjustment of the parameters is simultaneously made with this sensing procedure. Hence, the parameter adjustment scheme must obey the following requirements.

(1) The sensing procedure is not interrupted by the parameter adjusting procedure.
(2) The operation of the parameter adjustment is stable and robust enough for the changes in environmental conditions.
(3) The response is sufficiently fast for real-time operation.

3. EXAMPLES OF ADAPTIVE SENSOR SYSTEMS

The study of adaptive sensors has been made in many fields. One of the most extensive developments of adaptive sensor systems has taken place for the application of the video camera. The quality of the image taken by a video camera suffers from the imperfections of the imaging optics. For example, an out-of-focus imaging lens results in a deterioration of the spatial resolution. Image saturation will occur if its intensity exceeds the dynamic range of the solid-state image sensor. In addition, vibration of the camera may blur the image. If even one of these conditions is violated, no clear image can be obtained from the video camera. In order to overcome these problems, many procedures for the automatic adjustment of imaging optics have been developed. There are several automatic focusing methods that use a feedback circuit from the obtained image to the motor-driven lens [1]. The control of the image intensity can be achieved by changing the iris diameter of the imaging lens or the integration time of the image sensor. To compensate for vibration, a variable angle prism filled with high refractive-index fluid can be used [2].

In the field of computer vision, much attention has been paid to the adaptive control of the motion of the video camera. Attention should be given to the fact that active eye motion is essential for human perception. It follows that the active motion of a video camera, such as the rotation of the camera and the variation of the imaging magnitude, causes changes in images from which useful information about the object may be obtained. In the past decade, extensive study has been conducted toward this strategy, called active vision [3–5], and considerable progress has been made to realize many functions, such as perception of the structure and tracking of the object motion.

An enormous amount of work has gone into another type of adaptive sensing scheme in astronomy, called 'adaptive optics' [6–10]. This scheme was developed to improve the resolution of a telescope. The resolution of a conventional large-aperture telescope is mainly limited by the restriction due to atmospherically induced wave-front distortion. In a telescope possessing adaptive optics, the wave-front deformation is first extracted from the light wave arriving at the telescope by a Shack-Hartmann wave-front sensor [11] or an interferometric wave-front sensor [12]. The wave-front computer then combines these measurements in such a way to yield an estimate of the wave-front errors. The deformable mirror in the telescope is driven in real-time to compensate for the estimated distortions, allowing us to obtain a diffraction-limited image from the telescope. More than ten observatories have constructed or have been constructing telescopes using adaptive optics [13–15].

4. ADAPTIVE SPATIAL FILTER FOR VELOCIMETRY

The remainder of this report is devoted to a description of a recent study on an adaptive velocity sensing system based on the spatial filtering technique. The spatial filtering technique has played an important role in the field of noncontact velocimetry [16–19]. This technique has the advantages of its simple structure and its ability for real-time operation. Nevertheless, it has the disadvantage that the detected signal is heavily dependent on the characteristics of the object pattern. With this point in mind, a novel type of adaptive spatial filtering system has been developed [20–23].

4.1. Problems of the conventional spatial filtering technique

Before describing the working principle of the adaptive spatial filtering system, we must first consider the problems of the conventional technique. Its basic scheme is illustrated in Figure 2. The illuminating light is scattered by a moving object with a random surface pattern. Using lens L_1, the scattered light forms an image of the object on the parallel slit reticle possessing a spatially periodic transmittance distribution with a period p. The light passing through the parallel slit reticle is collected by lens L_2 and generates a photocurrent $i(t)$ at the photodetector. Cartesian coordinates are taken on the parallel slit reticle in such a way that the x axis is set perpendicular to the slit direction. If $f(x, y)$ and $h(x, y)$ are, respectively, the intensity distribution of the object image and the transmittance distribution of the parallel slit reticle, then the photocurrent $i(t)$ can be written

$$i(t) = \iint f(x - Mut, y - Mvt)h(x, y)dxdy. \tag{1}$$

where u and v are the x and y components of the velocity of the moving object and M is the magnification of the imaging lens L_1. Since the parallel slit reticle has spatially periodic transmittance distribution along the x axis, the reticle works as a narrow bandpass filter, which picks up the spatial frequency component at $1/p$ for the moving image. The measurement of the center frequency results in the velocity along the x axis

$$u = \frac{p}{M} f. \tag{2}$$

In this configuration, spatial multiplication and integration are carried out simultaneously by means of the optical system. Therefore, this method is remarkable in its simple structure and rapid response. In spite of these advantages, however, this method has the following problems.

(1) No signal is detected if the characteristics frequency component is not contained in the object pattern.
(2) Since the sensitivity is primarily determined by the fixed reticle pitch p, the dynamic range is restricted by the band width of the electronic circuit.

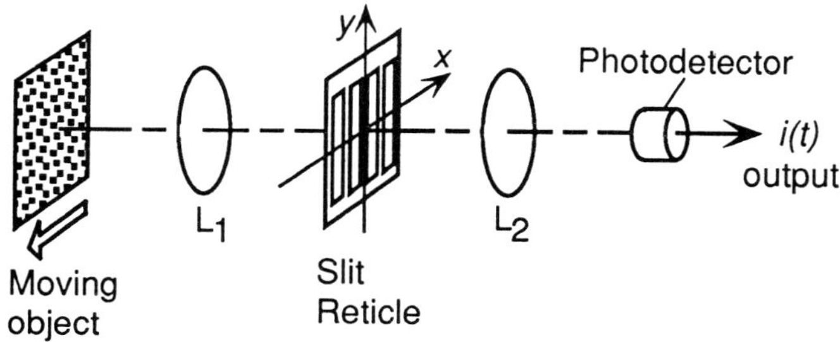

Figure 2. Schematic of the conventional spatial filtering technique.

(3) The sensor output is insensitive to the direction of the velocity.

Problem (1) originates from the fact that the spatial filter acts as a narrow bandpass filter around the characteristic frequency $1/p$. In order to solve this problem, adjustment of the reticle pitch is necessary when the characteristic component is not sufficiently contained in the object pattern. Problem (2) can also be solved by changing the reticle pitch according to the speed of the object. The direction of the object movement can be identified by translating the slit reticle. Nevertheless, the adjustment and the translation of the slit reticle of the conventional spatial filter requires mechanically movable elements, which may introduce other problems, such as misalignment of the optics and isolation of the vibration. With this point in mind, a novel spatial filtering method has been developed for the flexible operation of velocity sensing.

4.2. Flexible spatial filter built with an image sensor

The performance principle of the flexible spatial filter built with a solid-state image sensor is illustrated in Figure 3 [20,21]. The moving object is imaged on a solid-state image sensor. The obtained image is scanned to give a sequential video signal, which is further converted into digital codes by the A/D converter. This video signal is modulated by a weighting function corresponding to the parallel slit reticle and integrated for every frame interval of the image sensor by means of an accumulator and a latch. It follows that the output from the latch corresponds with the instantaneous output of the photodetector in the conventional spatial filter. In this system, the multiplication of the slit reticle and the spatial integration are carried out by the digital electronic circuit. Although the merit of the optical parallel signal processing is nullified, greater flexibility is obtained in its place. The pitch p as well as the waveform of the slit reticle can easily be tuned by changing the digitally coded weighting function.

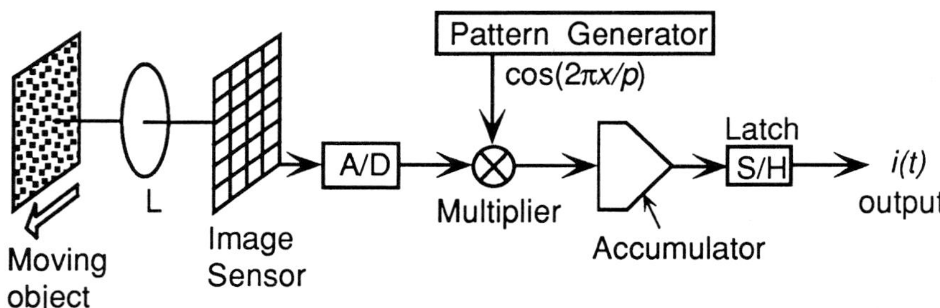

Figure 3. Schematic of the flexible spatial filter built with an image sensor.

4.3. Adaptive velocity sensing system with flexible spatial filter

The adaptive velocity sensing system is fabricated using the flexible spatial filtering method described in the preceding section. Its configuration is schematically shown in

Figure 4 [22,23]. The digitally coded sequential video signal emerging from the A/D converter is simultaneously fed into a pair of spatial filtering circuits consisting of a multiplier, an accumulator, and a latch. The sinusoidal weighting functions for these spatial filters have identical spatial periods, but are in quadrature with each other. The outputs from these spatial filters, referred to as $i_c(t)$ and $i_s(t)$, can be written

$$\begin{aligned} i_c(t) &= \iint f(x - Mut, y - Mvt)\cos(2\pi x/p)dxdy \\ &= |F(1/p)|\cos(2\pi Mut/p - \arg[F(1/p)]), \end{aligned} \tag{3}$$

$$i_s(t) = |F(1/p)|\sin(2\pi Mut/p - \arg[F(1/p)]), \tag{4}$$

where $F(\mu)$ is the one-dimensional Fourier spectrum of the object image in the spatial frequency domain and $\arg[Z]$ designates the argument of Z. The speed of the moving object can be determined from the center frequency of $i_c(t)$ or $i_s(t)$. Since $i_c(t)$ and $i_s(t)$ are in quadrature with each other, the instantaneous frequency can be obtained

$$f_i(t) = \frac{Mu(t)}{p} = \frac{i_c(t) \cdot i_s'(t) - i_s(t) \cdot i_c'(t)}{2\pi\{i_c^2(t) + i_s^2(t)\}}, \tag{5}$$

where $i_c'(t)$ and $i_s'(t)$ are the time derivatives of $i_c(t)$ and $i_s(t)$, respectively. The direction of motion is also identified by its polarity.

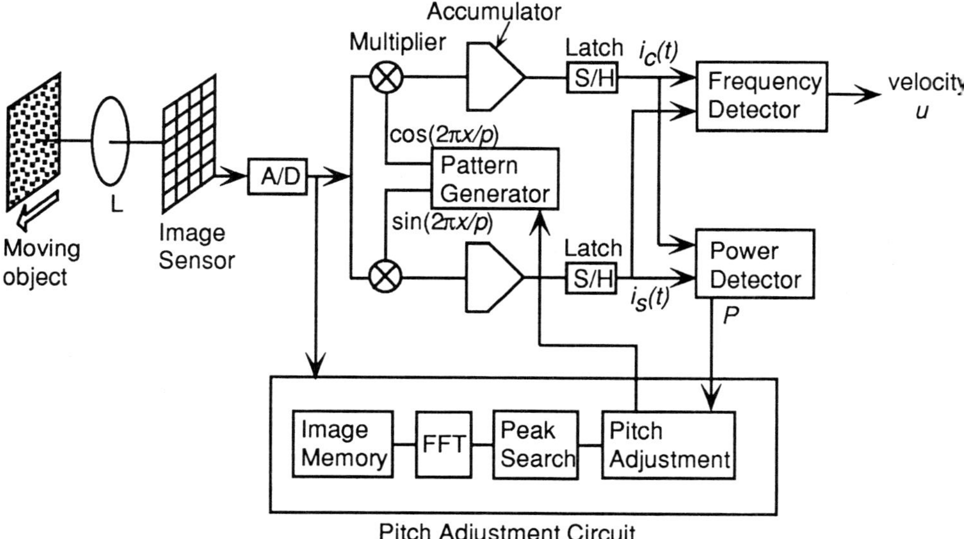

Figure 4. Configuration of the adaptive spatial filter.

We can now consider the adaptive adjustment of the reticle pitch. The power of the characteristic frequency components of the object image can be calculated from the sensor outputs as

$$P = i_c^2(t) + i_s^2(t) = |F(1/p)|^2. \tag{6}$$

The calculation of the signal power can be made simultaneously with the speed measurement. When the signal power decreases to a level that is too weak for speed measurement, it is necessary to adjust the pitch p so that sufficient signal power can be obtained. For this purpose, the video signal from the A/D converter is also connected to the pitch adjustment circuit (see Figure 4). In this circuit, the object image at a certain instant of time is stored in the image memory and its one-dimensional spatial power spectrum is computed using the fast Fourier transformation (FFT) method. The peak frequency of the spectrum is then determined to give the optimum pitch. Note that the reticle pitch should also be changed when the frequencies of $i_c(t)$ and $i_s(t)$ exceed the band width of the frequency detector.

A demonstrative experiment on the adaptive spatial filter was carried out using an MOS type of solid-state image sensor with 320 × 244 pixels. The surface pattern of the moving object and its spatial power spectrum are shown in Figure 5. In the first step of the experiment, the pitch adjustment circuit was disconnected from the spatial filtering circuit, so that the reticle could be set at an inadequate pitch. The variations of the sensor outputs $i_c(t)$ and $i_s(t)$ and the power spectrum of $i_c(t)$ obtained with the inadequate reticle pitch are shown in Figure 6a and b; the spatial frequency corresponding to the reticle pitch is indicated by (A) in Figure 5b. The variations of the sensor outputs $i_c(t)$ and $i_s(t)$ are shown by the solid lines in Figure 6a, whereas the dotted line shows their amplitude $\sqrt{P} = \sqrt{i_c^2(t) + i_s^2(t)}$. The sensor outputs have many harmonic components as

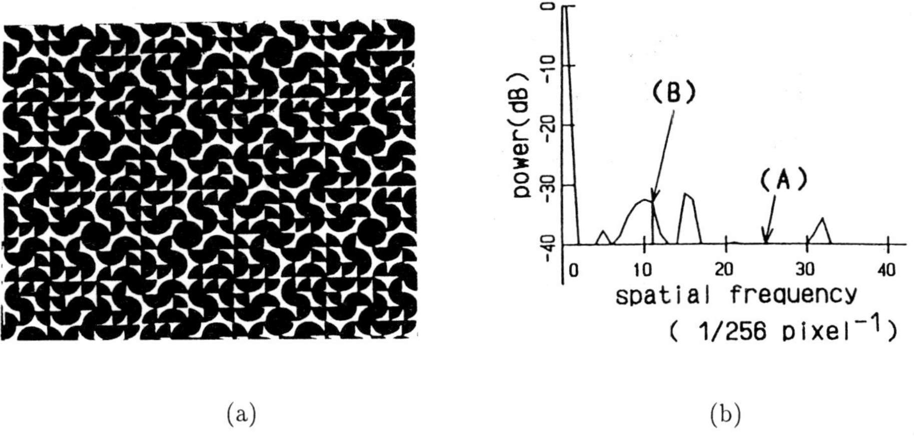

Figure 5. (a) Surface pattern of the object and (b) its spatial power spectrum.

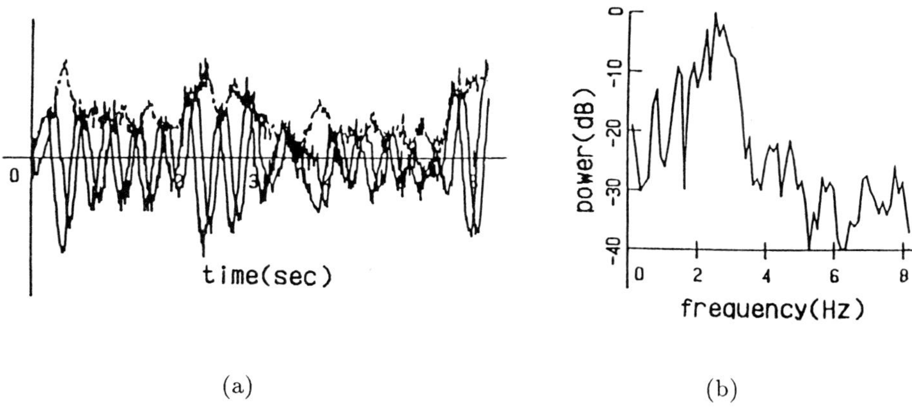

(a) (b)

Figure 6. (a) Variations in the sensor outputs $i_c(t)$ and $i_s(t)$ and (b) the power spectrum of $i_c(t)$ obtained with the inadequate reticle pitch.

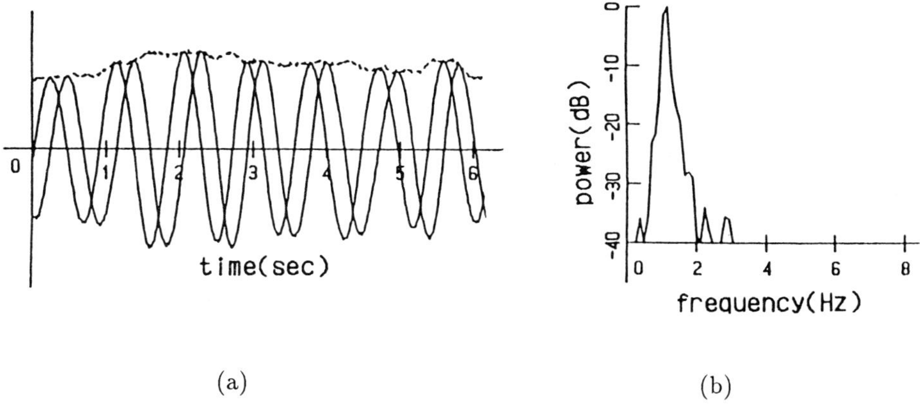

(a) (b)

Figure 7. (a) Variations in the sensor outputs $i_c(t)$ and $i_s(t)$ and (b) the power spectrum of $i_c(t)$ obtained with the optimum reticle pitch.

well as noise components because of the mismatching of the reticle pitch. In the second step, the pitch adjustment circuit was connected to the spatial filtering circuit. Then, the optimum reticle pitch was selected to extract the spatial frequency component at (B) in Figure 5b. The results obtained with the optimum reticle pitch are shown in Figure 7. As may be seen from the figure, the signal quality is successfully improved by the reticle pitch adjustment.

5. ACKNOWLEDGEMENTS

The author is grateful to Professor Hiro Yamasaki and Dr. Wataru Mitsuhashi for their fruitful discussions and many useful comments. The useful suggestions for active vision given by Professor Masatoshi Ishikawa are greatly appreciated. The author is also indebted to Professor Yoshihiro Ohtsuka for his continuous encouragement.

REFERENCES

1. K. Suzuki, Jap. J. Opt., 18 (1989) 604.
2. M. Toyama, Jap. J. Opt., 22 (1993) 101.
3. J. Aloimonos, I. Weiss, and A. Bandyopadhyay, Int. J. Comput. Vis., 1 (1988) 333.
4. A. Blake and A. Yuille, Active Vision, MIT Press, Cambridge, 1992.
5. T. Hamada, J. Soc. Instrum. Control. Eng., 32 (1993) 759.
6. H.W. Babcock, J. Opt. Soc. Am., 48 (1958) 48.
7. J.W. Hardy, Proc. IEEE, 66 (1978) 6.
8. H.W. Babcock, Science, 249 (1990) 253.
9. R.K. Tyson and J. Schulte in den Bäumen, eds., Adaptive Optics and Optical Structures, Proc. SPIE, 1271 (1990).
10. R.K. Tyson, Principles of Adaptive Optics, Academic Press, San Diego, 1991.
11. P.Y. Madec, M. Sechaud, G. Rousset, V. Michau, and J.C. Fontanella, Proc. SPIE, 1114 (1989) 43.
12. R. Roddier, Appl. Opt., 27 (1988) 1223.
13. L.A. Thompson and C.S. Gardner, Nature, 328 (1987) 229.
14. L. Goad and J. Beckers, Proc. SPIE, 1114 (1989) 73.
15. G. Rousset, J.C. Fontanella, P. Kern, P. Gigan, P. Lena, C. Boyer, P. Jagourel, J.P. Gaffard, and F. Merke, Astron. Astrophys., 230 (1990) L29.
16. J.T. Ator, J. Opt. Soc. Am., 53 (1963) 1416.
17. A. Kobayashi and M. Naito, Trans. Soc. Instrum. Control. Eng., 5 (1969) 142.
18. A. Kobayashi, J. Soc. Instrum. Control. Eng., 19 (1980) 409 and 571.
19. Y. Aizu and T. Asakura, Appl. Phys. B, 43 (1987) 209.
20. W. Mitsuhashi and H. Mochizuki, Trans. Inst. Electron. Commun. Eng., 65-C (1982) 578.
21. W. Mitsuhashi, K. Oka, and H. Yamasaki, Trans. Soc. Instrum. & Control Eng., 24 (1988) 1111.
22. H. Yamasaki, K. Oka, and W. Mitsuhashi, Digest of Tech. Papers of TRANSDUCERS '87, (1987) 137.
23. K. Oka, W. Mitsuhashi, and H. Yamasaki, Trans. Soc. Instrum. & Control Eng., 25 (1989) 271.

Intelligent Sensors
H. Yamasaki (Editor)

Micro Actuators

Hiroyuki Fujita

Institute of Industrial Science, The University of Tokyo
7-22-1 Roppongi, Minato-ku, Tokyo 106 Japan

Abstract

This paper gives a concise review of various types of microactuators fabricated by IC-based micromachining. A system architecture oriented to MEMS(micro electro-mechanical systems) is proposed. Promising fields of application are briefly overviewed. As an example, an integrated tunneling current unit which was fully fabricated by micromachining technology is described.

1.INTRODUCTION

Microactuators fabricated by IC-based micromachining[1,2] are reviewed. The microactuators are driven by various forces that are suitable at a micro level. There are two types of microactuators: one uses an electric or magnetic field for actuation and the other uses the properties of materials directly. Of the former type, a typical example is an electrostatic actuator[3,4] and, of latter, a shape-memory-alloy (SMA) actuator is representative. Table 1 shows some recent examples of IC-fabricated microactuators. Various processes for both structural and functional materials are to be developed especially for the latter type (numbers 4-6 in Table 1). However, the advantages of IC-compatible fabrication should be maintained.

Table 2 campares MEMS manufactured by micromachining and miniaturized machines made by conventional mechanical machining. Unlike miniaturized machines in which the three-dimensional structure, assembled in various shapes, is tightly associated with its function, the limitation of a typical IC-based fabrication process for MEMS only allows us to make planar structures[1,2] and micromotors[3,4], folded structures of thin poly-silicon films[5,6] or a projected image of two-dimensional mask patterns in deep resist[7,8]. Therefore it is difficult to realize various functions only by changing the shape of the machine. However full use must be made of the advantages that many structures can be obtained simultaneously by pre-assembly and batch processing and that integration with electronic circuits and sensors is possible. Various functions should be realized by using logic circuits and software in them. We can have many complicated modules in IC-based MEMS since many micromodules with sensors, actuators and electronic circuits can be made with exactly the same effort as takes to make just one module.

The friction on sliding surfaces is one of the major problems in rotational micromotors. The reason why friction cause trouble with MEMS is that the frictional force obeys an unfavorable scaling law in micro domains. When the characteristic dimension, L, is decreased, the frictional force is proportional to L^2, while the inertial force is proportional to L^3. The frictional force dominates over the inertial force and prevents micro gears or rotors from moving smoothly, if they move at all. Friction and tribology in the micro domain are under investigation. A couple of ways of reducing the friction are proposed: (1) using rolling contact rather than sliding contact between moving and stationary parts[9,14-17]; (2) supporting moving parts elastically[10] and (3) levitating moving parts[18].

Table 1
Microactuators fabricated by micromachining technology

	contact	size	speed	material	input	reference
1 electrostatic motor	sliding	60-120 μm ϕ	500 rpm	poly-Si	60-400 V	[4]
2 electrostatic motor	sliding	100 μm ϕ	15000 rpm	poly-Si	50-300 V	[9]
3 electrostatic motor	rolling	100 μm ϕ	300 rpm	poly-Si	26-105 V	[9]
4 electrostatic resonator	elastic	5x100x100μm	10-100 kHz	poly-Si	$40V_{DC}+10V_{AC}$	[10]
5 piezoelectric cantilever	elastic	8x200x1000μm	N.A.	Si+ZnO	30 V	[11]
6 shape memory alloy	(elastic)	2x30x2000μm	20 Hz	TiNi	40 V, 2 mA	[12]
7 thermal bimorph	elastic	6x100x500μm	10 Hz	Si+Au	130 mW	[13]

Table 2
Comparison between micromachined MEMS and miniaturized machines

	micromachined MEMS	miniaturized machines
assembly and adjustment	pre-assembly	part-by-part
integration of many elements	possible	difficult
combination with electronics	integration with the same process	wire connection
dimension	2.5 D	3 D

2. MEMS ORIENTED ARCHITECTURE

The limitation of the process, which was discussed earlier in this article, must be overcome by the design of micro electro-mechanical systems (MEMS), which would be completely different from the simple miniaturization of macro machines. A possible architecture oriented to MEMS is a system composed of many smart micromodules; this module has microactuator, sensors and electrical circuits integrated in itself[19]. Figure 1 is a schematic representation of such a system. As an example, a ciliary motion system and its actuators are explained. The mechanisms in the various organs of animals, insects and microscopic organisms help us to have innovative ideas. The ciliary motion is based on the motion of ciliates, which are microscopic organisms having many hairlike protrusions(cilia) on the surfaces of their cells. They accomplish locomotion by vibrating cilia cooperatively. This method of locomotion in a ciliate can be adapted in a device to convey objects. The modules of the ciliary motion system comprise an actuator (such as a cantilever-type actuator) and a oscillator. Adequate interconnection is provided and external signals can be synchronized. When the frequencies of the oscillations are synchronized and the fixed phase difference between adjacent oscillators is uniformaly maintained, each actuator runs cooperatively. Cantilevers propagate a wave and carry objects like balls. When a plate is carried, the required logic circuits are as simple as shift registers(Fig.2).

This system is a one-dimensional system and is composed of exactly similar modules. The motions of the actuators in the modules are very simple and can be easily realized by microactuators[20]. The present actuator[21] consists of a metal microheater sandwiched between two layers of polyimides which have different thermal expansion coefficients. Figure 3 shows the SEM photogragh of arrayed actuators. When the current in the heater is turned on and off, the actuator moves up and down. The dimensions of the cantilever are 500 μm in length, 110 μm in width. A vertical displacement of 130 μm and a horizontal displacement of 60 μm were obtained with 40 mA in the heater. The cut-off frequency (3 dB lower in amplitude) was 8 Hz for a sinusoidal current.

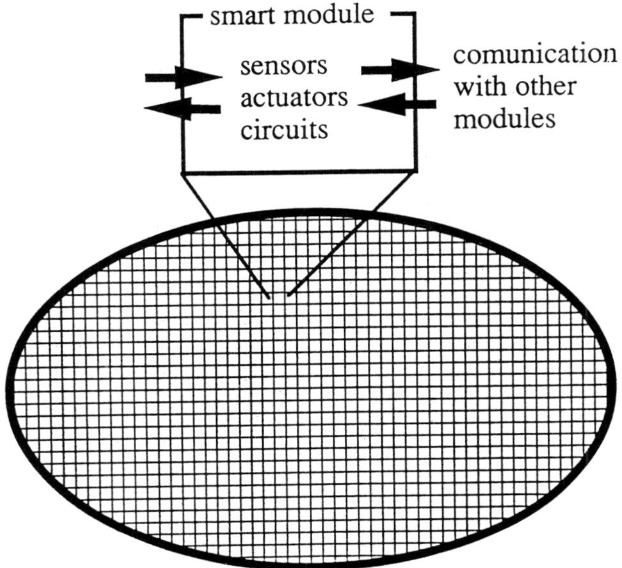

Fig.1 A schematic representation of MEMS composed of many smart modules.

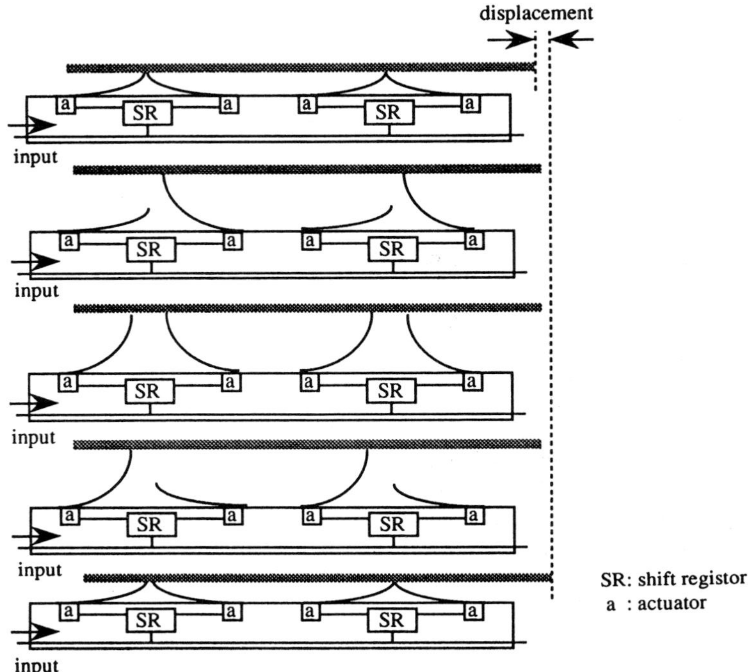

Fig.2 An example of a ciliary motion system made by silicon micromachining[19].

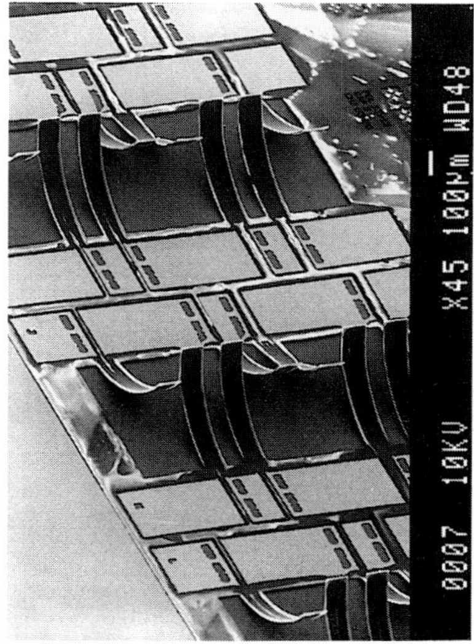

Fig.3 .SEM photograph of polyimide thermal actuators. The size is 500 μm x 110 μm. Cantilevers curl up at room temperature owing to intentionally introduced residual stress[21].

3. APPLICATIONS

Promising applications of MEMS for the near future will be in optics[22-26], magnetic and optical heads[27], fluidics[28-29], OA apparatus[30], the handling of cells and macro molecules[31,32], and microscopy with microprobes[11, 33-36] such as STMs (scanning tunneling microscopes) and AFMs (atomic force microscopes). These applications have a common feature in that only very light objects such as mirrors, heads, valves, cells and microprobes are manipulated and that little physical interaction with the external environment is necessary.

An example is a lateral tunneling unit (LTU) driven by an electrostatic linear actuator[33]. Micromachined tunneling units[11,34] and a STM (scanning tunneling microscope) [35] have been reported. Some of them require assembly and coarse adjustment of the opposing surface[11,34]. In an other case, the tip is opposed to an protrusion from the substrate[35]. The lateral configuration of our tunneling unit has the following advantages:

(1) Simple surface micromachining process with only one mask.
(2) The lateral electrostatic actuator has a large operating range and is easy to control.
(3) Integration with other microstructures such as AFM tips.
(4) Surfaces of the tip and the opposing wall can be covered by a variety of conductive materials.

The fabrication process of the LTU is a simple one-mask process. We started with a wafer covered with a 2.5 μm-thick oxide layer and a 4 μm-thick polysilicon layer. The polysilicon was patterned by RIE using a nickel mask which was vacuum coated and wet etched. The sacrificial oxide was half etched by straight HF; oxide just beneath the structure remained unresolved. After tungsten was spattered on to the surface, the structure was released by removing the oxide completely. Relatively large features were not fully undercut and still fixed to the substrate. Figure 4 shows the fabricated LTU. The tip and the comb-drive are suspended by four double folded beams. The distance between the tip and the opposing wall is determined by photolithography to within a few μm; the distance can be covered by the actuator with an applied voltage of 40-100 V. In operation the voltage was gradually increased by the control circuit until the tunneling current (0.1-100 nA) was detected. A small voltage was superimposed to keep the current the same as the reference value.

Fig.4 SEM photograph of the LTU

The result is shown in Figure 5. The upper trace is the tunneling current; the lower trace is the applied voltage which followed the reference input of a triangular wave (only the superimposed component is shown). Since this component is much smaller than the DC bias of a few tens of volts, it can be regarded as being proportional to the tip displacement. The current, I, and the distance between the tip and the wall, d, have the following relation:

$$I \sim \exp(-Kd), \qquad \text{where K is a constant.}$$

The upper trace of Figure 5 clearly shows the nonlinear dependence.

The LTU can be used as an extremely sensitive displacement sensor. One may make a very sensitive accelerometer. Another application is the detector for the AFM tip. A piezoresistive cantilever has been proposed for the AFM tip[36] but the sensitivity is still low. The LTU offers the possibility of making a very sensitive detector. Figure 6 shows a SEM photograph of the LTU fabricated together with the suspended sharp tip of the AFM. The integrated LTU/AFM tip allows us to scan the tip on the sample, while the conventional detection scheme using the light lever method only allows the sample to move. We are working towards an experimental verification of the LTU/AFM tip and the development of a process to achieve an overhung[37] AFM tip.

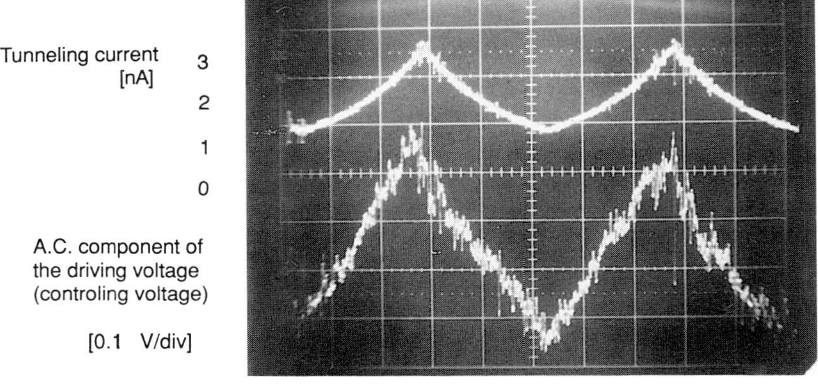

Fig.5 The tunneling current (upper trace) and the displacement of the tip (lower trace)

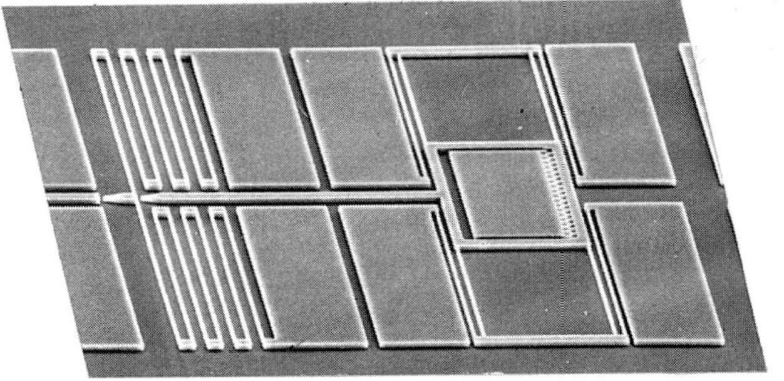

Fig.6 SEM photograph of the integrated LTU/AFM tip

CONCLUSION

Microactuators fabricated by IC-compatible micromachining have been reviewed. The feasibility of both fabrication and operation of basic devices has been demonstrated by many researchers. Further refinement based on detailed specifications will follow in a few years provided that practical applications are indicated. I hope that this article introduces this useful technology to those looking for suitable solutions to their applicatinal problems.

REFERENCES

[1] L.-S. Fan, Y.-C. Tai, and R. S. Muller,"Integrated Movable Micromechanical Structures for Sensors and Actuators" IEEE Trans. on Electron devices, ED-35, (1988) pp. 724-730.

[2] M. Meheregany, K. J. Gabriel, and W. S. Trimmer,"Integrated Fabrication of Polysilicon Micromechanisms" , IEEE Trans. on Electron devices, ED-35, (1988) pp. 719-723.

[3] L.-S. Fan, Y.-C. Tai, and R. S. Muller,"IC-processed electrostatic micromotors" Sensors & Actuators 20 (1989) pp.41-48.

[4] Y.-C. Tai and R. S. Muller,"IC-processed electrostatic synchronous micromotors" Sensors & Actuators 20 (1989) pp.49-56.

[5] K.S.J.Pister, M.W.Judy, S.R.Burgett and R.S.Fearing,"Microfabricated Hinges", Sensors & Actuators (A), 33, (1992) pp. 249-256.

[6] K.Suzuki, I.Shimoyama, H.Miura and Y.Ezura,"Creation of an Insect-Based Microrobot with an External Skelton and Elastic Joints", Proc. 5th IEEE Workshop on Micro Electro Mechanical Systems, Travemünde, Germany, February 4-7 (1992) pp. 190-195.

[7] W. Menz, W. Bacher, M. Hermening, and A. Michel,"The LIGA Technique -- a Novel Concept for Microstructures and the Combination with Si-Technologies by Injection Molding" Proc. 4th IEEE Workshop on Micro Electro Mechanical Systems, Nara, Japan, January 30-February 2 (1991) pp. 69-73.

[8] H. Guckel, K. J. Skrobis, T. R. Christenson, J. Klein, S. Han, B. Choi, and E. G. Lovell,"Fabrications of Assembled Micromechanical Components via Deep X-ray Lithography" Proc. 4th IEEE Workshop on Micro Electro Mechanical Systems, Nara, Japan, January 30-February 2 (1991) pp. 74-79.

[9] M. Meheregany, P. Nagarkar, S. D. Senturia, and J. H. Lang,"Operation of Microfabricated Harmonic and Ordinary Side-Drive Motors" Proc. 3rd. IEEE Workshop on Micro Electro Mechanical Systems, Napa Valley, CA, Feb.(1990) pp.1-8.

[10] W. C. Tang, T.-C. H. Nguyen, and R. T. Howe,"Laterally driven polysilicon resonant microstructures" Sensors & Actuators, 20 (1989) pp.25-32.

[11] S. Akamine, T. R. Albrecht M. J. Zdeblick, and C. F. Quate,"A Planar Process for Microfabrication of Integrated Scanning Tunneling Microscopes" Sensors and Actuators, A21-A23 (1990) pp. 964-970.

[12] J. A. Walker and K. J. Gabriel,"Thin-Film Processing of TiNi Shape Memory Alloy " Sensors and Actuators, A21-A23 (1990) pp.243-246.

[13] W. Riethmüller, W. Benecke, "Thermally Excited Silicon Microactuators", IEEE Trans on Electron Devices, ED-35 (1988) pp.758-763 .

[14] S. C. Jacobsen, R. H. Price, J. E. Wood, T. H. Rytting, and M. Rafaelof, "A design overview of an eccentric-motion electrostatic microactuator" Sensors & Actuators, 20 (1989) pp. 1-16.

[15] W. S. N. Trimmer and R. Jebens,"Harmonic Electrostatic Motors" Sensors & Actuators, 20 (1989) pp. 17-24.

[16] H. Fujita and A. Omodaka,"Fabrication of an Electrostatic Linear Actuator by Silicon Micromachining" IEEE Trans. on Electron devices, ED-35, (1988) pp. 731-734.

[17] M. Sakata, Y. Hatazawa, A. Omodaka, T. Kudoh, and H. Fujita,"An Electrostatic Top Motor and its Characteristics" Sensors & Actuators, A21-A23, (1990) pp. 168-172.

[18] Y.-K. Kim, M. Katsurai and H. Fujita,"A Levitation-Type Linear Synchronous Microactuator Using the Meissner Effect of High-Tc Superconductors", Sensors & Actuators A, 29(1991) pp.143-150.

[19] N. Takeshima and H. Fujita,"Design and Control of Systems with Microactuator Arrays" in Recent Advances in Motion Control (ed. by K.Ohnishi, et al.), Nikkan Kogyo Shimbun Ltd., (1990) pp. 125-130.

[20] W. Benecke and W. Riethmüller,"Applications of Sillicon-Microactuators Based on Bimorph Structrues" Proc. 2nd IEEE Workshop on Micro Electro Mechanical Systems, Salt Lake City, UT, February 20-22 (1989) pp. 116-120.

[21] N. Takeshima and H. Fujita,"Polyimide Bimorph Actuators for a Ciliary Motion System" ASME DSC-Vol. 32 (1991) pp.203-209.

[22] K. E. Petersen : Appl. Phys. Let. 31, 8, (1977) p.521

[23] R. Jebens, W. Trimmer, and J. Walker,"Microactuators for aligning optical fibers", Sensors & Actuators 20, (1989) pp. 65-73.

[24] S. Nagaoka,"Micro-Magnetic Alloy Tubes for Switching and Splicing Single-Mode Fibers" Proc. 4th IEEE Workshop on Micro Electro Mechanical Systems, Nara, Japan, January 30-February 2 (1991) pp. 86-91.

[25] J. H. Jerman, D. J. Clift, S. R. Mallinson,"A Miniature Fabry-Perot Interferometer with a Corrugated Diaphragm Support", Tech. Digest IEEE Solid-State Sensor and Actuator Workshop, Hilton Head Island, SC, June 4-7 (1990) pp. 140-144.

[26] R. Sawada, H. Tanaka, O. Ohguchi, J. Shimada, and S. Hara,"Fabrication of Active Integrated Optical Micro-Encoder" Proc. 4th IEEE Workshop on Micro Electro Mechanical Systems, Nara, Japan, January 30-February 2 (1991) pp. 233-238.

[27] M. G. Lim, et al."Design and Fabrication of a Linear Micromotor with Potential Application to Magnetic Disk File Systems" Presented at 1989 ASME Winter Annual Meeting, San Francisco, CA, Dec. (1989).

[28] S. Nakagawa, et al.,"Integrate Fluid Control Systems on a Silicon Wafer", in Micro System Technologies 90 (ed. H.Reichl), Springer-Verlag(1990), p.793.

[29] F. C. M. Van De Pol, et al.,"Micro Liquid-Handing Devices - A Review", in Micro System Technologies 90 (ed. H.Reichl), Springer-Verlag(1990), p.799.

[30] M. Shibata, et al.,"Bubble Jet Printing Elements", Proc. of 1987 Annual Meeting of IEE Japan, paper S.7-3-4(in Japanese).

[31] M. Washizu,"Electrostatic Manipulation of Biological Objects in Microfabricated Structures" in Integrated Micro motion Systems (ed. F. Harashima), Elsevier Science Publ., (1990) pp.417-432.

[32] G. Fuhr, R. Hagedorn, and T. Müller,"Linear Motion of Dielectric Particles and Living Cells in Microfabricated Structures Induced by Traveling Electric Fields" Proc. 4th IEEE Workshop on Micro Electro Mechanical Systems, Nara, Japan, January 30-February 2 (1991) pp. 259-264.

[33] D.Kobayashi, T.Hirano, T.Furuhata and H.Fujita,"An Integrated Lateral Tunneling Unit", Proc. 5th IEEE Workshop on Micro Electro Mechanical Systems, Travemünde, Germany, February 4-7 (1992) pp. 214-219.

[34] T.W. Kenny, S.B. Waltman, J.K. Reynolds and W.J. Kaiser, Appl. Phys. Lett. 58 (1991) p. 100.

[35] J.Jason Yao, S.C. Arney, and N.C. MacDonald, "Fabrication of High Frequency Two-Dimensional Nanoactuators for Scanned Probe Devices" J. Microelectromechanical Systems, 1 (1992) pp. 14-22.

[36] M.Tortonese, et al., "Atomic force microscope using a piezoresistive cantilever", Proc. of the 6th International Conference on Solid-state Sensors and Actuators, San Francisco, June 23-27 (1991) pp.448-451.

[37] C.-J. Kim, A.P.Pisano, R.S.Muller, "Silicon-Process Overhanging Microgripper", IEEE/ASME Jour. of Microelectromechanical Systems, 1 (1992) pp. 31-36.

Intelligent Sensors
H. Yamasaki (Editor)

Intelligent Three-Dimensional World Sensor with Eyes and Ears

Shigeru ANDO

Department of Mathematical Engineering and Information Physics
Faculty of Engineering, The University of Tokyo
7-3-1 Hongo, Bunkyo-ku, Tokyo 113, JAPAN

Man perceives the three-dimensional environment and the detailed shape of objects almost inattentively, and uses them for recognition and action. Knowing the mechanisms and adapting them into a technological system are the main goals of intelligent sensors. In this paper, we outline our approach to an intelligent three-dimensional vision sensor with an integrated architecture, new measurement principles, and a unified algorithm. This sensor not only forms both expanded and detailed three-dimensional sensations, but also has great similarity to man's binocular vision systems and binaural auditory systems, which work incorporatively, actively, and autonomously.

1 Introduction

Among many three-dimensional (3-D) measurement technology as shown in Table 1, many scientists are very interested in advanced research on a binocular stereo method by which completely passive and non-invasive measurements are possible by only aiming two TV cameras at the object. The binocular stereo method is also suitable for environment-free, general-purpose 3-D perception for robot vision, etc., because it requires less assumption and prior information on the object.

The best model of the binocular stereo method is the human vision system. Classical binocular stereo methods have been developed in the field of photographic measurement. Main subject of it was the automatic reconstruction of a precise 3-D profile using computers. However, this system is very different from human binocular vision system. There will be several reasons for this. We should consider, for example, many channels for 3-D information such as shape from parallaxes, shape from shadows, and shape from movement, etc. [1–4]. But the most important difference is in the sensory architecture between the artificial one and the highly integrated human one. Not only does man's binocular vision system acquire 3-D information as a single depth map [4], but it also perceives multilayered depth like a semi-transparent object, luster on the surface, material feeling (texture), etc., as shown in the next section. In addition, the human binocular vision system is integrated with many sensory motor functions for finding and pursuing an object,

Table 1. Classification and Comparison of 3-D Measurement Methods.

Method	Applications		Related articles
	3-D profile	3-D range	
Mechanical	very good	n.a.	[28],
Interferometic	excellent	n.a.	[13], [14], [15], [16]
Passive stereo (binocular vision)	applicable	good	[20], [12], [23], [22], [2], [3], [21],
Active stereo (contrived lighting)	applicable	good	[17], [18] [19]
Time-of-flight (laser, ultrasound)	n.a.	very good	[25], [26], [27], [28]
Other visual cues	n.a.	applicable	[24], [30], [29]

by which the overall external world is understood efficiently utilizing his own motion. The most important goal of our intelligent 3-D vision sensor is, therefore, to create such *an integrated sensing system* of human binocular vision, especially with *a unified architecture* of it.

2 Principle of Binocular Stereo Vision

Before describing our sensor, let us outline some properties of the stereo ranging method the human vision systems.

2.1 Two Roles: Range Perception and Shape Perception

Because of target difference, the 3-D measurement system is divided into: (1) the section that does not have to obtain a distance and position information of the object, but does require the accurate measurement of ruggedness and the detailed shape of it, and (2) the section that does not have to know the shape of an individual object, but should acquire the spatial arrangement of objects with respect to the observer. Many of the 3-D measurement devices so far correspond to the first category. Robot vision systems and a system of triangular measurement are included in the second category.

The 3-D measurement principle described above is classified and listed in Table 1 according to their suitableness for these two purposes. So far, the binocular stereo method has not been thought to turn to the measurement of highly accurate shape profile because the triangulation method is believed to be the measurement principle of distance. However, it must be kept in mind that the human binocular vision system achieves both of the above mentioned objectives of the 3-D measurements: man does not only understand the wide-ranging 3-D environment from his adjacent area to the distant circumference, but he also perceives the detailed solid profile of the object. Thus, what kind of mechanism is working in man's binocular vision system to achieve this?

2.2 Geometry of Binocular Stereo Vision

In the human binocular vision system, absolute depth and the big difference in depth are perceived by the angle between the eyeballs (vergence angle) at fusion (state that both eyes catch the same object at the center of FOV, FOV:field of view) or by the angle change during saccadic eye movement (rapid rotation of the eyeball for changing FOV). This means the perception is sequential. This movement is activated mostly by intention and occasionally by some sensory events. Small differences in depth and detailed solid profile, on the other hand, are perceived from the position disparity of the retinal images (See Figure 1 [8]). This means the perception is mostly parallel.

It will be very interesting to consider the difference in human 3-D perception ability with relating to the abovementioned differences. For instance, the absolute depth perception from the angle of the eyeball hardly exists. This can be clearly understood by the experiment of seeing one point source of light in a darkroom. You will feel as if it is located 1 or 2 meters from you, regardless of its actual distance. On the other hand, the distinction in the small depth difference by the retinal image disparity is very high. However, a relation between the amount of disparity and the scale of perceived solid profile depends on the absolute feeling of depth at which the object is located. This is understood when we see the stereogram like Figure 2: the dependency of solid profile feeling on the presentation method even if the retinal image disparity is kept constant.

Anyway, to achieve a 3-D environmental acquisition and a detailed solid profile feeling like with man's vision system, it is understood that 1) a mechanism for FOV movement and fixation of both eyes, 2) a mechanism for disparity detection of retinal images under fixation are needed. The former one may be sequential but must be highly intelligent. The latter one should be systematic, fast, and parallel.

Now, why and how does the human binocular vision system make use of perception information sources? The concept "Homeostasis of shape perception" is useful for considering this. Suppose the depth of the object is T, the vergence angle is θ, and a separation of both eyes be D, as shown in Figure 3. It will be clear that the accuracy of the obtained depth information will satisfy the relation

$$d\theta = -\frac{D}{R^2}dR. \tag{1}$$

This means that its resolution worsens as the distance coordinates increase in proportion to the second power of the depth D, and also the information capacity of the detailed shape profile decreases in proportion to D. Notice this conclusion contradicts the scale invariance of the shape perception (invariance of perceived shape to a similar transformation in three dimensions) and the homeostasis of the shape perception (the same object is perceived as having the same shape regardless of the distance) [8]. Since the scale invariance of the

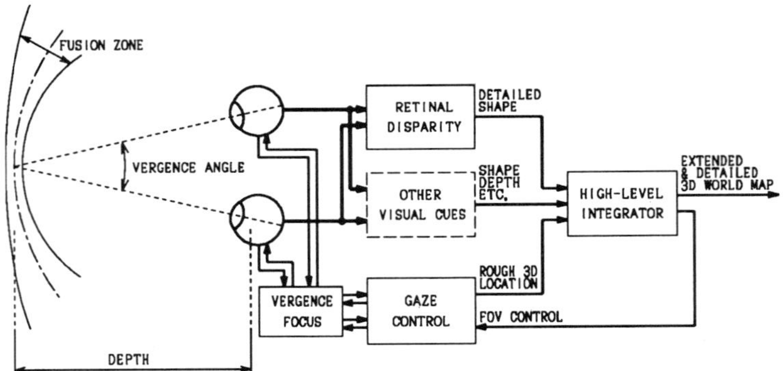

Figure 1. A simplified model of the human binocular vision system. Roughly speaking, an aspect movement mechanism by eyeball rotation and a detection mechanism of the disparity between left and right retinal images is equipped for binocular 3-D perception. A state in which the same point is seen stationary at the center of FOV of both eyes is called fixation or gaze, an angle difference between both eyes in this state is called a vergence angle, a rapid FOV movement from one fixation state to another is called a saccade, and a small vibrative eye motion under fixation is called a microsaccade or microvibration of fixation.

shape is always maintained as valid, it is natural to believe that the vergence angle is perceived according to the logarithmic scale

$$\frac{d\theta}{\theta} = d(\log \theta) = \frac{dR}{R} = d(\log R). \tag{2}$$

Then, because the range coordinate also becomes a logarithm scale, it is indeed consistent with the relative homeostasis.

Then, what kind of benefit and convenience exists besides homeostasis in such a three dimension perception framework of the logarithmically linear depth scale? The first ben-

Figure 2. A stereogram by which the existence of low-level binocular solid perception of man are recognized. This random dot stereogram (reprint from Julesz 1964, SCIENCE 145 356/362) shows that the solid feeling is perceivable only by fusing both eyes, regardless whether any solid figures are not recognized as an image itself. A square area will dash out at the center when the fusion is complete.

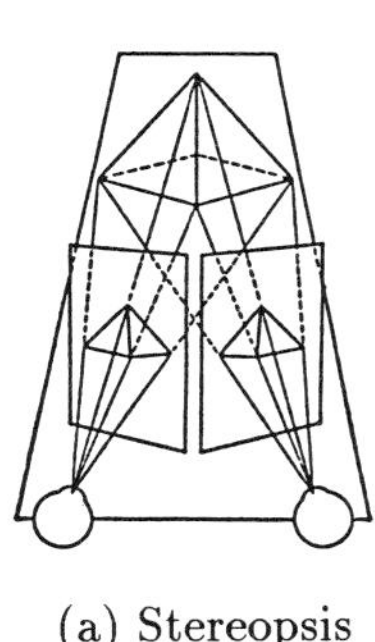

(a) Stereopsis

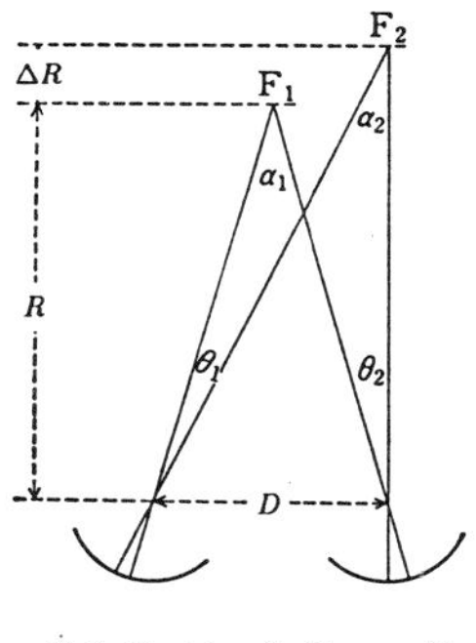

(b) Retinal disparity

Figure 3. A simplified geometry of binocular vision. A distance is detected by a vergence angle, and a detailed profile is detected by retinal disparities, but the relations between them are nonlinear, the sensitivity is quite different and position-dependent.

efit will be in the cooperation. That is, the shape detection system by the retinal image disparity is highly accurate and has patterning functions but it also has narrow rangeability. The distance detection system by the vergence angle, on the contrary, has a very wide rangeability but it is not so accurate. The second benefit is that we can grasp our exterior world in a logarithmically linear resolution distribution. With regard to the degree of urgency of approaching danger, it is natural for living things that the nearby external world is more important. The function to allocate the limited processing performance to a more important part has already been built in an early stage of the sensor. The 3-D sensing methods which use coherent waves have a contrasting character in this. Neither the "chirp" radars nor the synthetic aperture systems decrease the resolving capability of objects in the far distance. However, if we want to understand the entire external world by this, a necessary information processing performance must occur at the speed of a third power of R. The binocular vision system of man is indeed clever even in this respect.

3 Depth Perception Mechanisms: Binocular Correspondence and FOV Movement

As shown in the previous section, acquisition of a 3-D environment and detailed solid profile similar to man's vision system requires mechanisms for the FOV movement and binocular fixation of both eyes while detecting the vergence angle and its changes. An algorithmic foundation of this method is given by the stereo matching method of computer vision.

The most fundamental stereo matching system discovers a lot of correspondent image point pairs from a right and left image, and then forms a depth map using the binocular disparity (misregistration between both eye images) between them. With regard to detailed descriptions of this method, please see textbooks on the computer vision [9,10]. One basic procedure of binocular correspondence is that some feature points are determined properly in one image, and then the positions of the feature points in the other image are sought. By using the property that the correspondence point only exists only on a so-called epipolar line (a curvilinear trajectory on one image by mapping the straight line penetrating the feature point in the other image and the lens center) based on the geometric optical constraint, the correspondence search can be sped up considerably.

In the human vision system, the first half of the above processing will also be done to find the new destination of the FOV movement. That is, although the foreground processing system is paying attention to a certain point in the binocular fusion state, the background processing system is searching unconsciously for the new correspondence candidates in the marginal view of right and left images. When it becomes unnecessary to continue staring or the event which requires the concentration of attention occurs, the most important point is selected from the candidates and the FOV is moved to that point. The differences from the stereo matching method are:

1. only a single point is selected as the correspondence point,
2. it moves the view centers of both eyes to the actual point and completes the agreement between them,
3. events other than the image features take part in the dynamic selection of the correspondence candidate.

We describe first a system and an algorithm in which the FOV is controlled electronically to establish a stable binocular correspondence.

3.1 Extracting efficient feature points

The processing begins with the extraction of the feature points from both right and left images. Because of the assumption of the saccadic FOV movement system, the search must be performed anywhere far beyond the fusion zone if FOV movement is possible. In addition, disparity can hardly be expected because the feature points are usually new for the vision system. Since, the use of low-level image features such as gradients or zero crosses results in an enormous number of candidate points, it becomes impossible to perform the correspondence in real time.

Let us first show that it is very effective to use some classified features with category attributes for this situation. Let the total numbers of left and right images be N_L and N_R. Then a number of possible correspondence between all of them is $N_L N_R$. Howerver, if they are classified into M categories with numbers $N_{L1}, N_{L2}, \cdots, N_{LM}$ and $N_{R1}, N_{R2}, \cdots, N_{RM}$, respectively, and correspondence are sought between each class independently, we obtain a greatly reduced number of combinations

$$\sum_{i=1}^{M} N_{Li} N_{Ri} \ll N_L N_R, \tag{3}$$

where

$$\sum_{i=1}^{M} N_{Li} = N_L, \quad \sum_{i=1}^{M} N_{Li} = N_L. \tag{4}$$

For an example, if we could classify $N_L, N_R = 100$ features into $M = 10$ categories equally, then the number of correspondence reduces from 10000 to 1000, which is actually a 1/10 reduction.

We designed several feature extracting cells to find this type of correspondence points according to the newly developed differential geometric operations. Here, we describe the method very briefly. Let the brightness be f and its spatial gradients be f_x, f_y. Suppose we measure these quantities in some small spatiotemporal observing window, say $X \times Y$ pixel wide region $[X \times Y]$. Then, how are they distributed in a 3-D space (f, f_x, f_y)? Our method is based on the fact that the degree of degeneration is strictly related to a functional property of the brightness pattern in $[X \times Y]$ if both X and Y are small. To describe this more concretely, we must first construct a sample covariance matrix

$$\begin{aligned} \mathbf{S}_{.XY} &\equiv \iint_{[X \times Y]} \begin{bmatrix} f \\ f_x \\ f_y \end{bmatrix} \begin{bmatrix} f & f_x & f_y \end{bmatrix} dxdydt \\ &= \begin{bmatrix} S & S_x & S_y \\ S_x & S_{xx} & S_{xy} \\ S_y & S_{xy} & S_{yy} \end{bmatrix}. \end{aligned} \tag{5}$$

Applying some algebraic operations to the elements, such as

$$P_{EG} \equiv \frac{(S_{xx} - S_{yy})^2 + 4S_{xy}^2}{(S_{xx} + S_{yy})^2} \qquad \text{(Any directional edge)} \tag{6}$$

to extract edge features, and

$$P_{HP} \equiv \frac{\det[\mathbf{S}_{.XY}]}{S(S_{xx}S_{yy} - S_{xy}^2)} \qquad \text{(Heterogeneus peak)} \tag{7}$$

to extract blob features by discriminating its degree of degeneration while removing noise etc [39–41]. From the outputs of these cells, the brightness pattern in a small area $[X \times Y]$ is categorized into edge classes, ridge classes, and blob classes. For each category we obtain some quantities which describes a degree of membership to it, orientation attributes, etc. Outputs of several cells to some brightness ridge and peak patterns are shown in Figure 4. The meanings and some additional attributes of the most important three classes are as follows:

1. Edge feature and its strength and orientation.

2. Ridge feature and its sign, strength, and orientation.

3. Blob (point, peak, saddle, etc.) feature and its sign, strength, and orientation.

The correspondence is established using the abovementioned pattern 3 (blobs) of a right and left image as primary candidates. Evaluating and ranking the degrees of vertical misregistration, horizontal disparity, and parametric agreement between attributes based on the criterion, we can determine an optimum correspondence very stably. If there is no correspondence by which enough of the criterion is obtained, the search is repeated using 1 (edges) and then 2 (ridges) of the abovementioned attributes as secondary candidates.

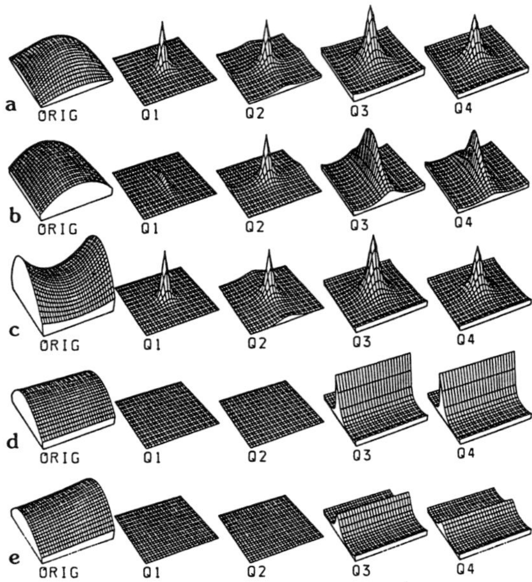

Figure 4. Feature extraction using differential geometric categorization of local brightness pattern. The upper row is the typical brightness patterns which are displayed by taking brightness as a vertical axis. The lower rows show the output of the categorized feature extraction operator to each brightness pattern. Not only the difference of the reaction but also the sharpness of the detection response will be clear.

3.1.1 Electronic FOV movement

Another distinct characteristic of our system is in the movement of the center of the FOV to a point of best correspondence with the result of the optimum search. Usual methods for the movement of view are the rotation of each TV camera [35]. There was a problem, however, of generating a different distortion between right and left images by the movement. In the method described here, the time difference is given between right and left synchronous signals of TV cameras, as shown in Figure 5, so that the picture positions on the frame memory suffer an equivalent horizontal shift. Such a composition is used even in a recent hand-swinging-prevention video cameras. In the FOV movement, no different distortion is caused as long as the optics system can be regarded as the pinhole camera. The fault of this composition, however, is that the movement of FOV is limited within a considerably narrow range. Therefore, the range must be extended by some other mechanical means such as rotation of the mounted head.

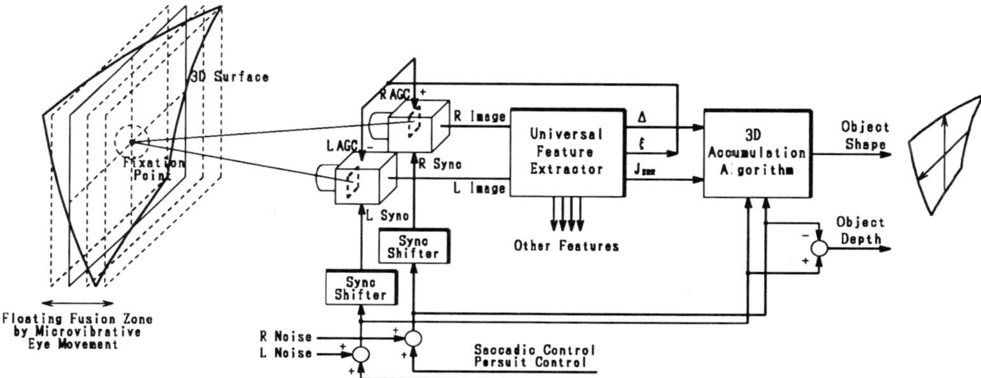

Figure 5. A block diagram of an electronic FOV movement and a binocular fixation mechanism. Correspondence is taken between the features of both eyes category by category, and the amount of FOV movement of each eye to the point with the highest degree of correspondence and the highest degree of significance (interest) is sent to a synchronous signal shifting circuit. As a result, the center of FOV of right and left eyes moves to the correspondence point, and rough correspondence is achieved there between both eyes. The amount of small shift in the center of FOV is fed back to the shift circuit of synchronous signals, so that sight is locked to the object, and the system enters into the state of binocular fixation (gaze).

Several merits achieved by the addition of FOV movement capability in a sensor stage are as follows:

1. Subpixel level movement and correspondence of left and right FOV is possible.

2. Modulation or spatial-to-temporal conversion of pixel data using vibrative movement of FOV.

3. Randomization of the relation between a pixel and an object point (Removal of sampling artifacts and pixel noise).

4. Vast network for image shift becomes unnecessary when parallel network image processing is used.

5. It is suitable for space-variant processing hardware, e.g. dual processors for central and marginal FOVs or processing in the complex logarithmic coordinate.

On the other hand, not much attention is paid to the abovementioned advantages because the movement of view is performed only by moving the pointer of the image array in the sequential image processing on the computer. However, in the real-time and parallel binocular vision systems like the human vision system, we can easily think about a very significant use. Actually, this system supplies the observed image data to the differential binocular vision system described in the next section. Using this construction, the disparities between left and right images are kept enclosed in the fusion zone, so the reconstruction of the profile is assured in the differential binocular vision system. Another function of this system is to give small FOV vibrations like microsaccadic movements of human eye. This enables the differential binocular vision system to improve the accuracy of the solid profile.

4 Changing Mechanisms of Visual Attention

In this section, we consider some methods to determine where the correspondence should be established. The autonomous sensor must decide it himself and concentrate its limited sensing power on it.

4.1 Auditory Localization

The quad-aural 3-D sound source localization sensor introduced here as an auditory sensor is based on the same principle as the differential binocular vision [31], the binocular luster extraction [33] and the active incremental solid profile acquisition [35], described in the next section.

4.1.1 Principles

The sensor consists of four adjacent microphones on a square, as shown in Figure 6. They measure a set of spatiotemporal differentiations f, f_t, f_x, f_y of the sound pressure field imaginarily at its center by the difference and differentiation of the microphones. Then, what information is included in the spatiotemporal gradients observed at a single point?

Let a Cartesian coordinate with an origin at the measuring point be x, y, z. In the front half space $z > 0$, there are multiple uncorrelated sound sources S_i $(i = 1, 2, \cdots)$ whose position is (x_i, y_i, z_i). Let the sound velocity be C, the distance from a sound source S_i to a position (x, y, z) of each sensor be $R_i = \sqrt{(x - x_i)^2 + (y - y_i)^2 + (z - z_i)^2}$, the source sound be $g^i(t)$, and the sound pressure generated by each source be $f^i(t)$. Then, the total pressure f is expressed by the sum of the spherical wave emitted from the sources as

$$f = \sum_i f^i(t) = \sum_i \frac{1}{R_i} g^i(t - \frac{R_i}{C}). \tag{8}$$

The x, y and t gradients of the sound field at the measuring point can then be written as

$$f_x = -\sum_i \{\xi_x^i f^i(t) + \tau_x^i f_t^i(t)\} \tag{9}$$

$$f_y = -\sum_i \{\xi_y^i f^i(t) + \tau_y^i f_t^i(t)\} \tag{10}$$

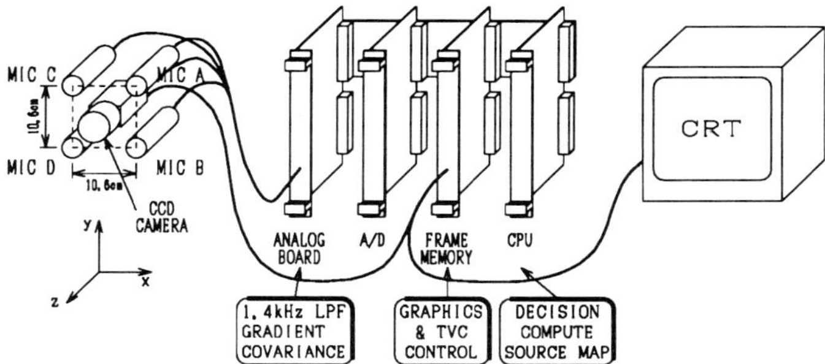

Figure 6. A block diagram of the quad-aural 3-D sound source localization sensor based on the spatiotemporal gradient method. Amplitude, spatial gradients, and a temporal gradient of the sound pressure field are measured with four microphones. From them, an instantaneous covariance matrix is formed by an analog circuit. The categorization and location information of multiple sound sources are calculated from this matrix using generalized correlation coefficients and the least squares method in a computer.

$$f_t = \sum_i f_t^i(t) \tag{11}$$

using an x, y directional phase (time) gradient τ_x^i, τ_y^i and an amplitude gradient ξ_x^i, ξ_y^i, where $f_t^i(t)$ is the time derivative of $f^i(t)$.

It is then understand that the trajectories of this vector (multidimensional "Lissajous figure") momentarily show some degeneration. From $(f_x, f_y, f_t, f)^T$ in a short interval $[t-T, t]$, we can obtain instantaneously the following sound source arrangement information. According to the number of sound sources, we can summarize it as follows:

1. The azimuth and distance of one sound source — When only a single sound source is active, the trajectory becomes two dimensional. In this case, we can localize the 3-D position of the source.

2. The azimuth and distance of the curve which passes two sound sources — If two sound sources are located at an equal distance, or if they are in the vertical plane which contains the measurement point, the trajectory becomes 3-D, and it can be judged that they are lying on an intersection with a plane parallel to the axis and a sphere which contains two sound sources. In this case, obtainable localization information is the intersection line.

To describe this more concretely, the covariance matrix is constructed using each differential signal within a short temporal observation window $[t-T, t]$ (The actual used window in the following experiments is an exponentially decaying window with a 5 ms

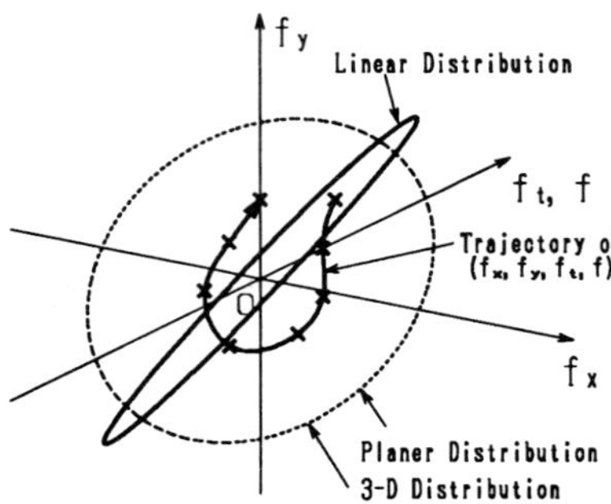

(a) Time locus of $(f_x, f_y, f_t, f)^T$.

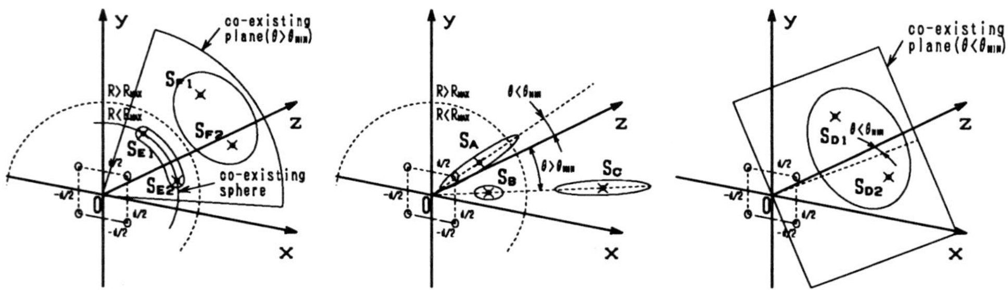

(b) Three distribution types of sound sources.

Figure 7. Geometrical relations between the sound source arrangement and spatial differentiation of sound field. (a) Trajectories of a measurand $(f_x, f_y, f_t, f)^T$ in a gradient space. (b) Correspondent distribution of sound sources according to the degree of degeneration.

time constant) as follows:

$$\begin{aligned} \mathbf{S}_{XYT} &= \int_{t-T}^{t} \begin{bmatrix} f_x \\ f_y \\ f_t \\ f \end{bmatrix} \begin{bmatrix} f_x & f_y & f_t & f \end{bmatrix} dt \\ &= \begin{bmatrix} S_{xx} & S_{xy} & S_{xt} & S_x \\ S_{xy} & S_{yy} & S_{yt} & S_y \\ S_{xt} & S_{yt} & S_{tt} & S_t \\ S_x & S_y & S_t & S \end{bmatrix}. \end{aligned} \tag{12}$$

To compute this matrix, an electronic circuit is easily applicable as the instantaneous correlator array among all the differential signals. From these outputs, we can compute generalized correlation coefficients such as

$$Q_{1S} = \frac{(S_{xt}S_y - S_xS_{yt})^2}{(S_{xx}S_{yy} - S_{xy}^2)(S_{tt}S - S_t^2)} \qquad \text{(Single source)} \tag{13}$$

and judge the degree of degeneration by thresholding it. For each judgement, we can obtain localization information such as

$$\tau_x = \frac{S_tS_{xt} + S_{tt}S_x}{S_{tt}S - S_t^2}, \quad \tau_y = \frac{S_tS_{yt} + S_{tt}S_y}{S_{tt}S - S_t^2} \qquad \text{(Interaural temporal disparity)} \tag{14}$$

$$\xi_x = \frac{S_{ff}S_x - S_tS_x}{S_{tt}S - S_t^2}, \quad \xi_y = \frac{S_{ff}S_y - S_tS_y}{S_{tt}S - S_t^2} \qquad \text{(Interaural intensity disparity)} \tag{15}$$

For a detail discussion of this algorithm, see [46].

4.1.2 Preliminary experiments

Figure 8 shows the experimental example confirming the performance of this system in the real environment. The result for a single source is shown in (a). In it, the results of the abovementioned judgement and parameter extraction are displayed by a mark '+' near the speaker's mouth. In case of double sources, a higher order model is selected automatically, as shown in (c), and a corresponding measureable parameter, i.e., a co-existing plane of the two sources, is displayed as a straight line which connects the mouths of the two speakers.

4.2 Motion and Accretion Detection

As is well known, the existent or nonexistent correlations between f_t and f_x and f_y supplies us the motion information in the visual fields [6]. We show a method to use this information for the visual attention change.

From Euler's relation [5] in the moving image field (optical flow), an equation holds such that

$$v_xf_x + v_yf_y + f_t = 0 \quad (v_x, v_y : x, y \text{ velocity}), \tag{16}$$

which is equivalent to saying that the distribution of f_x, f_y, f_t shows perfect degeneration if motion exists there [32]. If degeneration does not occur although the image changes, we know that some events other than motion, e.g., dynamic occlusion/accretion or changing illumination, occur [39].

To use these properties for changing attention of the 3-D vision sensors, we observe the correlation matrix

$$\begin{aligned}
\mathbf{S}_{XYT} &\equiv \iiint_{[X\times Y\times T]} \begin{bmatrix} f_x \\ f_y \\ f_t \end{bmatrix} \begin{bmatrix} f_x & f_y & f_t \end{bmatrix} dxdydt \\
&= \begin{bmatrix} S_{xx} & S_{xy} & S_{xt} \\ S_{xy} & S_{yy} & S_{yt} \\ S_{xt} & S_{yt} & S_{tt} \end{bmatrix}.
\end{aligned} \tag{17}$$

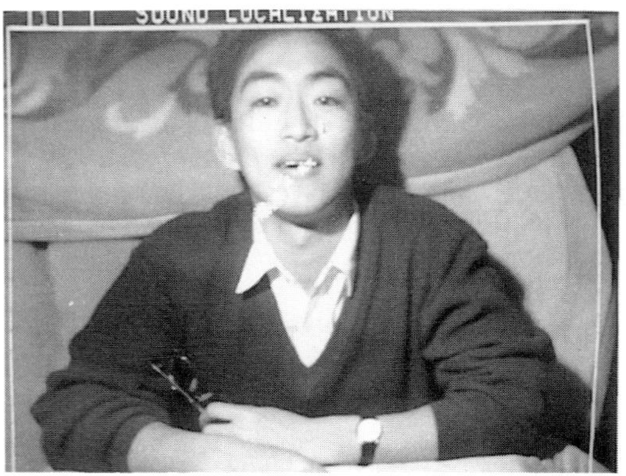

(a) One speaker

(b) One moving source

(c) Two speakers

Figure 8. Experimental results of quad-aural sound localization sensor. (a) Localization of a speech sound from a single speaker. (b) Localization of a moving white noise source. (c) Localization of speech sounds from two speakers.

for the left and right added image anywhere in FOV. From this matrix, we can easily derive optical flow velocity

$$v_x = \frac{S_{yy}S_{xt} - S_{xy}S_{yt}}{S_{xx}S_{yy} - S_{xy}^2} \qquad (x\text{-velocity}) \tag{18}$$

$$v_y = \frac{S_{xx}S_{yt} - S_{xy}S_{xt}}{S_{xx}S_{yy} - S_{xy}^2} \qquad (y\text{-velocity}), \tag{19}$$

and some dynamical image features such as

$$P_{DA} \equiv \frac{\det[\mathbf{S}_{XYT}]}{S_{tt}(S_{xx}S_{yy} - S_{xy}^2)} \qquad \text{(Dynamic accretion)} \tag{20}$$

to find dynamic occlusion of image segments.

We use them for changing visual attention as follows. Usually, the energy of brightness change S_{tt} is small because the sensor and its surrounding environment are stationary.

If some strong/large values appear, we obtain their position information (direction) from the image coordinates, and, according to the rank and the corresponding eigenvectors, we know whether 1) there is a moving object with a velocity v_x, v_y (rank=1), or 2) brightness changes possibly because an object appears or disappears suddenly (rank$\geq$ 2).

Figure 9(a),(b) shows experimental results of these algorithms [32]. The target was a water surface (with aluminium powder) which was stirred by a small plate. Owing to a special hardware for computing the correlation matrix, measurement of motion on 32×32 points requires only 0.15 second. Additional features to determine measurement reliability are displayed by squares, dots, or none at the measuring points.

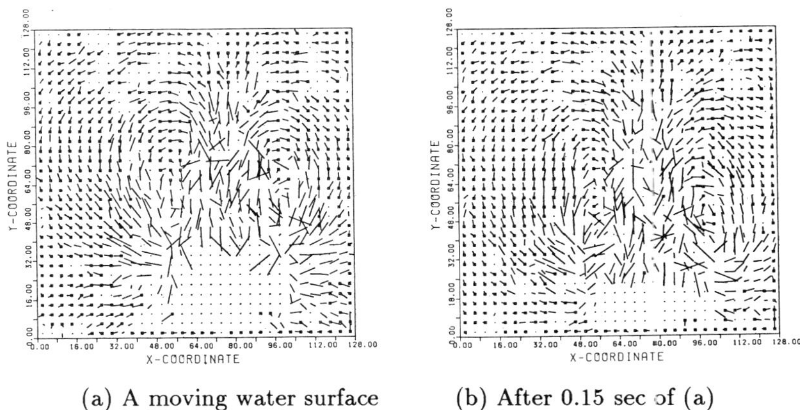

(a) A moving water surface (b) After 0.15 sec of (a)

Figure 9. Experimental results of optic flow detection. The target was a water surface (with aluminium powder) which was stirred by a small plate. Input of image sequence from TV camera, generation of the correlation matrix by hardware, and computation of motion and image features on 32×32 points requires only 0.15 second. Image features to determine measurement reliability are displayed by squares, dots, or none at the measuring points.

5 Shape Perception Mechanisms: Differential Binocular Vision Method

One of the most interesting algorithms of binocular stereo methods is the differential method of disparity analysis [5–7]. In this algorithm, the image pair obtained from both eyes is regarded as an image field, and the gradient covariance relations, which were satisfied locally in this field [31], are introduced to judge the establishment of binocular fusion and to determine the binocular disparity for reconstructing the solid profile. We call this algorithm a differential binocular vision method. The most outstanding trait of the differential binocular vision method is its subpixel accuracy for depth measurement and its easy implementation by the hierarchical parallel processing hardware. Using them, some 3-D image measurement systems have been developed with gradient correlation circuits and multiple processors [31–33]. Introducing this algorithm into the active 3-D binocular vision sensor by making the best use of the real-time high-speed performance is a purpose in this section.

So far, the differential binocular vision method has been regarded as having a fundamental weakness in narrowness of the solid profile measurement range. Here, we should remember that acquisition of both the wide-range 3-D environment and the detailed solid feeling obtained in human vision system is achieved by two separate mechanisms: detection of the retinal image disparity and detection of the vergence angle under fixation and FOV movement. Moreover, even in the human system, the range of solid feeling (fusion zone) with a latter mechanism is very narrow, at most 10 cm in the front [8]. If the differential binocular vision method is introduced as the second mechanism of the integrated vision system with the former as a preprocessing mechanism, the weakness of narrow rangeability diminishes, and its advantage in accuracy and processing speed will be used best.

In this section, we will first describe the principle of binocular vision methods which is extended to extract the interocular intensity disparity as luster feeling. Next, we will introduce to this method the criterion functions to judge the validity of the measured profile (relative altitude with respect to a base plane) and pick out a profile slice (a surface within a measureable range of depth), excluding the areas deviating from the fusion zone. Based on these, we will present an active binocular 3-D measurement algorithm by which a wide and accurate solid profile is acquired incrementally using a vibrative or microsaccadic FOV movement.

5.1 Differential Binocular Vision Method

First, we will describe a principle of the differential binocular vision method briefly. Picture a binocular camera with an interocular separation D and a distance H ($H \gg D$) to its base plane (Figure 10). Let the imaginary image observed at the center of the cameras be $f(x,y)$. The use of $f(x,y)$ is only for the simplicity of expression, as it will be vanished later. By using $f(x,y)$, we can easily describe a left and right image $f_L(x,y)$ and $f_R(x,y)$ respectively as a small symmetric intensity change $1 \pm \xi$ and a small symmetric positional shift $\pm\Delta$ with respect to $f(x,y)$ as

$$f_R(x,y) = (1+\xi)f(x+\Delta,y) \tag{21}$$

$$f_L(x,y) = (1-\xi)f(x-\Delta,y) \tag{22}$$

where Δ is expressed as

$$\Delta \equiv \frac{Dh(x,y)}{2(H+h(x,y))} \sim \frac{D}{2H}h(x,y) \tag{23}$$

by using a relative height $h(x,y)$ with respect to the base plane. If the left and right images show satisfactory agreement, we can assume that

$$h(0,0) = 0 \quad \text{(origin: center of FOV)}, \tag{24}$$

and the base plane coincides with a plane being parallel to the image plane and intersecting the object point at a center of FOV. Let the addition and subtraction of the left and right image be $f_+(x,y)$ and $f_-(x,y)$, respectively. Applying first order approximation of Δ and ξ to these two images $f_+(x,y), f_-(x,y)$, and then adding and subtracting them so that $f(x,y)$ is cancelled, we reach the relation

$$f_-(x,y) = \xi f_+(x,y) + \Delta f_x(x,y), \tag{25}$$

where $f_x(x,y)$ is an x-directional derivative of the addition image $f_+(x,y)$. This relation is almost satisfied whenever both a positional and intensity difference between right and left images are small, hence the first order approximation is still valid. Conversely speaking, it is possible to interpret this relation such that the area where the difference is small forms the fusion zone in which we can perceive solid feeling.

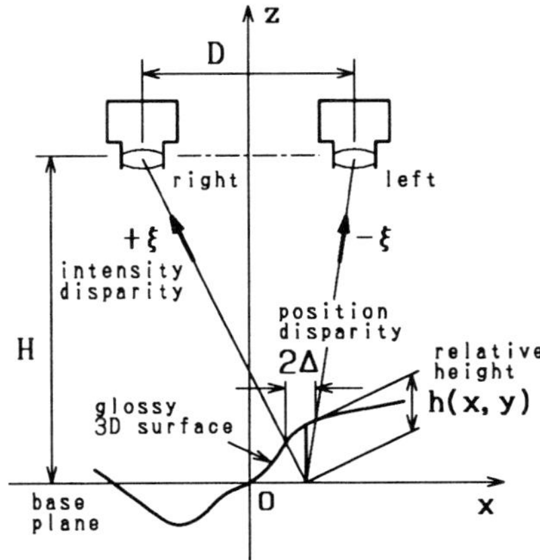

Figure 10. A geometry of the differential binocular vision system which the interocular intensity difference is considered. A parallel shift construction of the binocular correspondence is assumed. A right and left image contains an intensity difference $(1 \pm \xi)$ owing to the difference in reflection direction and a positional difference 2Δ owing to a relative altitude $h(x)$ with respect to the base plane. Although both of them appear as brightness differences between right and left images, it is possible to separate these two causes by using the gradient covariance relations included in a binocular image field.

Then, how can we obtain a binocular position disparity Δ and a binocular intensity disparity ξ using this relation? The simplest way is to solve Eq. (25) by the least squares method in a small window $[X \times Y]$ (e.g. 3×3 or 5×5 pixel area is used in the experiments described later) as

$$J \equiv \iint_{[X \times Y]} (\Delta f_x + \xi f_+ - f_-)^2 dxdy = \text{minimum}. \tag{26}$$

As a result, we obtain the position disparity Δ and the intensity disparity ξ as follows

$$\xi = \frac{S_{xx}S_{+-} - S_{x+}S_{x-}}{S_{xx}S_{++} - S_{x+}^2} \tag{27}$$

$$\Delta = \frac{S_{++}S_{x-} - S_{x+}S_{+-}}{S_{xx}S_{++} - S_{x+}^2}, \tag{28}$$

where

$$\mathbf{S}_{X+-} \equiv \iint_{[X\times Y]} \begin{bmatrix} f_x \\ f_+ \\ f_- \end{bmatrix} \begin{bmatrix} f_x & f_+ & f_- \end{bmatrix} dxdy \tag{29}$$

$$= \begin{bmatrix} S_{xx} & S_{x+} & S_{x-} \\ S_{x+} & S_{++} & S_{+-} \\ S_{x-} & S_{+-} & S_{--} \end{bmatrix}. \tag{30}$$

Each value of Eq. (30) means the sum of products in the small observing window $[X \times Y]$ of the addition image, the subtraction image, and the x differential addition image. By substituting Δ into Eq. (23), a relative height $h(x, y)$ with respect to the height of FOV center is obtained.

Because Eq. (23) includes the absolute distance H, the scale of the relative altitude $h(x, y)$ varies according to the given value of H. This is also true for man's solid feeling described above.

5.2 Perceiving Luster from Interocular Intensity Difference

In the situation when we cannot assume a completely diffused reflection or an illumination by a spatially uniform light source, one perceives a luster on the object surface. This perception occurs very strongly, especially when the position of the bright part shifts between both eyes, and is generally called binocular luster. Binocular luster occurs in the fusion state with the solid feeling when there is difference (interocular intensity disparity) between right and left brightness. It can also be perceived in the stereogram in Figure 11 where the sign of brightness is reversed right and left [8]. In this case, it will be seen like a glossy gray solid with luster. By using the abovementioned method, we can detect a small interocular intensity disparity owing to the looking angle difference on the surface, and relate it to the luster feeling, that is, a direction dependence of surface reflectance, although such an extreme case like Figure 13 cannot be treated. According to a detailed analysis [33], the distribution of the interocular intensity disparity obtained for a uniform ray being incident from a single angle has been shown to agree with a parallax-directional differentiation of the logarithm reflectance function [33] with the departure angle as an independent variable.

5.3 Preliminary Experiments

Here, we will show an example of an experimental system and its results using the simple differential binocular vision method without the mechanism of FOV movement and gaze control. This experimental system is shown in Figure 12. It is using two CCD cameras (RETICON MC5128, 128 × 128 resolution) by distance H = 110 cm and a separation $D = 12$ cm. A base plane (perfect correspondence plane of both eyes) is placed at about

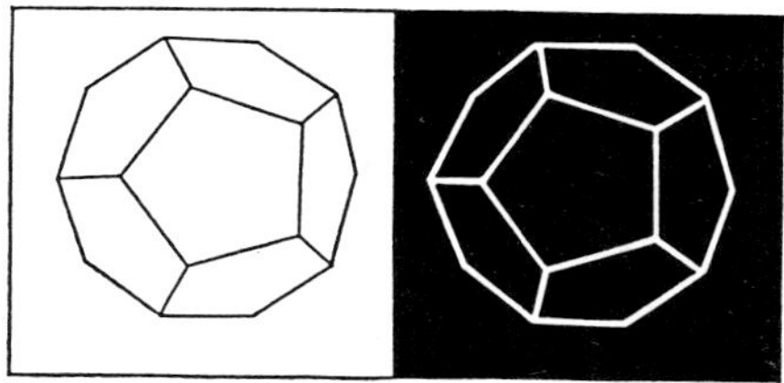

Figure 11. A brightness reversal stereogram (reprinted from Tazaki et al. [8]). We feel a black-gray lustrous solid from this although the level of fusion weakens a little by the contradictory phenomenon (struggle) in the visual field. The luster feeling caused by the brightness difference in the right and left image under fusion like this is called binocular luster.

the central depth of the object profile. After quantizing the video signals by 8 bits, it calculates rapidly the quantities in Eq. (30) by using the gradient covariance hardware which made the best use of the hierarchic structure of the algorithm. The processings hereafter are done in software by the computer.

(a) 3-D object

(b) binocular cameras

Figure 12. A photograph of the experimental system of the differential binocular vision system. The interval between two CCD cameras is D = 12 cm, and the distance between the camera and the object is H = 110 cm. The base plane (perfect correspondence plane of both eyes) is taken at about the central depth of the object profile. The right and left video signals quantized in 8 bit were processed at a 0.15 sec/frame speed using special gradient covariance hardware and software.

A) Experiment on Profile Reconstruction [31]

The object was the surface of a cone-type loudspeaker. To give correspondence clues, a random dot light pattern was projected on the surface after painting it white. Figure 13 displays the detected height map of the solid profile as a 3-D wireframe. As shown in the results, accuracy of profile, if it is in the fusion zone, is very good and spatial resolution is also sufficiently high, although it requires only 0.15 second for measurement and computation.

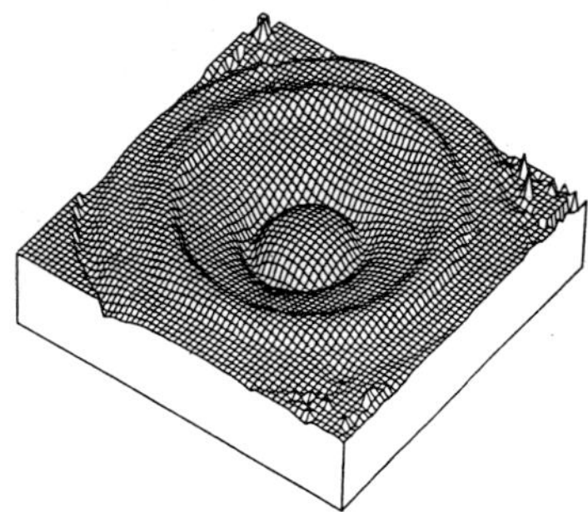

Figure 13. An example of solid profile measurement by the differential binocular vision system. The object is a surface of a cone-type loudspeaker. In this example, we obtained 64 x 64 spatial resolution, about 2 cm of the rangeability, about 1mm of the profile measurement accuracy in a smooth area (the accumulation is 30 times).

B) Experiment on Luster Perception [33]
The object was a metallic can of photographic film (a cylinder about 10 cm in diameter and about 4 cm in thickness), and its surface is somewhat glossy and slightly rough. A brand seal made of paper was pasted on one part of its lid. A right and left image, a measured distribution of position disparity Δ, and intensity disparity ξ as brightness maps are shown in Figure 14 (a) to (d). The intensity disparity map (d) looks bright almost everywhere on the object. That this is the effect of luster will be understood well when the brand seal (with no luster) is observed. It has a uniform gray tone (intensity disparity=0). This experiment shows that the binocular intensity disparity provides enough information to recognize and classify the condition of the surface.

C) Measurment of Differential Logarithmic Reflectance Function [33]
Under quite the same conditions as described in the preceding paragraph, we used no-reflection glass with black paper to exclude the back surface reflection as a target. This glass has a uniform diffusing surface, and was used to suppress surface reflection when taking a picture of the copy. Figure 15 is a 3-D plot of the obtained distribution of intensity disparity ξ. It is forecast that the surface has the Gaussian distribution type reflectance function, and if so, its logarithm is an upward convex parabolic surface and its directional derivative becomes an inclined plane. Its tendency can be read from Figure 15. Because the reflection light becomes weak in the surrounding part and the intensity disparity is not obtainable there, it looks like a plane with only the center part cut out.

6 Fusing Multiple Views Using Microsaccadic Eye Movement

In this section, we will consider the tight connection between the grasp mechanism of a 3-D environment using the FOV movement and fixation mechanism and the detailed

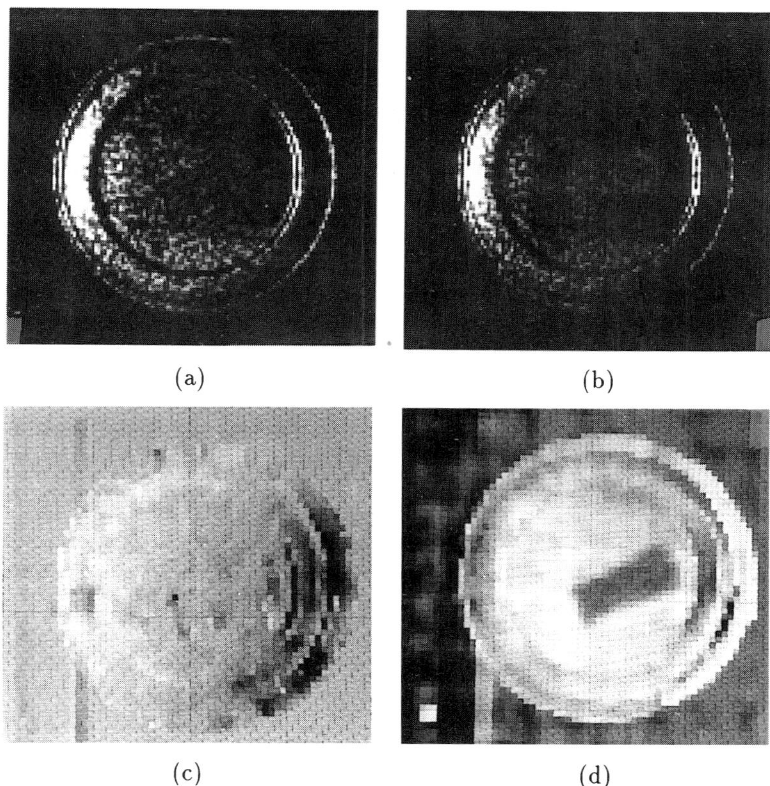

(a) (b)

(c) (d)

Figure 14. An example of luster perception. The object is a metallic can of a photographic film. Its right and left images are shown in (a) and (b). In (c), a height map of the solid profile is shown by the brightness (brighter means higher), and in (d) a map of intensity difference is shown by the brightness (brighter means brighter right image). The shade part at the center indicates no luster. It is actually in the brand seal made of paper.

solid profile perception mechanism based on the disparity of retinal image. We call such a system, which is sophisticated, active, and mimicks human vision system, intelligent 3-D vision sensor.

The first problem that occurs is the pixel-by-pixel judgment of binocular fusion. In this case, the newly demanded capabilities for the differential binocular vision system are:

1. the ability to exclude the consistent data which come from the surface beyond the fusion zone so that meaningless and harmful outputs are prevented, and
2. the ability to move its base plane and/or FOV to acquire novel information from the image point where it could not fuse or it has never seen before.

To realize such capabilities, we developed first the self-evaluation criteria of the measurement validity and reliability, which are derived inference-theoretically in relation to the

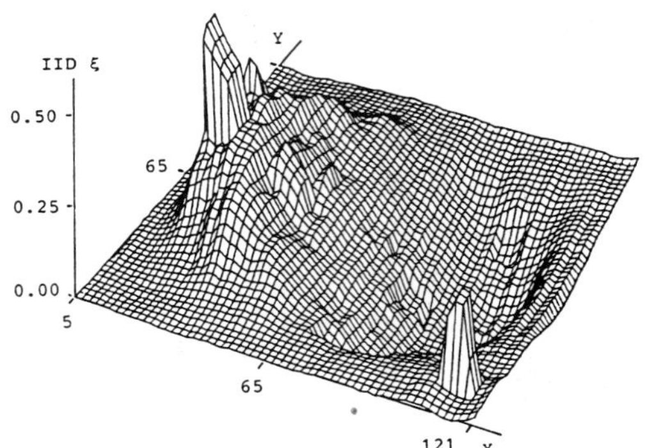

Figure 15. An example of measuring a differential logarithmic reflectance function. The object is "No-reflection glass" for making photographic copies of pictures. Because it has a high-diffusion surface, binocular intensity disparity distribution becomes an inclination plane like this figure.

least squares method [32].
i) Measureability measure
The determinant of the normal equation Eq. (26) is expressed as

$$J_{DET} \equiv S_{xx}S_{++} - S_{x+}^2, \tag{31}$$

and means the amount of measureability at the point under consideration.
ii) Validity measure
Substituting Eqs. (27,28) into the least squares criterion J, we obtain a sum of squared error J_{RES} under an optimum solution

$$J_{RES} \equiv S_{--} - \xi S_{+-} - \Delta S_{x-} \qquad (\xi, \Delta : \text{optimum solutions}). \tag{32}$$

Generally, J_{RES} is an indicator of the mixture of disturbance which was not expected in the model, and in this case judges the validity of differential equality.
iii) Error variance estimator
Most of the error in Eq. (25) originates from higher order Taylor coefficients which are ignored in the approximation. By moving this error term to the left-hand side, however, we can consider it as an additive pixel noise of which variance is estimated by the sum of squared residual J_{RES}. Generally, the larger the disparities, the larger this error will become. In order to use J_{RES} as an index of fusion, suppose the noise is white and uncorrelated with any pixel values and disparity values. Then, with the sum of the squares of errors p, q on the solution ξ, Δ is expressed approximately as

$$E[p^2 + q^2] = \frac{(S_{xx} + S_{++})J_{RES}L^{-2}}{S_{xx}S_{++} - S_{x+}^2} \equiv J_{ERR}. \tag{33}$$

Since we can estimate with J_{ERR} the amount of error variance involved in ξ, Δ without knowing their true values, it is most suitable for judging the degree of binocular fusion as the invariant with other measurement conditions like brightness.

6.1 Extraction and Classification of Profile Slices

The area where binocular fusion is established is limited within the narrow range of depth (fusion zone) surrounding a base plane. We call the part which consists of a measureable surface of the object in the fusion zone a profile slice. The measureable surface of the object is the surface with sufficient clues of light and shade pattern. Therefore, it is judged from the comparison of the criterion function J_{DET} for the measureability with J_{MIN} as

$$J_{DET} \leq J_{MIN} \sim \frac{\sqrt{3}\pi\sigma^2 L^2}{3}, \tag{34}$$

where J_{MIN} is the average of J_{DET} originating only from image noise. On the measureable surface of the object, the part where the measurement error variance is sufficiently small is the profile slice. Therefore, it can be extracted by Eq. (34) and

$$J_{ERR} \leq J_{MAX} \tag{35}$$

by giving a permissible limit of error variance J_{MAX} beforehand.

The combination of these criteria J_{ERR} and J_{DET} enables us the consistent judgement of binocular fusion as follows:

1. the area in which sufficient clues and reliable height are obtained (a surface in a fusion zone, profile slice),
2. the area in which sufficient clues exist but the measured height is unreliable (a surface outside the fusion zone, questionable area), or
3. the area in which no sufficient clues exist (a surface which can be any of the above, indefinite area).

6.2 A Method of Least Squares 3-D Accumulation

In this section, we will consider the active 3-D reconstruction scheme which combines the FOV movement mechanism and the differential binocular vision algorithm, and show a sensor system like the one shown in Figure 16(a) which forms an absolute and accurate height map of an extending 3-D solid. It puts the geometrical parameters of the TV cameras under its control, composes relative, fragmentary solid information obtained from the differential binocular vision algorithm, and forms incrementally the wider-ranging, more absolute 3-D information. This will have a significant analogy to the incremental acquisition mechanism of 3-D models by computers [4] and the man's ego-centric perceiving ability of the external world [11, 12].

Consider the ith profile slice. Let the spatial region (support) of it be ω_i, the relative height be $h_i(x, y)$, and the error variance be $v_i^2(x, y) \equiv 4H_i^2 J_{ERR}/D^2$. Let H_i be the distance from the camera to the ith base plane. Although the target value of the base plane distance sent to the vergence angle driver is G_i, it is predicted that the true value H_i (unknown) will involve some uncertainty with the variance σ_G^2, owing to disturbances. Suppose, until the ith accumulation, an absolute height map $H_i(x, y)$ is acquired in its

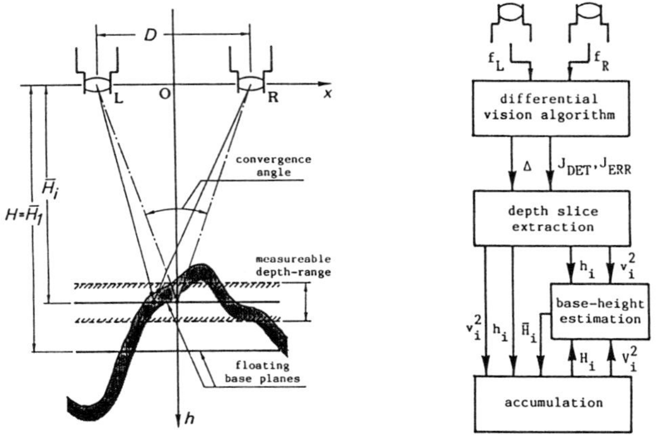

(a) Floating baseplane (b) Reconstruction system

Figure 16. A geometry and a block diagram of the differential binocular vision system which uses active vibrative control of vergence angle. (a) The change of vergence angle means back-and-forth floating of a base plane. By combining (accumulating) profile slices (relative altitude) included in each fusion zone, an entire 3-D shape is reconstructed incrementally.

support Ω_i, and its variance map is $V_i^2(x,y)$. Denoting the true height map as $H(x,y)$ and referring to Figure 16(b), we can describe this process using the following statistical model

$$h_i(x,y) = H(x,y) - H_i + \theta_i(x,y) \quad \text{in } \omega_i \tag{36}$$

$$H_i(x,y) = H(x,y) + \Theta_i(x,y) \quad \text{in } \Omega_i \tag{37}$$

$$H_i = G_i + \epsilon_i \tag{38}$$

where

$$E[\theta_i(x,y)^2] = v_i^2(x,y) \tag{39}$$

$$E[\Theta_i(x,y)^2] = V_i^2(x,y) \tag{40}$$

$$E[\epsilon_i^2] = \sigma_G^2 \tag{41}$$

and $\theta_i(x,y)$, $\Theta_i(x,y)$, and ϵ_i express noises involved in $h_i(x,y)$, $H_i(x,y)$, and H_i, respectively. Therefore, to estimate the unknown quantities $H(x,y)$ and H_i from the observed quantities $h_i(x,y)$ and G_i according to a statistical model described above becomes the next problem.

A suboptimal algorithm to this problem is given [35] as follows. First, we obtain a least squares estimation K_i of H_i by

$$K_i = \frac{\{\int_{\omega_i \bigcap \Omega_i} (V_i^2 + v_i^2)^{-1} (H_i - h_i) dxdy + \sigma_G^{-2} G_i\}^{-1}}{\{\int_{\omega_i \bigcap \Omega_i} (V_i^2 + v_i^2)^{-1} dxdy + \sigma_G^{-2}\}^{-1}}. \tag{42}$$

From the previous arguments, there are two ways to determine the current distance of a base plane, i.e., a visual one obtained from the difference between an internally accumulated absolute height and an instantaneously obtained relative height, and a mechanical one which is actually the target value to the movement system. Since Eq. (42) shows that the most reliable method to obtain the current distance is a weighted average of these two values, it expresses how the movement system and the vision system should be cooperatively combined.

Let the error variance involved in K_i be σ_K^2. Then, by using the above result, it can be written as

$$\sigma_K^2 \equiv \{\int_{\omega_i \bigcap \Omega_i} (V_i^2 + v_i^2)^{-1} dxdy + \sigma_G^{-2}\}^{-1}. \tag{43}$$

Therefore, we know a least squares estimation of $H_i(x,y)$, i.e., the ith absolute height map is

$$H_{i+1}(x,y) = \frac{V_i^{-2} H_i(x,y) + (v_i^2 + \sigma_K^2)^{-1}(h_i(x,y) + K_i)}{V_i^{-2} + (v_i^2 + \sigma_K^2)^{-1}} \quad \text{in } \omega_i \bigcap \Omega_i \tag{44}$$

$$H_{i+1}(x,y) = h_i(x,y) + K_i \quad \text{in } \omega_i \bigcap \overline{\Omega_i} \tag{45}$$

$$H_{i+1}(x,y) = H_i(x,y) \quad \text{in } \overline{\omega_i} \bigcap \overline{\Omega_i}, \tag{46}$$

and its error variance is

$$V_{i+1}^2(x,y) = [V_i^{-2}(x,y) + \{v_i^2(x,y) + \sigma_K^2\}^{-1}]^{-1} \quad \text{in } \omega_i \bigcap \Omega_i \tag{47}$$

$$V_{i+1}^2(x,y) = v_i^2(x,y) + \sigma_K^2 \quad \text{in } \omega_i \bigcap \overline{\Omega_i} \tag{48}$$

$$V_{i+1}^2(x,y) = V_i^2(x,y) \quad \text{in } \overline{\omega_i} \bigcap \overline{\Omega_i}. \tag{49}$$

The accumulation methods described in Eq. (42), Eqs (44) to (46), and Eqs (47) to (49) can be explained as follows. First, we estimate the base plane height H_i from the height difference within an overlap of the internally accumulated support Ω_i and the instantaneously obtained support ω_i. If no overlap exists, the target value G_i itself is used. An estimation K_i of H_i tell us where the current base plane is with respect to the internally accumulated height map. Therefore, by adding K_i to the relative height map $h_i(x,y)$, we can obtain an absolute one within ω_i, although it is instantaneous. If an overlap between ω_i and Ω_i exists, we can use it to improve the accuracy of the internally accumulated height map by averaging them with proper weights. This is the operation described in Eqs (44) and (47). If the area ω_i exists which is not included in Ω_i, we can use it to expand the support Ω_i of the internally accumulated height map. This is the operation described in Eqs (45) and (48). The method for renewing the support and its initial condition is

$$\Omega_{i+1} = \Omega_i \bigcup \omega_i, \quad \Omega_1 = \phi \text{ (empty set)}. \tag{50}$$

As a result, the support Ω_i of the absolute height map expands monotonically, and the error variance $V_i^2(x,y)$ of it decreases monotonically by this accumulation process. This means an incremental acquisition capability of the system; it continues expanding and improving its own knowledge both quantitatively and qualitatively as time advances.

6.3 Preliminary Experiments

Figure 5 shown before includes the experimental composition of the system based on the abovementioned principle. In this figure, the vergence angle change is done by shifting synchronous signals of two TV cameras reversely. The accumulation system is so designed that it can function even when the vergence angle varies randomly without any relation to the instruction from the system.

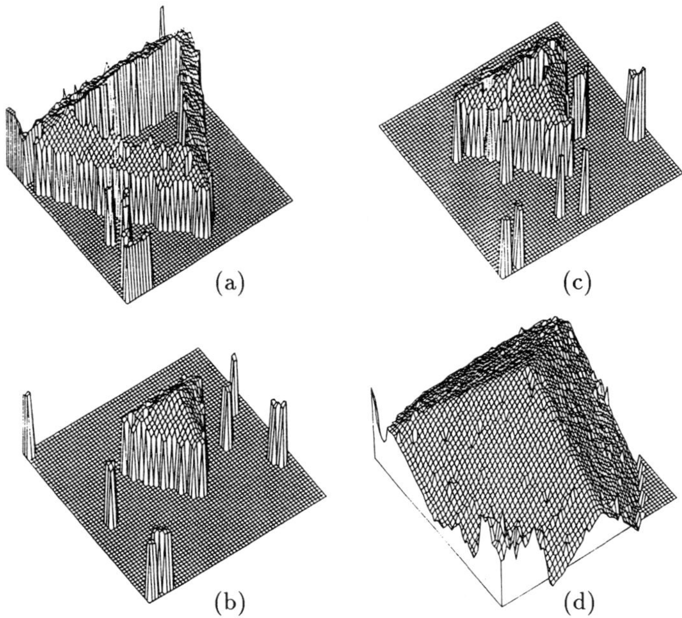

Figure 17. Example of the profile slice and the accumulation results of each step of the accumulation. The accumulation was done 20 times as a whole. (a) is the third result, (b) the fifth result, (c) the seventh result, and (d) the final result. The object was the cube whose vertex was turned to the binocular vision system side. The lowest flat part indicates a questionable area.

A) Reconstruction of simple solid [35]

The object is a gypsum cube about 10 cm on each side. Blob patterns 1-3mm large were put on the object surface at random to provide clear clues for the low resolution TV cameras (two 128×128 matrix cameras, RETICON MC5128). It can be predicted that the profile slice will be a triangular belt because the object was placed so that the vertex may turn to the measurement system. The base plane was placed manually about 5 cm below the vertex. Then, we took images were taken be moving the base plane forward 10 times at intervals of about 8mm, and profile slices were extracted from them. We used 2 mm^2 as the setting error variance σ_G^2 of a base plane, which was roughly equal to the backlash of the driving system. The thresholds for the profile slice extraction were assumed to be $J_{MAX} = 0.82$ and $J_{MIN} = 64.0$ from the results of preliminary experiments. Three-dimensional plots of the profile slice extracted from the 3rd, 5th, and the 7th images are shown in Figure 17 (a), (b), and (c). Both an indefinite area and a questionable area are displayed by the lowest plane. The surface (a triangular belt) shows the height value in its support, i.e., a profile slice. Some wrong isolated profile slices can be made harmless by the process of accumulation because J_{ERR} in that part is large enough when compared with the true ones. Figure 17 (d) shows the height distribution after the final accumulation. The corresponding error distribution is also stored in the computer.

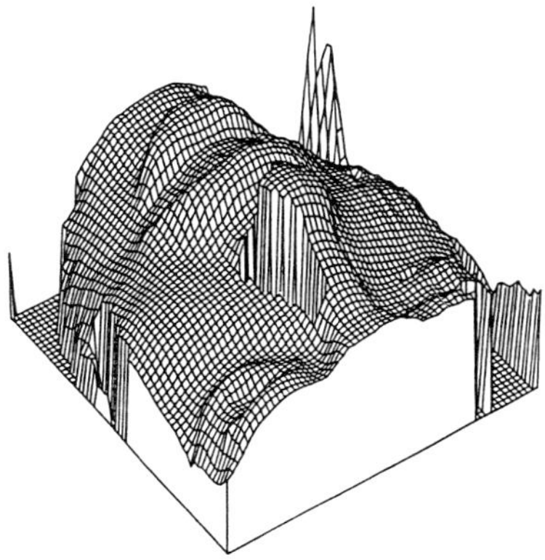

Figure 18. A measurement result of the face mask shown in Figure 14(a). While floating a base plane back and forth, 40 times accumulation was performed. A lack of profile at the side of the nose shows that it remains as an unmeasureable area. This is because the ruggedness change is too steep, therefore there is no right and left correspondence.

B) Application to face profile measurements

One of the promising applications of the intelligent 3-D vision sensor is surveillance and abnormal detection using a mobile robot. We tried a reconstruction of a human face (the face mask made of plastic seen in Figure 12). The conditions are almost the same as the experiment in the preceding paragraph. The face mask had a height difference of about 12 cm from nose top to the plane of both ears. We moved the base plane back and forth 40 times at intervals of about 4 mm. Figure 18 shows a 3-D plot of the accumulation result. Judgement of an indefinite area owing to a lack of matching cues is expressed by the lowermost plane. Without being influenced by the light and shade difference, etc., of each part of the face, an almost faithful solid reproduction is obtained. (The error in a smooth part is below 5%). A deep drop at the side of nose shows that the area remained questionable. This is because the profile change is too steep there, so that some occlusions occur. It does mean, that this system can moves the aspect in autonomy to acquire new information to eliminate such uncertain areas.

7 Intelligent World Sensor "SmartHead"

This section describes our newly developed intelligent sensor system "SmartHead" as shown in Fig.19, in which a binocular vision sensor and a quad-aural auditory sensor are mounted on an autonomously movable head. Both the vision and auditory sensors are ready to detect unusual events, i.e., visual motion/accretion and auditory localization of sound sources, in its surrounding environment as an early warning system. If such an event is found, then head motors and a high-speed saccadic eye movement system are activated and they quickly move the FOV of the vision sensor toward a direction of the object. A tracking and fixation system catches the most salient point in that direction and fixes the FOV on the object. Then a high-speed differential stereo vision sensor with microvibrative eye movements is activated, and it reconstructs a solid profile and extracts various image features of the object.

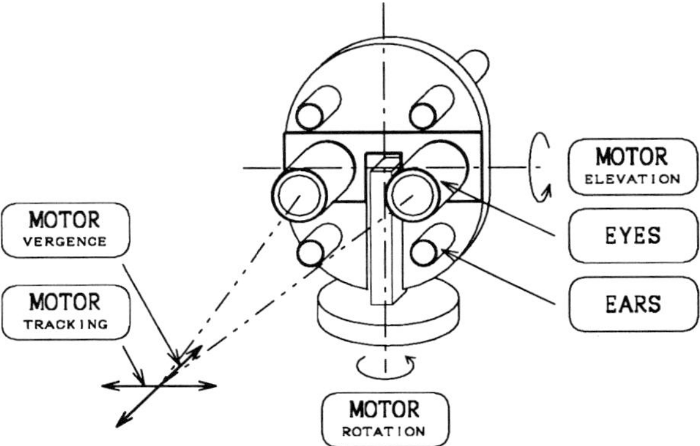

Figure 19. Robot head sensor "SmartHead" with a binocular vision sensor and a quad-aural auditory sensor on an autonomously movable head. All sensors and algorithms described above are installed with some motor functions controlled autonomously using the sensory information.

7.1 Gradient Correlation Sensing Principle

As shown in all the previous sections, our methods for vision sensing and auditory sensing are based on the gradient correlation sensing principle. Therefore, the vision section and the auditory section of our sensor "SmartHead" are designed so that it makes the best use of the unified sensing principles and architectures. The unified early processing architecture installed in the "SmartHead" is described schematically in Fig.20(a). In the first stage, a difference and/or differential of the incoming signals are generated by the sensing probes themselves or in the early stage of the circuits. Like any other differential measurement system, the selectivity for the difference (skew-symmetric) component is emphasized by the symmetric structure. But unlike the others, this architecture produces additive (symmetric or common mode) components simultaneously. The latter stage of this architecture produces sequences of cross correlations between the output of the gradient stage. Although it looks similar to correlation measurement systems, the biggest difference is that the correlations are multidimensional and of very short term. The output of this stage forms a temporal and/or spatial sequence correlation matrix.

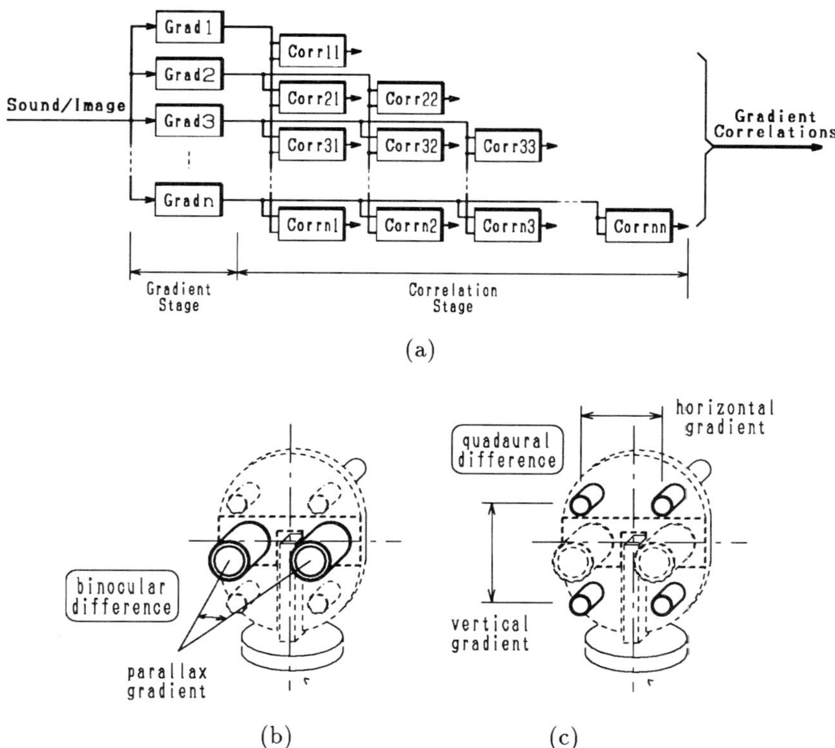

Figure 20. A schematic diagram of the gradient correlation based sensory architecture (a), and two gradient sensing probes used in the vision sensor and auditory sensor. The difference and/or differential of the incoming signals are constructed by the sensing element itself or in the early stage of the circuits. Then short-time and/or small-area correlations among them are generated so that they form a correlation matrix.

A) Architecture for vision section

Fig.21 shows the gradient stage of the architecture. The number of output components is five: The pair of binocular difference f_- and sum f, and three spatiotemporal gradients f_x, f_y, f_t of the sum f. Note that the binocular difference and sum are constructed simply by the subtraction and addition of binocular video signals, respectively. Symmetry of the binocular cameras and circuits is carefully considered. To maintain the symmetrical condition while changing FOV freely, precise control facilities for gain and vergence are installed in the cameras.

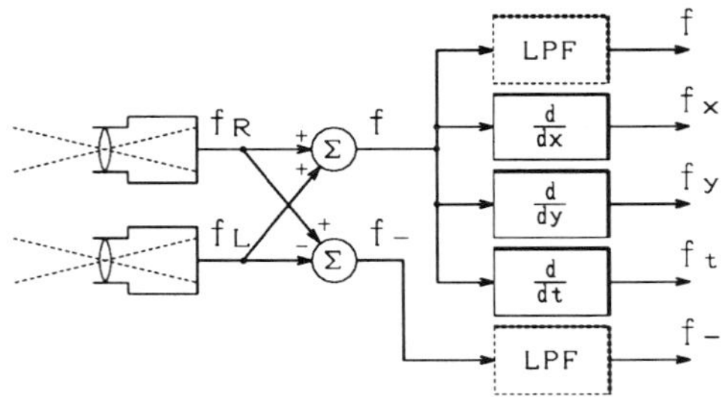

Figure 21. Gradient stage of the vision section. The number of output components is five: one left and right sum f, one left and right difference f_-, and three spatio-temporal gradients f_x, f_y, f_t of f. The left and right cameras followed by a subtraction circuit constitute a parallax differentiator.

B) Architecture for auditory section

Fig.22 shows the gradient stage of the auditory architecture. The number of output components is four: one quad-aural sum f and three spatiotemporal gradients f_x, f_y, f_t of f. Note that the quad-aural difference and sum are constructed simply by the subtraction and addition of four auditory signals, respectively. Symmetry of the microphones and circuits is carefully considered. To maintain the symmetrical condition and wide dynamic ranges, rough and precise gain control facilities are installed in the microphone amplifiers.

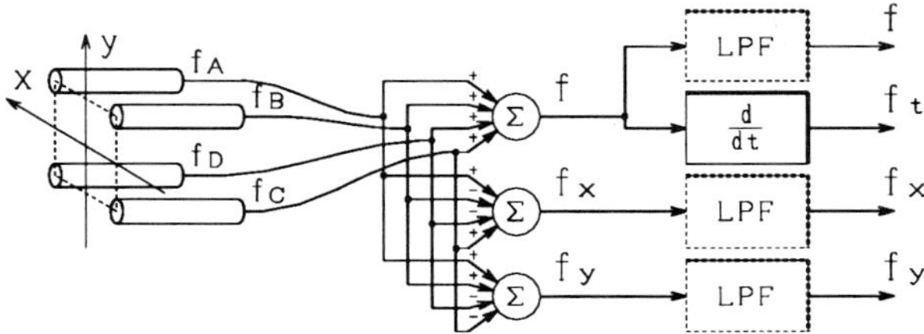

Figure 22. Gradient stage of the auditory section. The number of output components is four: one quad-aural sum f and three spatiotemporal gradients f_x, f_y, f_t of f. The quad-aural microphones followed by subtraction circuits constitute a sound field differentiator.

7.2 Gradient Correlation Hardware

Even though the gradient correlation architecture is systematic, computation for it requires tremendous time. Therefore we developed a circuit for computing the gradient covariance matrix (an extended matrix which involves $\mathbf{S}_{.XY}, \mathbf{S}_{XY}, \mathbf{S}_{XYT}$, and $\mathbf{S}_{X+-}$ as submatrices) for the vision section. For the auditory section, we used an analog circuit [43]. Fig.23 shows photographs of these circuits (single and double slot-wide VME board). It can generate whole elements of the gradient correlation matrix within a four pixel time of a TV camera. All succeeding processes are performed by conventional UNIX workstations (Solbourne 5/600 with 1 CPU, and a desktop SUN 4).

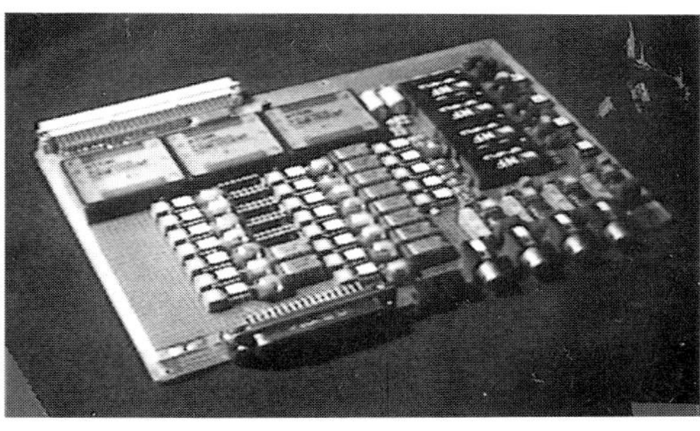

(a) auditory section

(b) vision section

Figure 23. A photograph of the gradient correlation hardware for the vision section and the auditory section of "SmartHead". Input of the hardware (b) is binocular TVC signals (8 bit/pixel each), and the output is the 5 × 5 gradient correlation matrix (32 bit/element within four pixel time).

7.3 Autonomous Sensing Algorithm

One of the most important objectives of the "SmartHead" sensor is acquiring a world image, i.e., a dynamical 3-D map of the surrounding environment, while performing humanlike actions naturally and autonomously. We realized such actions through the coordination of several processes. The tasks and algorithms of each process installed in "SmartHead" are as follows.

Binocular Correspondence Task — Establish the most reliable correspondence and trigger saccadic eye movement.

Tracking and Fixation Task — Compute binocular disparity and motion in the FOV center and pursue the object continuously.

3-D Feature Extraction Task — Fuse binocular images and extract 3-D shape and image features.

Object Segmentation Task — Identify an object figure from a background by using the shape and features.

Visual Early Warning Task — Examine marginal views and find salient or moving objects as candidates of a new FOV position.

Auditory Early Warning Task — Localize sound sources and find salient ones as candidates of a new FOV position.

These processes work as follows. Normally, "SmartHead" is widely examining his environment using its eyes and ears. In this case, both the visual early warning process and the auditory early warning process are active, and the other processes are waiting. When a new object appears, some motion is detected, or sound comes from somewhere, it moves its face and eyes toward it. In this case, the binocular correspondence process is activated by one of the early warning processes. Then it begins to pursue the object, if it is moving, and fixes its eyes on it. This is performed by the tracking and fixation process which is activated by the binocular correspondence process. If the tracking is unsuccessful, the binocular correspondence process is activated again. When the object stops, "SmartHead" begins to watch it very carefully. In this case, the 3-D feature extraction process is activated by the tracking and fixation process. This process extracts shape and image features for segmenting the object from the view. The extracted information is transferred to the object segmentation process which runs on another workstation connected by LAN. The extraction of the object figure from background requires higher level image processing and plenty of times. Since it is beyond a task of sensor, we do not describe it. Through the object segmentation process, a relational sensory information structure from the early warning source to the intentionally captured 3-D image of the object is established. Since the early warning processes are always active, the above processes can be suspended any time so that the attention is redirected to new more important events.

Fig.24 shows a photograph of when the sensor localized on a sound source and then caught by eyes a calling woman. Fig.25 shows examples of outputs of this sensor. The primary output is not a CRT monitor image such as is shown but arrays of digital data in which the shape and features in the processing region are stored.

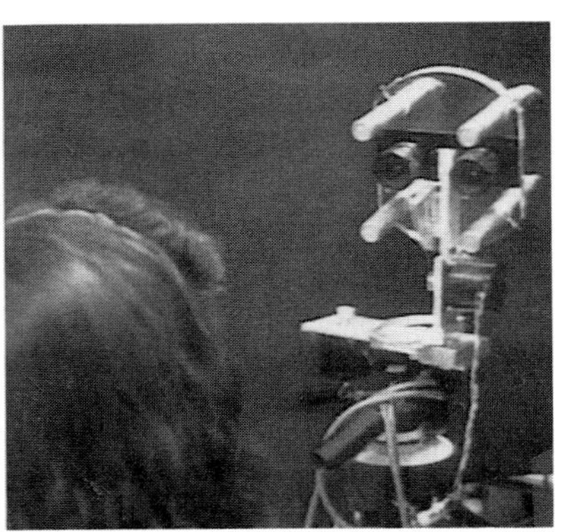

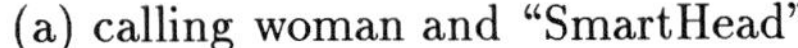

(a) calling woman and "SmartHead" (b) captured image

Figure 24. An addition image of both eyes immediately after completing the FOV movement and fixation. The vision of "SmartHead" correctly catches the mouth of the calling woman.

7.4 Performance

The performance of this sensor is summarized as follows.

1) A 3-D sound localization sensor with ± 2 deg (in anechoic room) and ± 5 deg (in usual room) accuracy, real-time processing speed (less than 1 ms delay time), and high temporal resolution (5 ms observation window time).

(a)

(b)

Figure 25. An example of the real-time shape reconstruction (a) and the image feature extraction (b). The central square of the monitor shows the processing region (64 × 64 measuring points in 128 × 128 area). In the region, a relative height map (a) with respect to the base plane and a strength map (b) of "ridgeness" feature is expressed by the degree of brightness.

2) A high speed saccadic eye movement system which responds quickly to motion and salient features over an entire FOV (several TV frame times from detection to matching).
3) A tracking and fixation system which pursues both the range motion and azimuth motion (feedback delay is less than 1 TV field time).
4) A high speed differential stereo vision sensor with microvibrative eye movements (from 0.1 sec to 0.5 sec for reconstructing solid on 64 × 64 measuring points).
5) A visual and auditory early warning system using real-time motion detection (0.5 sec for entire image search) and 3-D sound localization (1 ms/localization).

8 Summary

Man perceives both the 3-D arrangement of his surrounding environment and various detailed features of the objects in it. They are the most important information source for his activity. However, since much of this perception is performed unconsciously, man himself is hardly aware of the mechanisms underlying this preattentive perceiving activity. In order to achieve an autonomous machine such as robots, learning and incorporating these mechanisms in a technological system are definitely necessary. This is also the main goal for making the sensor intelligent. Although it is still a future goal that the intelligent vision sensor introduced in this section will fulfill its functions as a totally integrated sensor system like the human binocular vision, several basic elements, such as a structure to form an absolute solid feeling by moving the fixation point actively and autonomously, the perception of luster from interocular intensity disparity, etc., will provide a lot of fundamental architecture and measurement principles for developing the advanced flexible sensing system.

References

[1] H.C.Longuet-Higgins: A computer algorithm for reconstructing a scene from two perspective projections, Nature, 293, 133/135 (1981)

[2] S.T.Barnard and W.B.Thompson: Disparity analysis of images, IEEE Trans. Pattern Anal. Machine Intell., **PAMI-2**, 4, 333/340 (1980)

[3] W.E.L.Grimson: Computational experiments with a feature based stereo algorithm, IEEE Trans. Pattern Anal. Machine Intell., **PAMI-7**, 1, 17/34 (1985)

[4] M.Herman, T.Kanade and S.Kuroe: Incremental acquisition of a three dimensional scene model from images, IEEE Trans. Pattern Anal. Machine Intell., **PAMI-6**, 3, 331/340 (1984)

[5] B.K.P.Horn and B.G.Schunck: Determining optical flow, Artificial Intelligence, 17, 185/203 (1981)

[6] C.Cafforio and F.Rocca: Methods for measuring small displacement of television images, IEEE Trans. Information Theory, **IT-22**, 5, 573/579 (1976)

[7] B.D.Lucas and T.Kanade: An iterative image registration techniques with an application to stereo vision, Proc. 1981 Int. Jont Conf. Artificial Intell., 674/679 (1979)

[8] Tazaki,Ohyama,Hiwatashi(eds.): "*Visual Information Processing*", Asakura Pub. Co., 273/278 (1979) (in Japanese)

[9] E.L.Hall: *Computer Image Processing and Recognition*, Academic Press, NY, chap.8 (1979),

[10] Y.Shirai: "*Pattern Understanding*", OHM Pub. Co., chap.3 (1987) (in Japanese)

[11] H.A.Sedgwick: Environment-centered representation of spatial layout: available visual information from texture and perspective, Human and Machine vision, Academic Press, 425/458 (1983)

[12] D.Marr and T.Poggio: A theory of human stereo vision, Proc. Roy. Soc. London, B204, 187/217 (1980)

[13] K.Haines and B.P.Hildebrand: Coutour Generation by Wavefront Reconstruction, Phys. Letters, **19**, 1, 10/11 (1965)

[14] M.Takeda and K.Mutoh: Fourier transform profilometry for the automatic measurement of 3-D object surface, Applied Optics, **22**, 24, 3977/3982 (1983)

[15] J.C.Dainty (ed.): "*Laser speckle and related phenomena*," Topics in Applied Physics vol.9, Springer, Berlin (1983)

[16] M.Idesawa, T.Yatagai, and T.Soma: Scanning Moire method and automatic measurement of 3D shapes, Applied Optics, **16**, 2152/2162 (1977)

[17] Y.Shirai and M.Suwa: Recognition of polyhedrons with a range finder, Proc. 2nd Int. Conf. Artificial Intell., London, 80/87 (1971)

[18] T.Binford: Visual perception by computer, Proc. IEEE Int. Conf. System and Control, Miami (1971)

[19] P.Will and K.Pennington: Grid coding: A preprocessing technique for robot and machine vision, Proc. 2nd Int. Conf. Artificial Intell., London, 66/68 (1971)

[20] M.Levine, D.O'Handley, and G.Yagi: Computer determination of depth maps, Comput. Graph. Imag. Proc, **2**, 134/150 (1973)

[21] Y.Yakimovsky and R.Cunningham: A system for extracting 3-D measurements from a stereo pair of TV cameras, Comput. Graph. Imag. Proc., **7**, 195/210 (1978)

[22] H.Baker: Edge-based stereo correlation, Proc. ARPA Image Understanding Workshop, Maryland (1980)

[23] D.Marr and T.Poggio: Cooperative computation of stereo disparity, M.I.T. AI Lab., Memo. 364 (1976)

[24] R.Bajcsy and L.Lieberman: Texture gradient as a depth cue, Comput. Graph. Imag. Proc., **5**, 52/67 (1976)

[25] D.Nitzan, A.Brain, and R.Duda: The measurement and use of registered reflectance and range data in scene analysis, Proc. IEEE, **65**, 206/220 (1977)

[26] R.Lewis and A.Johnston: A scanning laser rangefinder for a robotic vehicle, Proc. 5th Int. Conf Artificial Intell., 762/768 (1977)

[27] R.Jarvis: A laser time-of-flight range scanner for robotic vision, IEEE Trans. Pattern Anal. Machine Intell., **PAMI-5**, 122/139 (1983)

[28] R.Jarvis: A perspective on range finding techniques for computer vision, IEEE Trans. Pattern Anal. Machine Intell., **PAMI-5**, 122/139 (1983)

[29] T.Williams: Depth from camera motion in a real world scene, IEEE Trans. Pattern Anal. Machine Intell., **PAMI-2**, 511/516 (1980)

[30] K.Prazdny: Motion and structure from optical flow, Proc. 6th Int. Conf. Artificial Intell., Tokyo, 702/704 (1979)

[31] S.Ando: Profile Recovery System Using a Differential Identity of Binocular Stereo Images, Trans. Soc. Instrumentation and Control Engineers (SICE), **23**, 4, 319/325 (1987) (in Japanese)

[32] S.Ando: A Velocity Vector Field Measurement System Based on Spatio-Temporal Image Derivatives, Trans. SICE, **22**, 12, 1330/1336 (1986) (in Japanese)

[33] S.Ando: Detection of Intensity Disparity in Differential Stereo Vision Systems with an Application to Binocular Luster Perception, Trans. SICE, **23**, 6, 619/624 (1987) (in Japanese)

[34] S.Ando: Two-Dimensional Velocity Measurement Using Gradient Correlation with a Comparison with Spatio-Temporal Gradient Method, Trans. SICE, **23**, 8, 856/858 (1987) (in Japanese)

[35] S.Ando and T.Tabei: Differential Stereo Vision System with Dynamical 3-D Reconstruction Scheme, Trans. Soc. Instrumentation and Control Engineers, **24**, 6, 628/634 (1988) (in Japanese)

[36] S.Ando: A Theory of Semi-Transparent 3-D Surface Perception in Differential Stereo Vision Systems, Trans. SICE, **24**, 9, 973/979 (1988) (in Japanese)

[37] S.Ando: Image feature extraction operators based on curvatures of correlation function, Trans. SICE, **24**, 10, 1016/1022 (1988) (in Japanese)

[38] S.Ando: Texton finders based on Gaussian curvature of correlation function, Proc.1988 IEEE Int.Conf.Syst.Man Cybern., Beijing/Shenyang, 25/28 (1988)

[39] S.Ando: Gradient-Based Feature Extraction Operators for the Classification of Dynamical Images, Trans. SICE, **25**, 4, 496/503 (1989) (in Japanese)

[40] S.Ando: Image Grayness Feature Extraction Based on Extended Gradient Covariances, Trans. SICE, **27**, 9, 982/989 (1991) (in Japanese)

[41] S.Ando and K.Nagao: Gradient-Based Feature Extraction Operators for the Segmentation of Image Curves, Trans. SICE, **26**,7,826/832 (1990) (in Japanese)

[42] H.Yamada and S.Ando: An Analog Electronic Image Motion Sensor, Trans. SICE, **27**,7,729/734 (1991) (in Japanese)

[43] S.Ando: An Analog Electronic Binaural Localization Sensor, Technical Digest of 8th Sensor Symposium, Tokyo, 131/134 (1989)

[44] H.Yamada and S.Ando: The Analog Electronic Motion Sensor, Technical Digest of 8th Sensor Symposium, Tokyo, 127/130 (1989)

[45] Peng Chen and S.Ando: A Generalized Theory of Waveform Feature Extraction Operators for High-Resolution Detection of Multiple Reflection of Sound, Annual Report of Engineering Research Institute of University of Tokyo, **50**, 167/174 (1991) (in Japanese)

[46] S.Ando, H.Shinoda, K.Ogawa, and S.Mitsuyama: A Three-Dimensional Sound Localization System Based on the Spatio-Temporal Gradient Method, Trans. SICE, **29**, 5, 520/528 (1993) (in Japanese)

[47] H.Shinoda, T.Geng, S.Yamashita, M.Ai, and S.Ando: A Gradient-Based Image Measurement System for Characterizing 3-D Deformation Under Pressures, Trans. SICE, **29**, 7, 963/972 (1993) (in Japanese)

[48] S. Ando: An Autonomous Three-Dimensional Vision Sensor with Ears, Proc. Int. Workshop Machine Vision Applications, Kawasaki, Japan, 417/422 (1994)

[49] S. Ando: An Intelligent Three-Dimensional Vision Sensor with Ears, Sensors and Materials, **7**, 3, 213/231 (1995)

[50] S. Ando and H. Shinoda: Ultrasonic Emission Tactile Sensing, IEEE Control Systems Magazine, **15**, 1, 61/69 (1995)

Intelligent Sensors
H. Yamasaki (Editor)

Auditory System

Kota TAKAHASHI

Department of Communications and System Engineering,
The University of Electro-Communications,
1-5-1 Chofugaoka, Chofu-shi, Tokyo 182, Japan

Abstract

This paper presents four approaches for the development of an intelligent sound receiver. Prior to the description of their systems, remarkable characteristics of the biological sensing system are described in order to suggest a final goal of auditory sensing technology. In the latter half of the paper, a sensor fusion method for an intelligent sound receiver is also described.

1. INTRODUCTION

Sound sensing techniques can be classified into three categories as shown in Figure 1. The basis for this classification is the abstraction level of the sensing system outputs.

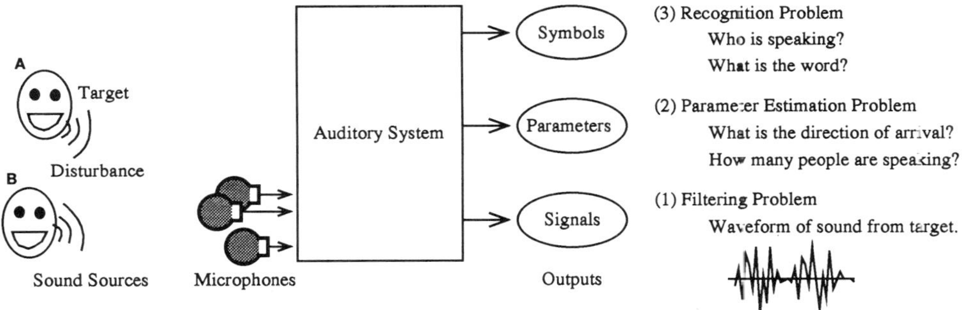

Figure 1. Three categories of outputs from a sound sensing system.

The first category, the lowest abstraction level, contains techniques for a filtering problem or a prediction problem. Outputs of the systems in this category are acoustical signals, whose abstraction level is equal to that of the inputs.

Parameter estimation problems, e.g., estimation of DOA (direction of arrival), estimation of number of sound sources, or estimation of spectrum, fall under the second category. The outputs of these systems are a few parameters, which are not acoustical signals.

The third category, the highest abstraction level, deals with recognition problems. The outputs are symbols or abstract information, which are no longer signals or numerical information. Speech recognition problems and speaker identification and verification problems are included in this category.

In this paper, sensing systems in the first category are mainly described. Thus, the function of the systems is to pick up or extract a target sound from a noisy environment. In other words, an intelligent sound receiver is the subject of this paper.

2. INTELLIGENCE OF AUDITORY SYSTEM

First of all, let us elucidate the remarkable functions of the biological auditory sensing system in order to formulate the final goal of auditory sensing technology.

Cocktail party effect: Human beings can catch important words, e.g., our own name or important keywords, even in a noisy environment such as a party room. In the human nervous system, important information is emphasized and disturbance noises are suppressed. Such adaptive information selectability is termed the *cocktail party effect* [1]. Although the cocktail party effect is originally a term used in psychology, it has found use in literature of sensing technology because many sensor engineers have been interested in this effect which describes well typical intelligent ability of human auditory sensing.

Auditory scene analysis: In the ears, all sound signals from different sound sources are added, thus we can receive only the combined sound signal. The difficulty of understanding sound environment is due to this situation. Nevertheless human beings have an ability to separate each sound signal and recognize acoustical environments. Such function is termed *auditory scene analysis.* This term became common after it was used as the title of a book by Bregman [2]. As yet there is no complete computational theory for auditory scene analysis; therefore, many engineers who are developing intelligent sound sensing systems are interested in this theory. If the computational theory is developed, it will be of great help in the study of intelligent sensing systems.

3. SOME APPROACHES

Sound sensing systems with the aim of realizing the cocktail party effect are described here. Two different approaches are taken to implement the cocktail party effect. One is constructing systems similar to the human nervous system using a computational algorithm; another is constructing systems different from the human nervous system using a novel algorithm. Sensing systems described in this paper fall under the latter approach, because computational theory of the cocktail party effect has not yet been developed

In this section, four approaches for the development of sound receivers are reviewed. Although some of them possess functions other than sound sensing, all of them can distinguish target sound from disturbance without the need for signals for learning or a priori knowledge about the positions of sound sources. The author chose these four approaches because such autonomous adaptability is one of the most remarkable characteristics of the cocktail party effect.

3.1. Widrow's adaptive noise canceler

Figure 2 shows an adaptive noise canceler [3] proposed by Widrow. Although microphones are utilized as sensors in this figure, Widrow's studies were not confined to acoustical sensing techniques. In this approach, sensors are divided into two groups: the *primary* sensor group and the *reference* sensor group. Usually one sensor is utilized as the primary sensor and the rest are utilized as reference sensors.

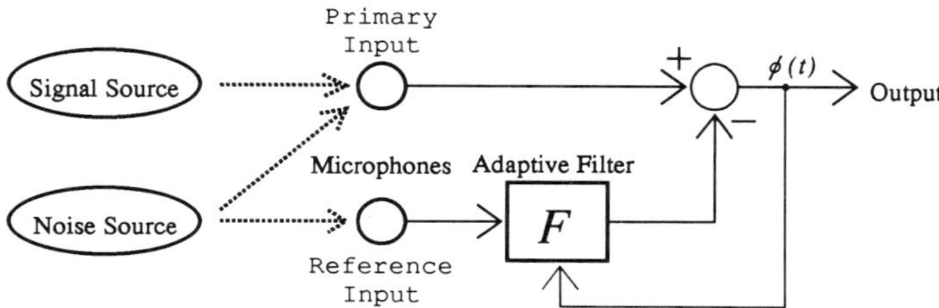

Figure 2. Adaptive noise canceler.

The main characteristic of this approach is the zero transfer functions between the signal source and the reference sensors. By using filters, noise components in the output are suppressed because the coefficients of the filter are adjusted in order to minimize the output power. In contrast, the signal component is not suppressed because of the zero transfer functions between the signal source and the reference sensors. Thus, only the target signal can be extracted. In this method, signal and noise are distinguished by the zero transfer function between the sound sources and the microphones.

3.2. Flanagan's array microphone system

Flanagan proposed an array microphone system [4] whose directivity can be controlled automatically. As shown in Figure 3, the system consists of sixty-three (7×9) microphones with their corresponding programmable delay elements and a microprocessor for controlling the delay elements.

The following are details of the operation. In the first step, directivity of the system is scanned throughout a room by the different delay time vectors. Next, the system judges whether a voice, or only the background noise, is included in the sound from each direction. For this judgement, the ratio of a long-term (150ms) average of sound power to a short-term (30ms) average is utilized. Finally, in order to receive a voice with optimum directivity, the sixty-three delay time vectors of the sound receiver are changed to the time vectors which give the maximum value of a short-term (30ms) average of the sound power.

3.3. Two-filter method

The two approaches mentioned above require a priori knowledge of the transfer functions or the relative positions of microphones. In contrast, the *two-filter method* [5] described here does not require such a priori knowledge. In other words, the system can adapt to sound environment using only received signals.

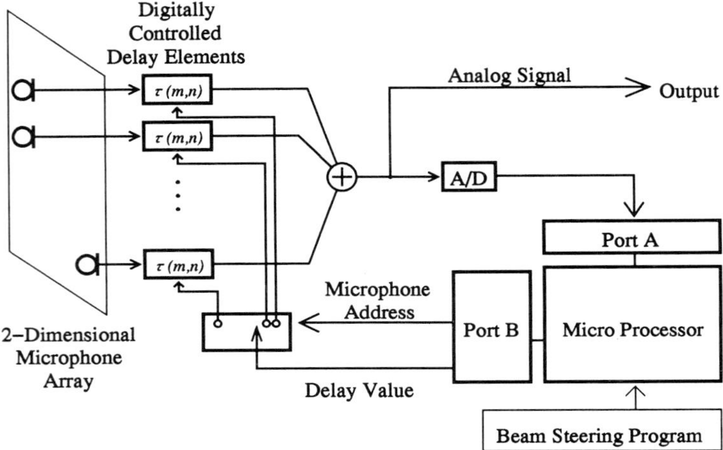

Figure 3. Flanagan's microphone array.

Figure 4 shows a block diagram of the two-filter method. Six channel inputs are converted to digital signals by A-D converters; then, six digital signals are filtered by two filter banks. Signal processor LSIs specific to a FIR filter are utilized to extract the target sound.

This system is designed to detect a target sound whose power fluctuates with time. Background noises are assumed to show no fluctuation. When these assumptions are satisfied, the inputs of the two latches in Figure 4 are described as

$$P_1(t_1) \simeq a_1 P_S(t_1) + b_1 P_N, \tag{1}$$
$$P_2(t_1) \simeq a_2 P_S(t_1) + b_2 P_N, \tag{2}$$
$$P_1(t_2) \simeq a_1 P_S(t_2) + b_1 P_N, \tag{3}$$
$$P_2(t_2) \simeq a_2 P_S(t_2) + b_2 P_N, \tag{4}$$

where t_1 and t_2 mean different times, $P_S(t_1)$ denotes the target sound power around t_1, P_N denotes the sum of the sound powers of disturbance noises, a_1 is an overall gain from the target to the output of digital filters of bank 1, a_2 is that of bank 2, b_1 is an overall gain from the noise sources to the output of digital filters of bank 1, and b_2 is that of bank 2.

Filter bank 1 is used to detect the target sound, whereas filter bank 2 is utilized to find better coefficients than those of filter bank 1. The comparison of the two filters is accomplished using the following evaluation value e:

$$e = \frac{P_1(t_1) + P_1(t_2)}{P_2(t_1) + P_2(t_2)} \cdot \frac{P_2(t_1) - P_2(t_2)}{P_1(t_1) - P_1(t_2)} \tag{5}$$
$$= \frac{P_S(t_1) + P_S(t_2) + 2\frac{b_1}{a1} P_N}{P_S(t_1) + P_S(t_2) + 2\frac{b_2}{a2} P_N}\ . \tag{6}$$

If bank 1 gives a better signal-to-noise ratio than that of bank 2, then $b_1/a_1 < b_2/a_2$, and $e < 1$. On the other hand, if bank 2 gives a better signal-to-noise ratio than that of

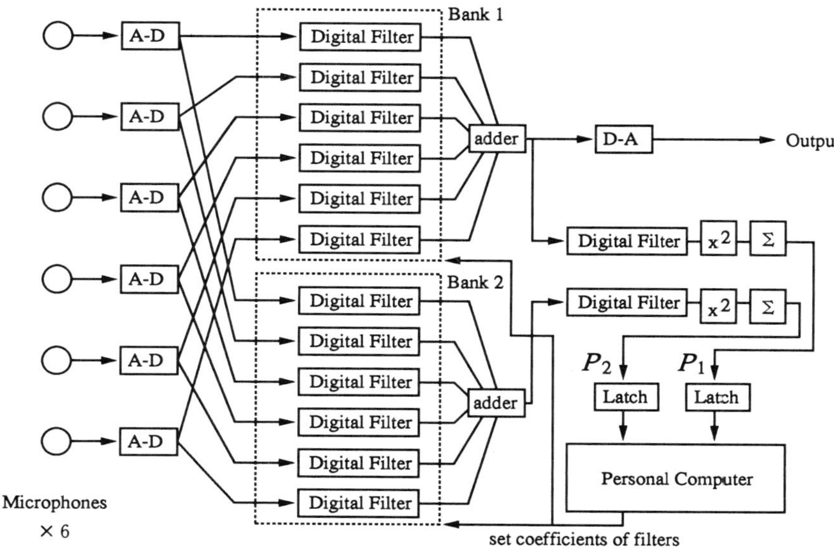

Figure 4. Block diagram of an intelligent receiver using the two-filter method.

bank 1, then $b_1/a_1 > b_2/a_2$, and $e > 1$. Therefore e can be used to judge which bank is better.

Figure 5 shows experimental results using the two-filter method. Here, weights (coefficients) of filter 1 ($f_1, f_2, \ldots f_6$) were fixed in one minute; weights of filter 2 were changed in order to find better weights than those of filter 1. If better weights were found, they were used as revised weights of filter 1 in the next one minute. Weights of filter 1 are plotted in this figure.

In the first half of the experiment, the disturbance sound was a white noise. In the second half, the disturbance sound was generated by a buzzer. When the noise source was changed to the buzzer, weights of filters changed to appropriate values in order to suppress the sound of the buzzer.

3.4. Cue signal method

We note that the algorithm of the two-filter method is restricted to the comparison of two filters. In other words, this method does not provide any instruction regarding the choice of the coefficients of filter 2. Thus, coefficients of filter 2 must scan all possible values or must change randomly. This means low efficiency of searching coefficients and longer adaptation time using the two-filter method.

The *cue signal method* [6–10] described here solves this problem by using an internal learning signal. An internal learning signal enables to adjust the coefficients using the steepest descent algorithm. The steepest descent algorithm allows more efficient search than random search. Figure 6 shows a block diagram of the cue signal method.

Details of the FIR filter part in Figure 6 are shown in Figure 7. In order to extract the target sound, a thirty-two-tap FIR filter is utilized for each microphone. Output of the

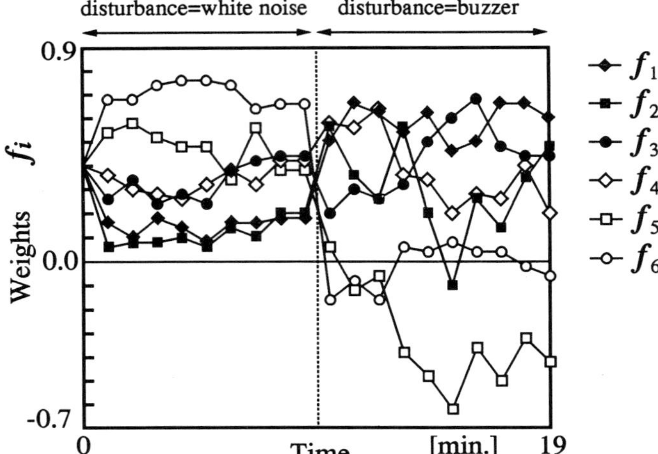

Figure 5. Adaptation of weights for six microphones.

filter is

$$\phi(t) = \sum_{n=1}^{N} f_n \, y_n(t), \tag{7}$$

where f_n are coefficients of the filter, $y_n(t)$ are values of whole taps, and N is the number of taps ($N = 192$). Let $d(t)$ be a learning signal and $e(t) = \phi(t) - d(t)$ be an error.

If the criterion for optimality is to minimize the time average of $e(t)^2$, the optimum solution for the coefficients can be described as the *normal equation*

$$\boldsymbol{p} = R\,\boldsymbol{f}, \tag{8}$$

where $\boldsymbol{p} = (p_n)$ is a correlation (time average) vector between $d(t)$ and $y_n(t)$, $R = (R_{n_1 n_2})$ is a correlation matrix of $y_n(t)$, and $\boldsymbol{f} = (f_n)$ is an optimum coefficient vector.

The main characteristic of the cue signal method is, as mentioned above, an internal learning signal. An internal learning signal $d(t)$ is obtained from the product of the cue signal $\alpha(t)$ and a delayed acoustical signal $\psi(t)$.

$$d(t) = \alpha(t)\,\psi(t). \tag{9}$$

For the explanation, let $\psi_S(t)$ be a target signal component in $\psi(t)$. Under the following four assumptions:

(1) target signal $s(t)$ can be separated into two factors: an envelope signal $a(t)$ and a stationary carrier signal $c(t)$; $s(t) = a(t)\,c(t)$;

(2) cue signal $\alpha(t)$ has no correlation with $c(t)$, $\psi(t)$, or $n_j(t)$, where $n_j(t)$ is interference noise of object j. In addition, target signal $s(t)$ has no correlation with $n_j(t)$;

(3) cue signal $\alpha(t)$ has a positive correlation with instantaneous target sound power $a(t)^2$, but no correlation with instantaneous power of interference;

(4) time average of cue signal $\alpha(t)$ equals zero;

the optimum coefficient vector with this internal learning signal is equal to the optimum coefficient vector with the learning signal $\psi_S(t)$, which is one of the most suitable learning signals because it does not contain a noise component [6,7]. As a result, a target signal

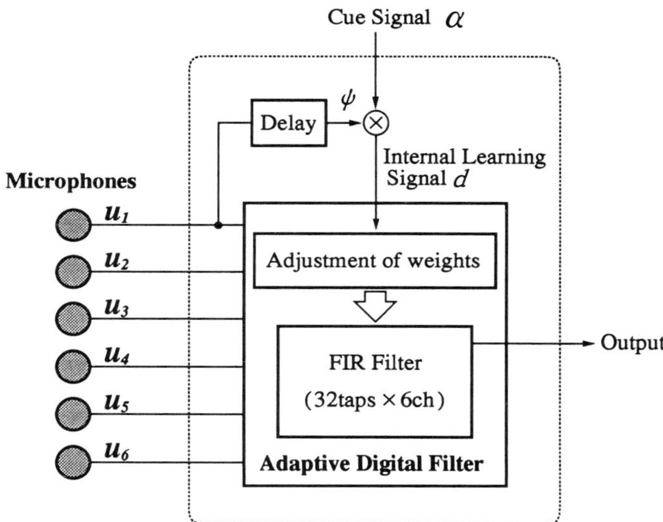

Figure 6. Block diagram of an intelligent receiver using the cue signal method.

can be estimated well by the cue signal method.

The equivalence between the internal learning signal $d(t)$ and the optimum learning signal $\psi_S(t)$ can be easily proven if two correlation vectors $\boldsymbol{p}$ for $d(t)$ and $\boldsymbol{p}$ for $\psi_S(t)$ are calculated, because both vectors are equivalent. Equivalent correlation vectors result in equivalent coefficient vectors from equation (8), since R is independent of learning signal.

In summary, if we can estimate the target sound power $a(t)^2$, the cue signal method can be used to extract the target sound. Furthermore, the cue signal method does not require precise estimation of the target sound power. In other words, the only requirement

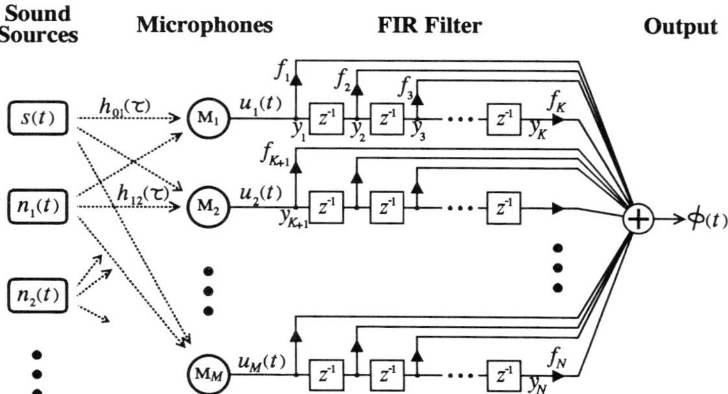

Figure 7. Sound signal flow in an intelligent receiver using the cue signal method.

is that the cue signal correlate with the sound level of the target but not with the sound level of the interference. Thus, many signals can be used as cue signals. For example, in the case of **3.3**, the power fluctuation signal from a microphone output can be used as a cue signal if the bias (DC component) of this fluctuation signal is eliminated.

4. INTELLIGENT RECEIVER USING SENSOR FUSION

Sensor fusion and integration techniques [11] have recently attracted much interest in the field of intelligent sensors. Fusion of different modal sensors is one of the most attractive topics for research. Many types of sensors are currently under development; thus many different cross-modal sensor fusion systems can be constructed. However, if we consider human sensory data, the most typical cross modality involves fusion of an auditory sensor (ear) and a visual sensor (eye). The McGurk effect [12,13] is one proof of the human ability to fuse audio-visual information. In addition, recent neurophysiological studies revealed the mechanism of fusion and integration of visual and auditory senses [14].

Using the cue signal method, an intelligent sound receiver using the sensor fusion technique can be constructed. The system described here distinguishes a target from interfering objects using visual information. The system monitors the environment with a video camera, and decides which object is the target. The system can be divided into two subsystems: a visual subsystem and an audio subsystem. These two subsystems are connected by a cue signal, which is an estimated signal of power fluctuation of a target sound.

The task of the visual subsystem is generation of the cue signal, and can be divided into two. One involves an automatic segmentation of an image taken by the video camera. The image contains not only the image of the target but also images of interfering objects. Therefore segmentation is necessary for capture and tracking of the target image only. The other task involves estimation of sound power of an object using the segmented image.

A simple example of the relationship between a sound power and the image is shown in Figure 8. The figure shows a cross-correlation function between the sound power of a speech and vertical velocity of lips. Since the velocity of lips can be calculated by measuring its optical flow, the sound power can be estimated visually. The estimated sound power of which bias component is eliminated can be utilized as a cue signal. For a more precise estimation, an internal model of the target is essential for generation of a cue signal.

We construct these two subsystems, i.e., visual subsystem and audio subsystem, using 71 digital signal processors (DSP) as shown in Figure 9.

The visual subsystem consists of 40 visual processing units (VPUs) with three frame memories and sets of high-speed DMA hardware which connect neighboring VPUs for communication of images. Automatic segmentation of an image and estimation of the sound power are carried out by the VPUs.

On the other hand, the audio subsystem consists of 31 visual processing units (APUs) with a local memory and common-bus structure. The calculations of FIR filtering and adjustment of the coefficients of the filter are carried out by this subsystem.

Figure 10 shows an example of a cue signal generator implemented in the VPUs. This

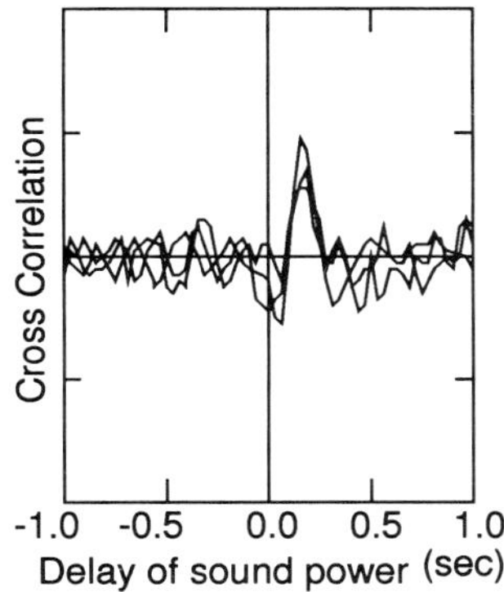

Figure 8. Correlation of velocity of lips with sound level.

method was developed in order to detect a voice from a noisy sound environment. The system knows the shape and color variation of faces, thus searching for and tracking of a face can be carried out. In this method, DOA was estimated using microphone signals, the results of which are also utilized to generate the cue signal.

5. SUMMARY

In this paper, two remarkable characteristics of the biological sensing system, the cocktail party effect and auditory scene analysis, were described in order to suggest a final goal of auditory sensing technology. Then four different approaches for the development of intelligent sound receivers were reviewed briefly. Although the gap between the biological sensing system and the artificial sensing system is not negligible, some of the intelligence of the biological sensing system has been realized in the artificial sensing systems.

In the last section, the sensor fusion techniques using the cue signal method were described. Other types of intelligent sound receivers will be developed in the near future, since sensor fusion techniques are currently areas of much interest in the field of intelligent sensors.

References

[1] E. C. Cherry, Some Experiments on the Recognition of Speech with One and with Two Ears, J. Acoust. Soc. Am., vol.25(5), 975–979 (1953)

[2] A. S. Bregman, Auditory Scene Analysis: The Perceptual Organization of Sound, The MIT Press, Cambridge, Mass. (1990)

[3] B. Widrow *et al.*, Adaptive Noise Cancelling: Principles and Applications, Proc. IEEE, vol.63(12), 1692–1716 (1975)

[4] J. L. Flanagan, J.D. Johnston, R. Zahn, and G. W. Elko, Computer-steered Microphone Arrays for Sound Transduction in Large Rooms, J. Acoust. Soc. Am., vol.78(5), 1508–1518 (1985)

[5] K. Takahashi, Auditory System, Computorol, No.21, 53–58, in Japanese (1988)

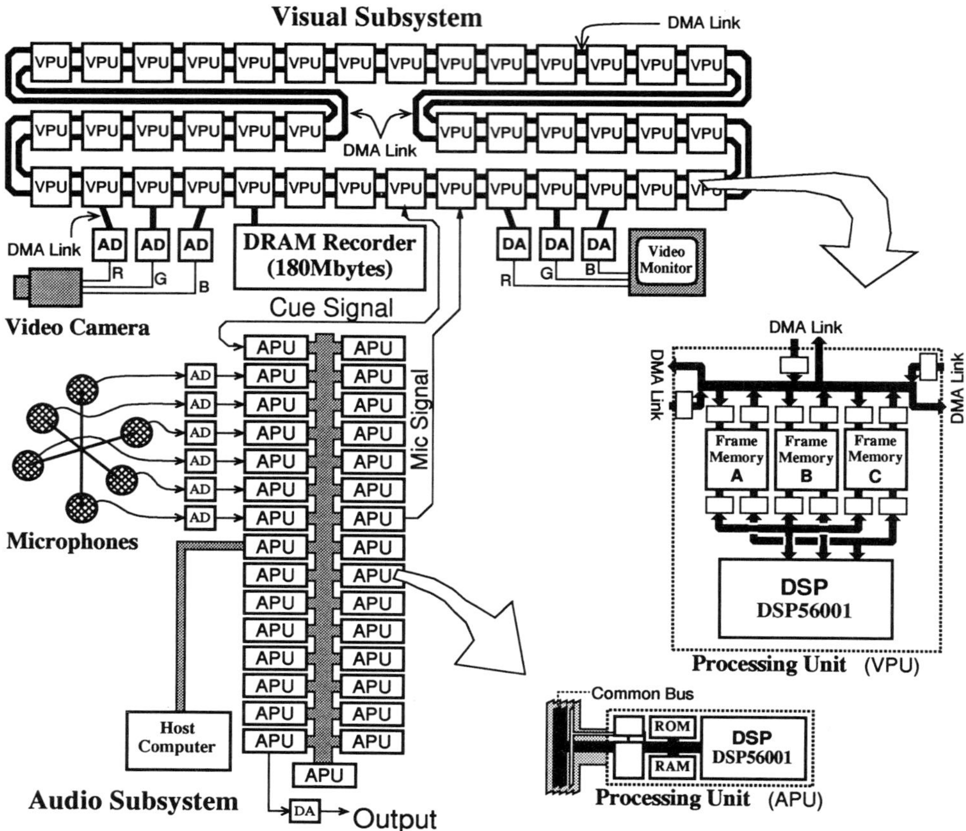

Figure 9. Real-time sensor fusion system.

[6] K. Takahashi and H. Yamasaki, Self-Adapting Multiple Microphone System, Sensors and Actuators, vol. A21-A23, 610–614 (1990)

[7] H. Yamasaki and K. Takahashi, An Intelligent Adaptive Sensing System, Integrated Micro-Motion Systems, F. Harashima(Editor), Elsevier, 257–277 (1990)

[8] H. Yamasaki and K. Takahashi, Advanced Intelligent Sensing System Using Sensor Fusion, International Conference on Industrial Electronics, Control, Instrumentation, and Automation (IECON '92), San Diego, USA, 1–8 (1992)

[9] K. Takahashi and H. Yamasaki, Real-Time Sensor Fusion System for Multiple Microphones and Video Camera, Proceedings of the Second International Symposium on Measurement and Control in Robotics (ISMCR '92), Tukuba Science City, Japan, 249–256 (1992)

[10] K. Takahashi and H. Yamasaki, Audio-Visual Sensor Fusion System for Intelligent

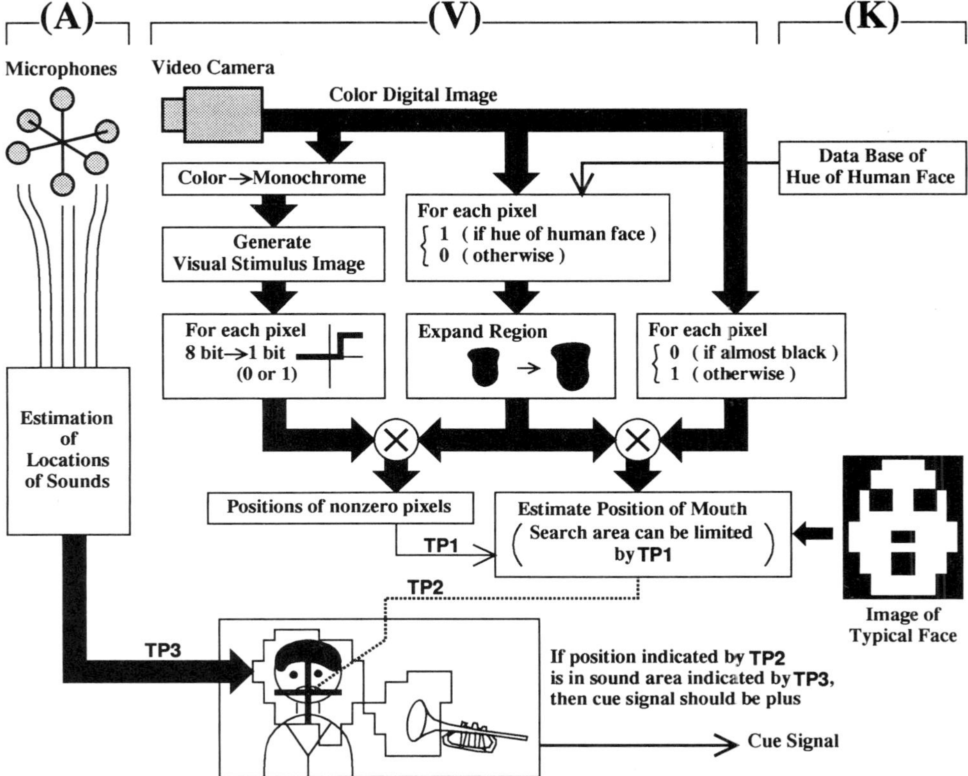

Figure 10. Block diagram of the cue signal generator. Audio information (A), visual information (K) and internal knowledge (K) are fused to estimate sound power of a target. Audio information is fused again with the result of the first fusion using the cue signal method.

Sound Sensing, Proceedings of the IEEE International Conference on Multisensor Fusion and Integration for Intelligent Systems (MFI '94), Las Vegas, USA, 493–500 (1994)

[11] R. C. Luo and M. G. Kay, Multisensor Integration and Fusion in Intelligent Systems, IEEE Trans. Sys. Man Cybern., vol.19(5), 901–931 (1989)

[12] H. McGurk and J. McDonald, Hearing Lips and Seeing Voices, Nature, vol.264, 746–748 (1976)

[13] J. McDonald and H. McGurk, Visual Influence on Speech Perception Process, Perception and Psychophysics, vol.24, 253–257 (1978)

[14] Meredith and Stein, Visual, Auditory, and Somatosensory Convergence on cells in Superior Collilculus Results in Multisensory Integration, J. Neurophysiol., vol.56, 640 (1986)

Intelligent Sensors
H. Yamasaki (Editor)

Tactile Systems

Masatoshi Ishikawa* and Makoto Shimojo**

*Faculty of Engineering, University of Tokyo, Bunkyo–ku, Tokyo, 113 Japan
**National Institute of Bioscience and Human–Technology, Tsukuba–shi, Ibaraki, 305 Japan

Abstract

Several types of tactile sensors and tactile information processing are described. There are some research subjects such as intelligent sensors using parallel processing, tactile imaging with video signal for tactile image processing using image processor, and active sensing for expanding capabilities of tactile sensors. In this paper, basic concepts on those subjects and examples are shown.

1. Introduction

Tactile sensors provide intelligent machines with a function comparable to the skin sensation. Visual sensors and auditory sensors i.e., detective devices such as video cameras and microphones, are already used in established applications. Techniques for image processing and speech recognition to be made using visual sensors or auditory sensors have also been developed. Compared with the development of the above types of sensors, that of tactile sensors has been slow and it is earnestly hoped that research and development work on tactile sensors will be promoted with respect to both devices and processing techniques.

Tactile sensors have special features different from those of other types of sensors. Attention is focused on such requirements as 1) large sensitive area, 2) flexible sensor material, and 3) high-density detection. These requirements have rarely been involved in the development of conventional types of solid state sensor devices mainly composed of elements made of silicon. To meet these requirements, it is necessary to urgently develop new sensor materials and structures.

From the viewpoint of processing techniques, the tactile sensors cannot be thought of separately from the movement of robot actuators. It is, therefore, necessary to develop techniques such as an active sensing technique in which the movement of robots is utilized for sensing and a sensory feedback technique in which the information detected by a tactile sensor is used for control. Another important research subject in the field of sensors is the sensor fusion to be realized to enable information integration, for example, between a tactile sensor and a visual sensor. These research subjects are shown in **Figure 1**.

2. Intelligent Tactile Sensors

2.1. Detection of Centroid

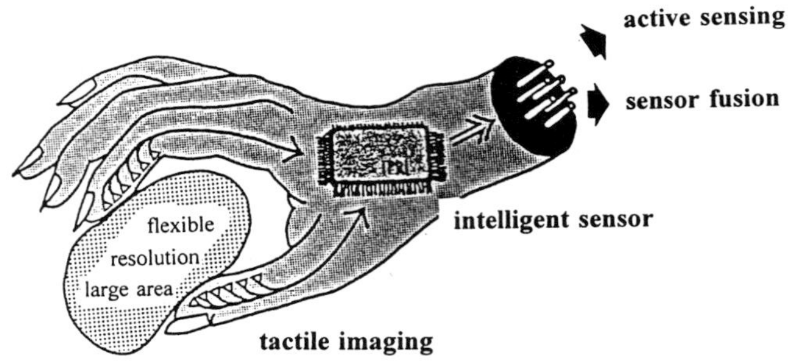

Figure 1 : Research subjects on tactile sensors

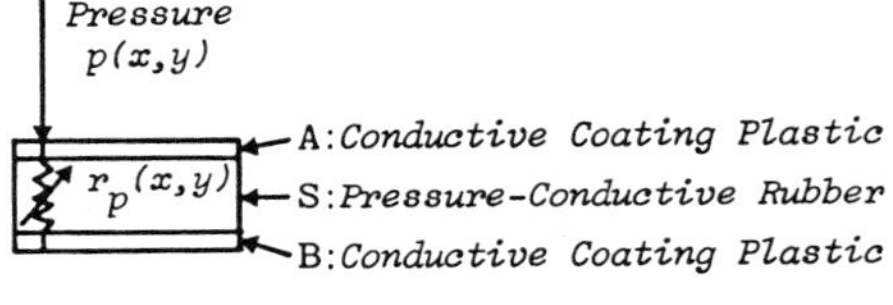

(a) Cross section

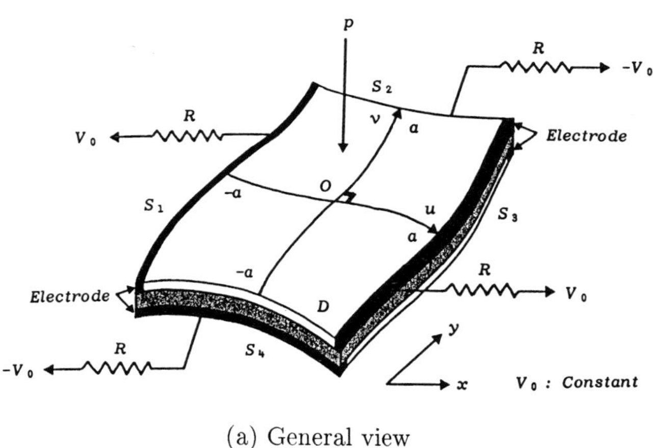

(a) General view

Figure 2 : Structure of centroid sensor

Pressure-conductive rubber is a pressure sensitive material with a softness like that of the skin of human beings. We have developed a sensor that detects the center position of a contact pressure distribution using a pressure-conductive rubber[1]. Its structure is illustrated in **Figure 2**. As illustrated, the sensor has a three-layer construction comprising a pressure-conductive rubber layer S sandwiched between two conductive-plastic layers A and B, with the electrodes formed on two opposing edges. This construction may be regarded as being like a two-dimensionally enlarged potentiometer. The materials making up the sensor are soft and can easily be obtained in various shapes with large surface areas.

Layer S is composed of a pressure conductive rubber in which the resistance r_p per unit area in the thickness direction changes according to the pressure distribution p. In layers A and B, voltage distributions $v_A(x, y)$ and $v_B(x, y)$ are produced, respectively, by the surface resistance r. A current distribution $i(x, y)$ is produced from layer A to B due to r_p, which changes according to the pressure. The basic equation for this sensor can be described as the Poisson equation for the current distribution i :

$$\nabla^2 v_A = ri, \qquad \nabla^2 v_B = -ri \tag{1}$$

An orthogonal coordinate (u, v) is set on the domain D on the sensor, with the four edges of the square being S_1, S_2, S_3, and S_4 (length of an edge being $2a$), with the origin at the center. The first-order moment of the current density $i(x, y)$ from layer A to B in regard to the coordinate u is calculated in the following:

$$I_u = \iint_D u(x, y)\, i(x, y)\, dxdy \tag{2}$$

The basic equation (1) is substituted into Eq. (2). Expanding using Green's theorem, we can write

$$I_u = \int_{\partial D} (u\frac{\partial v_A}{\partial n} - v_A\frac{\partial u}{\partial n})\, dm \tag{3}$$

where n denotes the normal direction of the boundary and m denotes the tangential direction[1].

Since the right-hand side of Eq. (2) is a line integral on the boundary, the boundary condition is substituted, and Eq. (2) is represented using only voltages $[v_A]_{S_1}$ and $[v_A]_{S_3}$ of the electrodes:

$$I_u = k\,([v_A]_{S_1} - [v_A]_{S_3}) \tag{4}$$

where k is a constant.

The sum I of the current flowing from layer A to B can be calculated from the current through resistor R. Combining these data with Eq. (4), the position of the centroid of the current density in regard to the coordinate u is obtained. Using a simple circuit, this value can be derived from the voltages of the electrodes $[v_A]_{S_1}$ and $[v_A]_{S_3}$. The same processing applies to layer B. The current density i can be described by the characteristic function $f(p)$ of the pressure-conductive rubber. Consequently, in practice, the centroid and the sum of $f(p)$ can be detected. The sensors with various shapes are shown in **Photo 1**. A white finger tip in **Photo 2** is a finger type of this sensor.

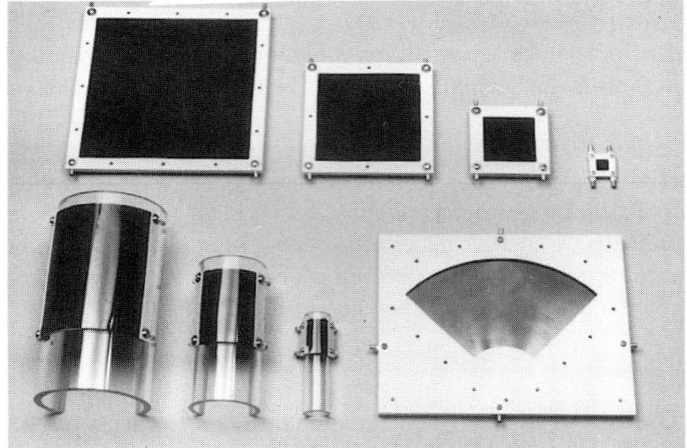

Photo 1 : Centroid sensors

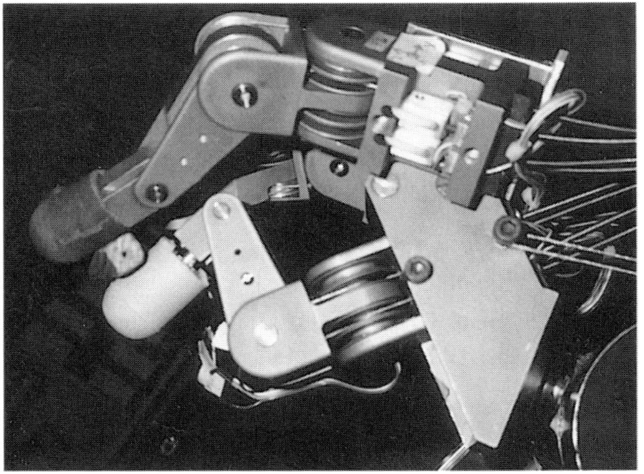

Photo 2 : Centroid sensor on a robot fingertip

2.2. Detection of n-th moment

As to the processing of signals from a set of sensors in array or matrix placement, several relatively well-organized processing methods have been proposed. For the extraction of the features of the Fourier transform, there exists the butterfly circuit (FFT), which is a multi-layer parallel processing system.

A method is also proposed which detects the moment features of the pattern using a multi-layer parallel processing circuit [2, 3]. This method is derived by considering the previous device structure for detecting the centroid and by transforming the distributed system described by the partial differential equation into the discrete described by the

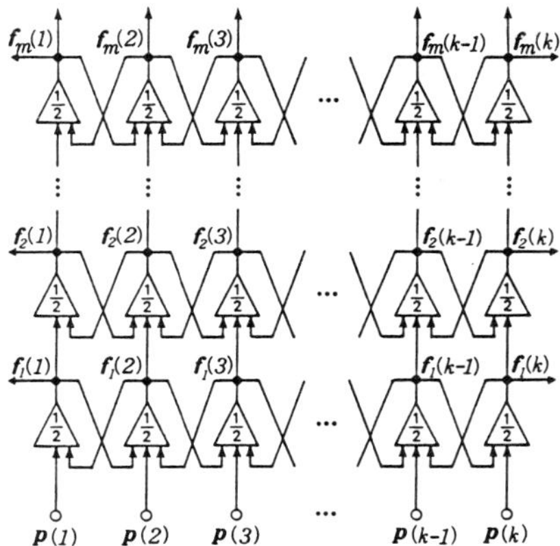

Figure 3 : Multi-layered parallel circuit for n-th moments

difference equation. The Poisson equation itself is not considered, but the basic equation such that the limit of convergence is the original Poisson equation.

By this method, the 0-th moment (sum) and the first moment (sum × position) can be detected by the parallel processing circuit for the set of sensors with array or matrix placement [2]. By constructing a multi-layer structure, the n-th moment can, in general, be detected [3]. As an example, **Figure 3** shows the processing circuit for the array sensor. As shown in this figure, the n-th moment of the array sensor information $p(i)\,(i = 1, 2, \cdots, k)$,

$$M_n = \sum_{i=1}^{k} x^n(i)\,p(i) \tag{5}$$

can be derived by a uniform multi-layer parallel processing circuit, by a simple operation using the boundary values $f_j(1)$ and $f_j(k)$ $(j = 1, 2, \cdots, m)$ of the processing circuits of $m = \mathrm{INT}(n/2) + 1$ layers ($\mathrm{INT}(a)$ indicates the maximum integer not exceeding a).

The zeroth to third moment, for example, can be detected as

$$\begin{aligned} M_0 &= f_1(k) + f_1(1) \\ M_1 &= f_1(k) - f_1(1) \\ M_2 &= f_1(k) + f_1(1) - 2\{f_2(k) + f_2(1)\} \\ M_3 &= f_1(k) - f_1(1) - 6\{f_2(k) - f_2(1)\}. \end{aligned} \tag{6}$$

2.3. VLSI Parallel Processing Tactile Sensor

To realize a versatile processing, it is effective to introduce some form of processors. As an example of such a processor, an LSI is developed for the parallel processing of a local pattern and applied to the processing in the tactile sensor [4].

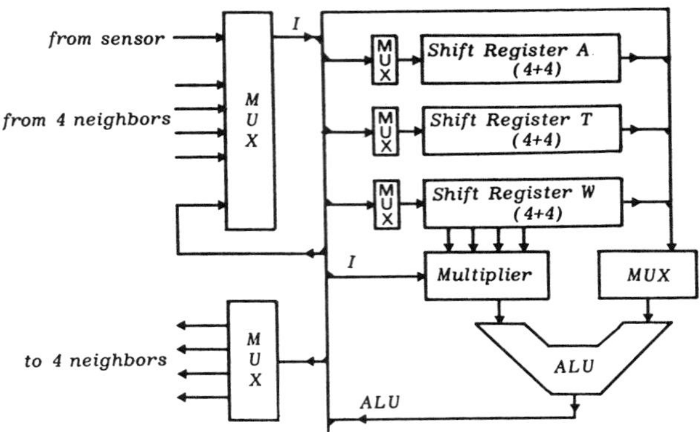

Figure 4 : Architecture of processing element

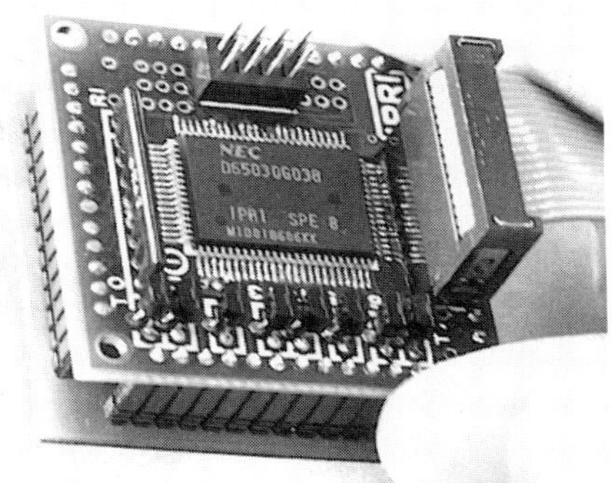

Photo 3 : Tactile sensor using parallel processing LSI

Figure 4 shows the structure of the processing element of the developed LSI. The information from sensor is directly inputed to each processing element. Only the 4 neighbor processing elements are mutually connected due to the constraints in the integration.

As is shown in the figure, various processings can be realized using ALU containing three registers (8 bits) and a multiplier (4 bits). This function is realized by a small number of gates, i.e., 337 gates per processing element using the bit-serial operation.

The processing elements are controlled by the common instruction (10 bits), realizing an SIMD-type parallel processing. The instruction may be provided from the firmware program memory or from the I/O of the upper-level computer. **Photo 3** shows an example of construction of the tactile sensor using this LSI. The sensor has 8×8 pressure-detecting points (substrate under the two layers of substrates) in a matrix form. By scanning in the detecting points in the row direction, the signals from a row are processed by a single processing element. When the edge is extracted by the logic operation from a 1-bit pattern, the detection, processing, and output are completed in 8.3 μs, indicating its high speed.

3. Tactile Imaging with Video Signal

In addition to the approach to realize a pattern processing in the sensor, an approach is considered where the pattern is processed in the upper level of the processing hierarchy, and the lower-level sensor produces an output which is the information in a form suited to the purpose of the processing.

One example which has been making progress recently is to utilize the image processing system as a pattern processing system. Several types of sensors are developed which includes the conversion process from the sensor output into the video signal as a part of the sensor processing mechanism [5, 6, 7, 8]. In these systems, the sensory information is the pressure distribution information. Since these sensors can essentially visualize the pressure information by a TV monitor, they are sometimes called the tactile imagers. **Figure 5** shows the basic system of tactile processing using this type of sensors.

In the sensor, the pressure-conductive rubber is used as the pressure-sensing material, and the change of the resistance is scanned and detected using electrodes. In such a circuit, especially the scanning circuit to detect the change of the resistance and capacitance, it is crucial to develop a faithful scanning circuit to cope with the high-speed video signal while preventing the crosstalk of the signal through the path other from the desired measuring point.

In the sensor of **Photo 4**, the completely independent circuit based on FFT switching is applied[5]. In principle, this method can realize a high-speed scanning without a crosstalk, but the integration is required since a large number of contact points is included. This is a table-mount sensor which can detect the pressure distribution in 64×64 matrix, and the output is processed by the image processor so that the edge or contour information of pressure distribution is extracted and displayed.

We also developed a tactile imaging sensor for measuring the body pressure measurement as shown in **Photo 5** where a diode is used at each detecting point to prevent the crosstalk [6]. As these types of scanning circuits are matched with wide area type of tactile imagers, we also developed a small size type of tactile imaging sensor using a circuit called zero-balancing as shown in **Photo 6** [7]. The tactile imaging sensor can be applied to measurement of pressure distribution of grasping as shown in **Photo 7** [8]. The sensor is useful as an input device for virtual reality systems.

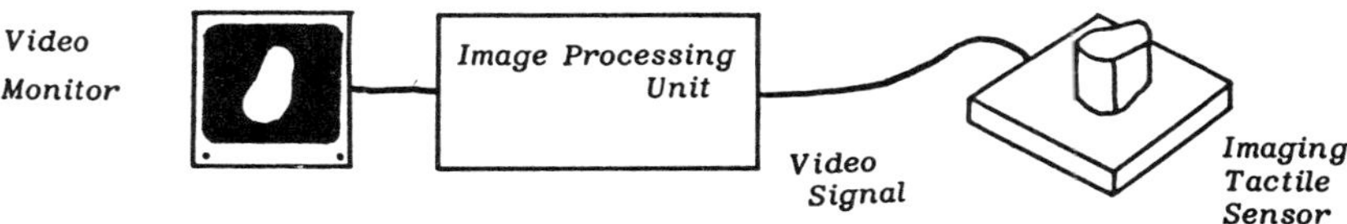

Figure 5 : Tactile pattern processing using image processor

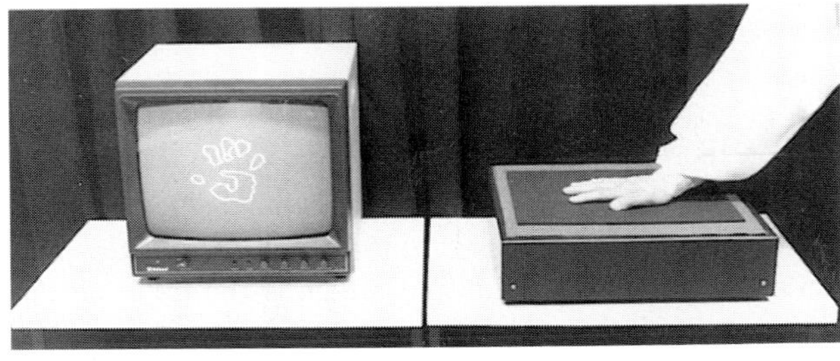

(a) Overview

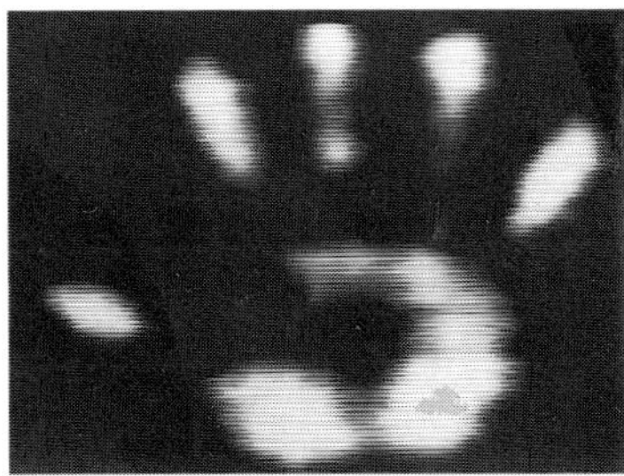

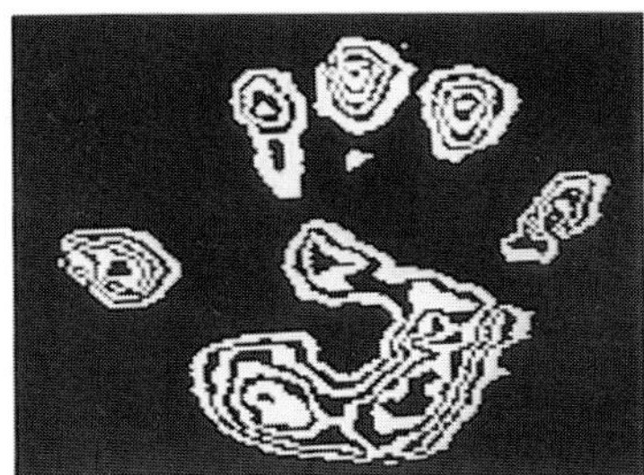

(b) Original tactile image

(c) Contour image

Photo 4 : Tactile pattern processing using tactile imaging sensor with video signal

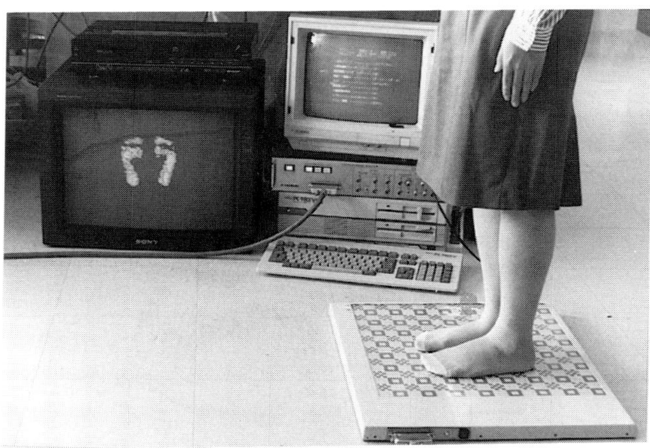

Photo 5 : Tactile imaging sensor for pressure distribution of footprints

Photo 6 : Flexible type of tactile imaging sensor with 64×64 detective points

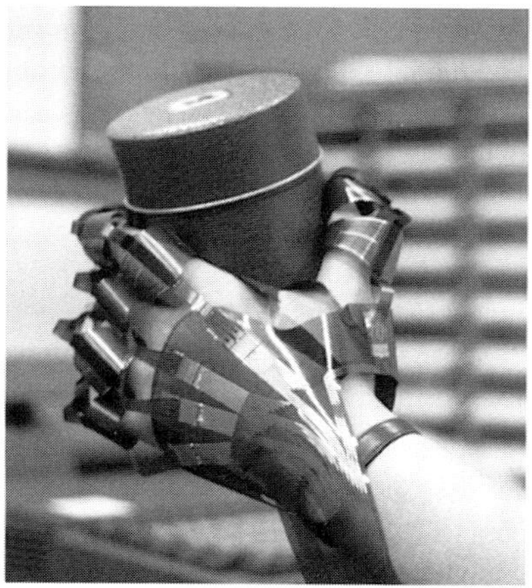

Photo 7 : Tactile imaging sensor for pressure distribution of grasping

4. Active Sensing

4.1. Basic structure of active sensing

The processing configuration of the sensing information is not restricted to the afferent type. There is a method to utilize the efferent information, i.e., a method to utilize the process of moving the sensor as a part of the sensing. This method is called active sensing and is utilized as a means to cope with the locality or uncertainty of the sensor.

In the case of the sensor, especially one installed on the actuator such as the tactile sensor, it is difficult to realize all of the desired sensing functions if sensing is considered simply as a problem of the detecting device. It is necessary to consider the sensing as the active process including the actuator system.

There are primarily three realization configurations for the active sensing.

The first configuration is the generation of measuring motion so that the sensor captures the object within the range of measurement. This operation is difficult if single sensory information is handled and often is

The second configuration is to prevent the localization of the sensory information and arrive at the whole image of the object. When the object of measurement is larger than the measuring range of the sensor, either the sensor or the object must be moved to see or touch the whole object.

The third configuration is to extract the microscopic structure of the object, especially the structure which is more detailed than the resolution of the sensor. In this case, the spatial pattern is transformed into the spatio-temporal pattern by moving the sensor so that the spatial resolution is improved.

4.2. Edge tracking system

An example of the second configuration of the active sensing is the sensory system which recognizes the contour of a still object using a local sensor [9]. **Figure 6** shows the configuration of the system. The system is to simulate the movement where the contour of the object is recognized by finger tip (called tactile movement). This realizes the active operation of the unknown object and the autonomous recognition of the whole image of the object.

More precisely, it is the system where the X-Y recorder is used as the actuator and the edge of the two dimensional object is searched and traced by a local sensor. The measuring operation is completely autonomous. In **Figure 6**, the future orbit (i.e., the object of measurement) shown by the dashed line is completely unknown. The efferent information is input to the internal model (called efferent copy). Then a prediction is made by the internal model using past data. The active movement is executed according to the result of prediction, and the sensory information also is corrected. In the processing of sensory information, the parallel processing discussed in section 2.3 is applied to extract the local edge from the object.

The following two ideas are shown as completely equivalent in the detection of the two-dimensional edge and executing the tracing by this model. The first is "edge detection" → "control of edge position and prediction of tangential direction" → "sensor movement." The second idea is to realize both "control of the edge position" and "constant movement." This is the case where the horizontal decomposition (corresponds to the latter) and the vertical decomposition (corresponds to the former) are completely equivalent, as was pointed out by Brooks [10].

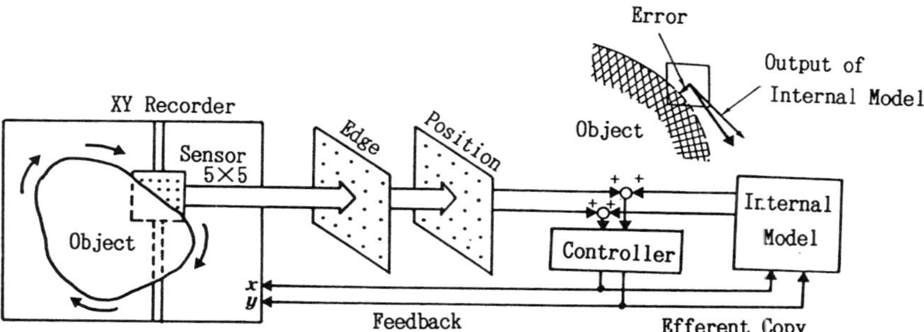

Figure 6 : Architecture of active sensing system

4.3. Use of spatial filter

An example of the third configuration of the active sensing is the sensor system which recognizes the tactile texture of the object surface using the spatial filter [11]. In this sensor, the outputs from the set of tactile sensors in array placement are input to the spatial filter where the response can be adjusted electrically. By executing the tactile motion and adapting the filter response, the coarseness of the object surface is recognized. A kind of parallel processing is realized by the spatial filter.

5. Conclusion

Various types of tactile sensors are described from the various viewpoints of intelligent sensor, tactile imaging and active sensing. Tactile sensing has quite distinctive features, which requires particular sensing processing as well as device materials. In such situation, intelligent sensing architecture plays an important roles to realize high performance tactile recognition.

References

[1] M. Ishikawa and M. Shimojo. A tactile sensor using pressure-conductive rubber. *Proceedings of THE 2nd SENSOR SYMPOSIUM*, pp. 189–192, 1982.

[2] M. Ishikawa. A method for measuring the center position and the total intensity of an output distribution of matrix positioned sensors. *Trans. SICE*, Vol. 19, No. 5, pp. 381–386, 1983 (in Japanese).

[3] M. Ishikawa and S. Yoshizawa. A method for detecting n-th moments using multi-layered parallel processing circuits. *Trans. SICE*, Vol. 25, No. 8, pp. 904–906, 1989 (in Japanese).

[4] M. Ishikawa. A parallel processing LSI for local pattern processing and its application to a tactile sensor. *Trans. SICE*, Vol. 24, No. 3, pp. 228–235, 1988 (in Japanese).

[5] M. Ishikawa and M. Shimojo. An imaging tactile sensor with video output and tactile image processing. *Trans. SICE*, Vol. 24, No. 7, pp. 662–669, 1988 (in Japanese).

[6] K. Kanaya, M. Ishikawa, and M. Shimojo. Tactile imaging system for body pressure distribution. *Proc. 11th Congress of Int. Ergonomics Association*, Vol. 2, pp. 1495–1497, 1991.

[7] M. Shimojo, M. Ishikawa, and K. Kanaya. A flexible high resolution tactile imager with video signal output. *Proc. IEEE Int. Conf. Robotics and Automation*, pp. 384–391, 1991.

[8] M. Shimojo, S. Sato, Y. Seki, and A. Takahashi. A system for simultaneous measuring grasping posture and pressure distribution. *Proc. IEEE Int. Conf. Robotics and Automation*, pp. 831–836, 1995.

[9] M. Ishikawa. Active sensor system using parallel processing circuits. *J. Robotics and Mechatronics*, Vol. 5, No. 5, pp. 31–37, 1993.

[10] R.A. Brooks. A robust layered control system for a mobile robot. *IEEE J. Robotics and Automation*, Vol. RA-2, No. 1, pp. 14–23, 1986.

[11] M. Shimojo and M. Ishikawa. An active touch sensing method using a spatial filtering tactile sensor. *Proc. IEEE Int. Conf. Robotics and Automation*, pp. 948–954, 1993.

Intelligent Sensors
H. Yamasaki (Editor)

Olfactory System

Masayoshi Kaneyasu

Hitachi Research Laboratory, Hitachi, Ltd.,
The Third Department of Systems Research
1-1, Omika-cho 7-chome, Hitachi-shi, Ibaraki-ken, 319-12 Japan

All living organisms are sophisticated sensing systems in which the nervous system handles multiple signals from sensory cells simultaneously or successively. The system is composed of various kinds of sensory cell groups and a signal treatment system centralized in the brain. This arrangement in living systems can be modeled with an electronic system composed of various kinds of sensor groups and a computer. Sensations in living systems, however, cannot be considered without also taking into account the intricate mechanism of signal treatment.

Studies have been carried out in our laboratories to obtain information of the same quality as the sensations in living systems utilizing an electronic system [1-5]. This chapter discusses, as an example of our research, the possibility of whether smells can be identified and quantified by an electronic system composed of an integrated sensor and a microcomputer with signal treatment algorithms in its memory modified to resemble those of living systems.

1. INTRODUCTION

The analytical method and experimental results of identifying and quantifying smells are described using an electronic system composed of an integrated sensor and a microcomputer. The integrated sensor with six different elements on an alumina substrate was fabricated with using thick-film techniques. The elements were kept at about 400℃ with a Pt heater mounted on the sensor's back. Since each element was made from different semiconductor oxides, they possessed different sensitivities to material odors. The integrated sensor was able to develop specific patterns corresponding to each odor as a histogram of conductance ratios for each element. The microcomputer identified the scent on the basis of similarities calculated by comparing standard patterns stored in the memory and a sample pattern developed by integrated sensor. The scent was then quantified using the sensor element with the highest sensitivity to the smell identified. The experimental results show that various smells can be successfully identified and quantified with the electronic system.

2. SMELL IDENTIFICATION METHOD

2.1. Identification in living systems

The mechanism by which vertebrates can identify smell is not yet fully understood. However the experimental results of Moulton et al. were extremely significant from the viewpoint of signal treatment for our research [6,7].

Figure 1 summarizes Moulton's experiments. Pulses were observed through electrodes implanted into the mitral cell layer of the olfactory bulb of a rabbit (Figure 1a). The pulses generated in the bulb at each respiration of a scent show different features depending on the material inhaled and the sites where the electrodes were inserted (Figure 1b). From this experiment, it was proposed that smell may be segregated by a spatio-temporal pattern of excitation in the bulb, shown in Figure 1c in the form of histograms. This indicates that a microcomputer should be able to identify smell, provided the sensor can generate a specific output corresponding to the scent.

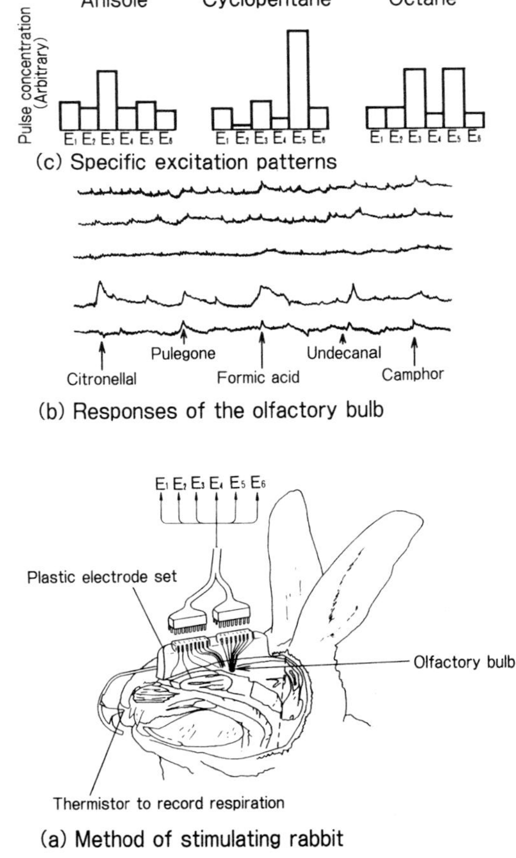

Figure 1. Outline of Moulton experiment

2.2. Identification using thick-film gas sensors

As shown in Table 1, it is known that semiconductor oxides change their conductance when exposed to reducing gases at temperatures around 400℃. However, the sensitivity of each oxide to gases shows a different value even when subjected to the same gas. This phenomenon indicates the possibility of obtaining a specific pattern corresponding to a smell by exposing a multiple number of sensor elements to the smell, since the majority of scents are reducing gases.

Table 1
Gas sensitivity of thick-film oxides

No.	Material	Resistivity ($\Omega \cdot cm$)	Conductance ratio			
			CH_4	CO	H_2	C_2H_5OH
1	ZnO	$6.3x10^3$	1	3.2	1.9	17
2	NiO	$1.1x10^4$	1	0.7	0.48	0.1
3	Co_3O_4	$4.5x10^2$	1	1	0.99	0.9
4	Fe_2O_3	$2.2x10^3$	1	1.1	1.1	2.0
5	TiO_2	$2.1x10^7$	1	0.97	0.71	0.6
6	ZrO_2	$5.1x10^8$	1	1	1.1	1.1
7	SnO_2	$1.0x10^5$	1.5	1.8	6.3	21
8	Ta_2O_5	$5.1x10^7$	1	1	3.6	2.3
9	WO_3	$1.0x10^4$	1	5.9	6.7	25
10	$LaNiO_3$	$2.2x10^2$	1	1	1	0.6

Material no.1, 1wt% Pt added ; Material nos. 2-9, 1wt% Pd added ;
Gas concentration : 1,000ppm
Sensor temperature : 400 ℃

3. EXPERIMENTS

3.1. Preparation of thick-film hybrid sensor

The hybrid integrated sensor was prepared as follows. Figure 2 shows the preparation process of the gas sensor pastes. Thick-film gas sensor paste was prepared by mixing together 89 wt% metal oxide powder, 1 wt% Pd(palladium) or Pt (platinum) powder, and 10 wt% binder glass powder for 1 hour in an automatic mixing machine. The mixture was then kneaded together with an organic vehicle (ethyl cellulose solution mixed with alpha-terpeneol) for 1 hour in the same mixing machine. The degree of polymerization, the concentration of ethyl cellulose, and the amount of the vehicle added were controlled in order to provide the pastes with an optimum printability.

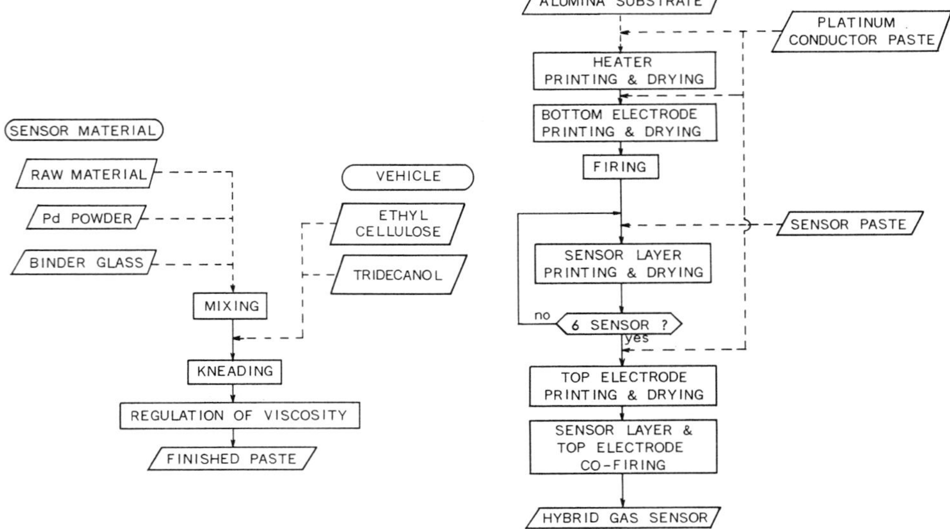

Figure 2. Preparation process of gas sensor pastes

Figure 3. Preparation process of thick-film hybrid gas sensor

The thick-film hybrid sensor was fabricated using thick-film processes. A meander type heater, which maintains sensor activity, was formed on the back side of an alumina substrate and the sensor elements were formed on the front. The heater and the bottom electrode of the sensor were printed with a 400-mesh screen and fired together at 1200℃ for 2 hours. A 165-mesh screen yielded the desired 40-45μm fired thickness for the double layer of gas-sensitive materials. The top electrode was printed with a 400-mesh screen and co-fired with the double gas-sensitive layer at 900℃ for 10 minutes. Figure 3 summarizes the hybrid sensor fabrication processes.

Figure 4 displays the thick-film integrated sensor used in this study. Six sensor elements were fabricated on an alumina substrate. The elements were fabricated using ZnO, ZnO doped with Pt, WO_3 , WO_3 doped with Pt, SnO_2, and SnO_2 doped with Pd as sensor materials and designated S1-S6, respectively.

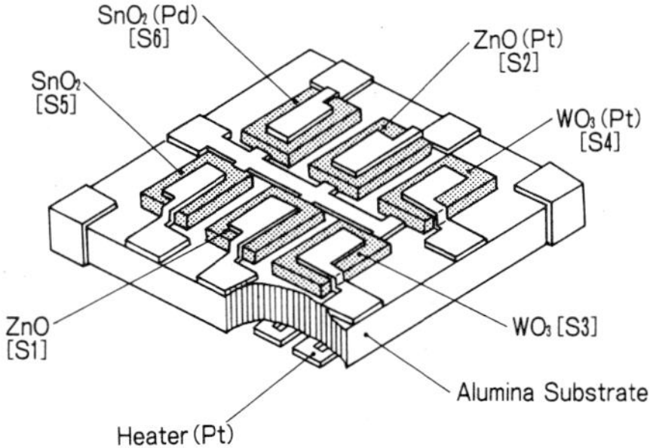

Figure 4. Schematic drawing of integrated sensor

3.2. Measurements

As shown in Figure 5, the measuring of sensing characteristics of each sensor element was carried out using a plastic chamber and a simple electrical measuring system. The smell materials were introduced into the chamber with a micro syringe. The gas concentration in the chamber was kept homogeneous during the measurement with a stirring fan. Ambient temperature and relative humidity in the chamber were kept at about 20℃ and 40~50%//RH, respectively. Six pick-up resistors (Rc) were connected to each sensor element in series. The value of Rc was adjusted to give the same electrical resistance the sensor element had before being exposed to the gas.

The variation in the electrical resistance of the sensor element is measured indirectly as a change in voltage appearing across the load resistor Rc. The analog voltage was received by a scanner and converted into digital voltage data by using a voltmeter. The digital data converted was then fed to a personal computer and became the conductance ratio G/Go described below. Go is the electrical conductivity when a specified gas is absent, and G is the conductivity when the gas flows in.

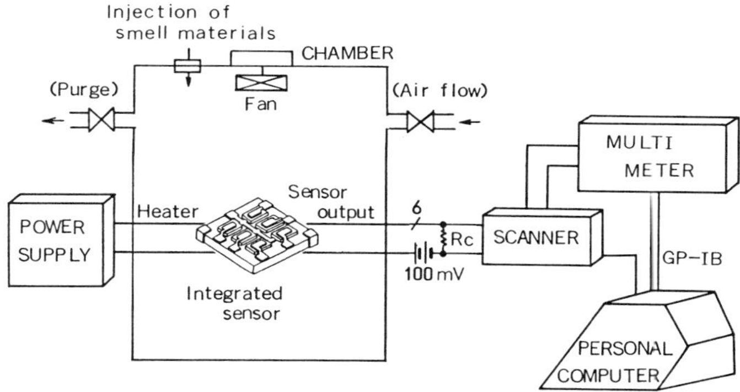

Figure 5. Experimental measuring system

4. EXPERIMENTAL RESULTS

4.1. Pattern generation using the hybrid sensors

The results of the integrated sensor exposed to vapors of anisole, cyclopentane, and octane at 400℃ were compared with those obtained by Moulton [6] and are shown in Figure 6. As is clear from the figure, a conductivity pattern specific enough to identify a smell can be obtained by the integrated sensor, although the pattern profile differs from that obtained with the rabbit.

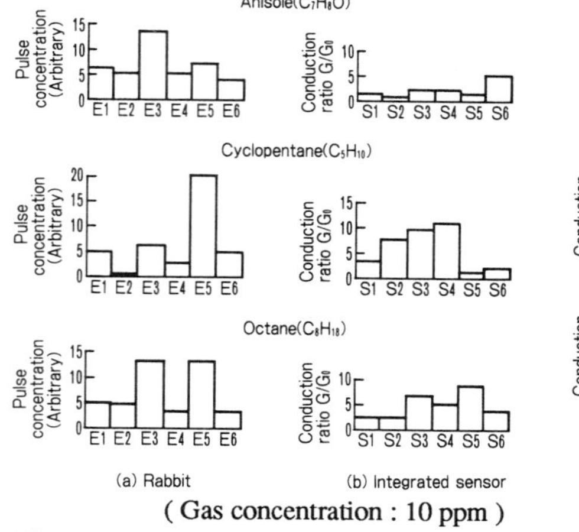

(Gas concentration : 10 ppm)

Figure 6. Comparison of the specific patterns generated by smell materials

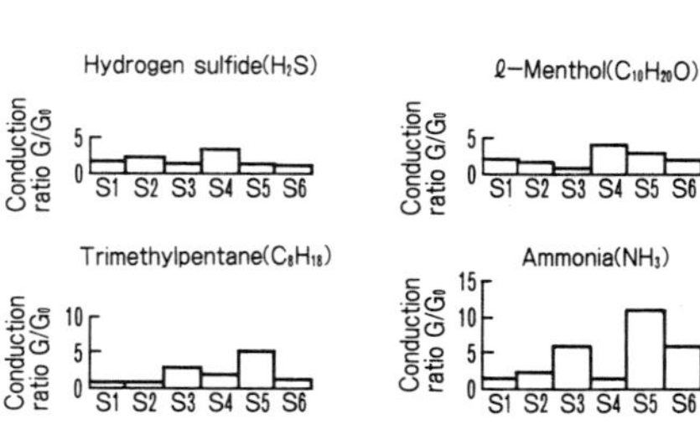

(Gas concentration : 10 ppm)

Figure 7. Specific patterns generated by typical smell materials

Figure 7 shows the specific patterns obtained with the integrated sensor that correspond with the typical smell materials listed in Table 2. These results confirm that the integrated sensor can develop specific patterns similar to those observed in the olfactory bulb of vertebrates such as rabbits.

Table 2
Smell materials dealt with in this research

Odorant	Odor
l-Menthol	Menthol
Hydrogen sulfide	Rancid smell
Ammonia	Intense smell
Anisole Octane	Organic solvent smell
Cyclopentane Trimethyl pentane	Standard fuel

4.2. Pattern recognition algorithm

Figure 8 shows gas sensitivities of various semiconductor oxides. The vertical axis shows the values for the conductivity ratio G/G_0. The ratio of conductivity of the sensor element is given on the ordinate in a logarithmic scale, while the gas concentration is given on the abscissa, also logarithmically. As can be seen, the conductivity ratio varies linearly as a function of the gas concentration. The linear variations have slopes which are noticeably different and dependent on the type of gas. Furthermore, the magnitude of the slope represents the sensitivity of the sensor element to the corresponding type of gases.

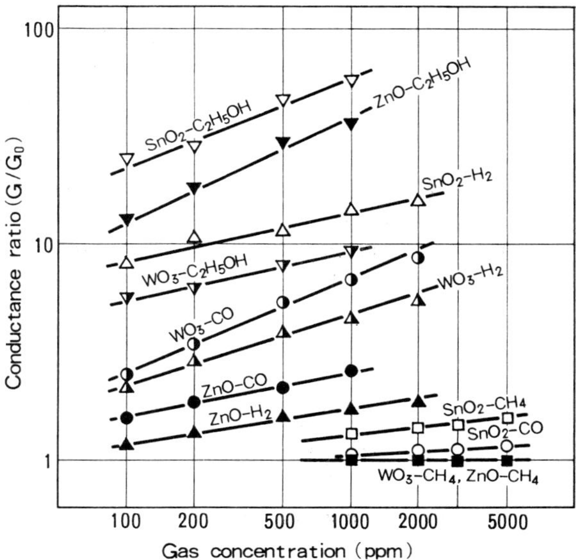

Figure 8. Gas sensitivity of various semiconductor oxides

The gas or smell detection characteristics of semiconductor oxides may be expressed empirically in the following general mathematical form:

$$y_j = (\alpha_i \cdot C + 1)^{m_{ij}}, \qquad (i = 1,2, \ldots ,M, \; j = 1,2, \ldots ,N) \tag{1}$$

In eq.(1), α_i represents the correction coefficient for the scent i to adjust it to the origin of the coordinate system. It may take different numerical values, depending on the gas, and is independent of the material or composition of the sensor element. m_{ij} represents the sensitivity value of the sensor element j for scent i and may differ depending on both the type of sensor material and the gas.

When both sides of eq.(1) are logarithmically rewritten, a numerical value p^i_j, which is independent of the concentration C, can be obtained:

$$\frac{\log y_j}{\sum_{j=1}^{N} \log y_j} = \frac{m_{ij} \cdot \log(\alpha_i \cdot C + 1)}{\log(\alpha_i \cdot C + 1) \cdot \sum_{j=1}^{N} m_{ij}} = \frac{m_{ij}}{\sum_{j=1}^{N} m_{ij}} \; (\equiv p^i_j), \; \sum_{j=1}^{N} p^i_j = 1 \tag{2}$$

The vector p^i_j thus obtained may be regarded as a standard pattern P^i with which the sample pattern X, obtained from the output signals of the N sensor elements, can be compared.

Our experiments, however, reveal a slight concentration dependence for P^i. In order to make smell identification more precise, the concept of a standard pattern class P^i_k was introduced. Eq.(3) is the formula for identifying the gas corresponding to the standard pattern. A schematic diagram for calculating the distances is shown in Figure 9. Each standard pattern class has seven patterns prepared with different concentrations of smell. To calculate the distance, we take the differences between these patterns and the sample pattern, and then multiply them by the weighting factor w. Namely, an arbitrary sample pattern X for an unknown scent is compared with each of the standard patterns P^i_k in order to obtain distance d^i from eq.(3). The formula represents the intraclass clustering procedure for transforming a set of patterns belonging to one class into a denser cluster:

$$d^i = \frac{1}{L^i} \sum_{k=1}^{L^i} \cdot \sum_{j=1}^{N} w^i_j \, (x_j - p^i_{kj})^2 \tag{3}$$

where L^i is the number of standard patterns which correlate with the different concentrations of smell, x_j and p^i_{kj} represent the jth elements of the patterns X and P^i_k, respectively, and w^i_j represents the weighting factor placed on the jth elements. The weight w^i_j is given as

$$w^i_j = \frac{1}{(\sigma^i_j)^2 \cdot \sum_{j=1}^{N} (\sigma^i_j)^{-2}} , \quad \sum_{j=1}^{N} w^i_j = 1 \tag{4}$$

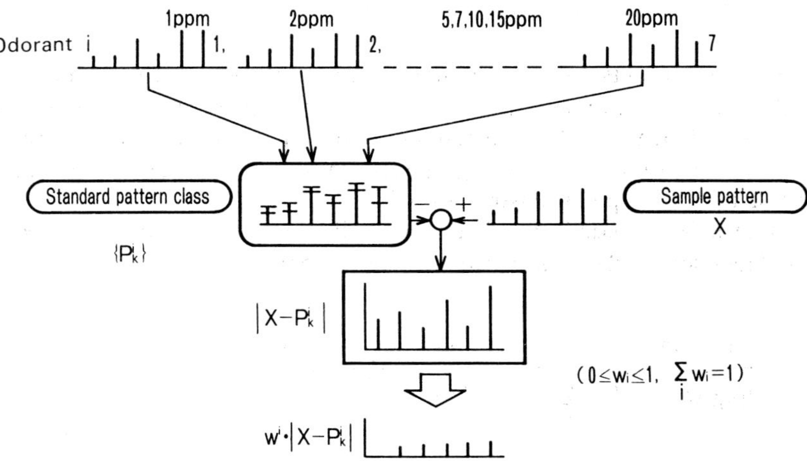

Figure 9. Calculation of distances

where σ^i_j represents deviation and is given as

$$(\sigma^i_j)^2 = \frac{1}{L^i}\sum_{k=1}^{L^i}(p^i_{kj})^2 - \left(\frac{1}{L^i}\sum_{k=1}^{L^i}p^i_{kj}\right)^2 \tag{5}$$

When i is determined to satisfy eq.(6)

$$d_{min} = \min\{d^i\} \tag{6}$$

the smell that comes in contact with the integrated sensor can be identified as the smell which belongs to the ith class. Figure 10 shows a pattern recognition algorithm. We obtained seven distances for each smell. They were then compared with a sample pattern and the smell was identified by finding a standard pattern from the minimum of distances.

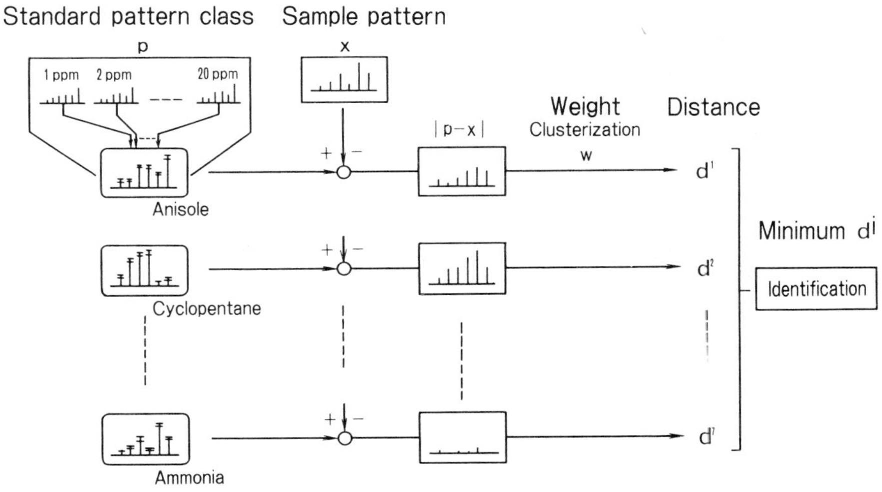

Figure 10. Pattern recognition algorithm

4.3. Identification and quantification

Figure 11 shows the smell detector developed for this study. The thick-film integrated sensor is covered with a stainless steel mesh cap (center left of the detector). The rectangular area with a round window on the top is an 8-bit single-chip microcomputer 6805. The printed circuit board mounted vertically between them is a related-circuit module that includes operational amplifiers. Analog signals put out from the integrated sensor are converted into digital signals which are sent to the microcomputer to calculate the distances. Then, the microcomputer identifies and quantifies the smell. The detector can generate the signals to sound an alarm or control a gas safety system as necessary. A basic flow chart for this process is shown in Figure 12.

Figure 11. The smell detector developed

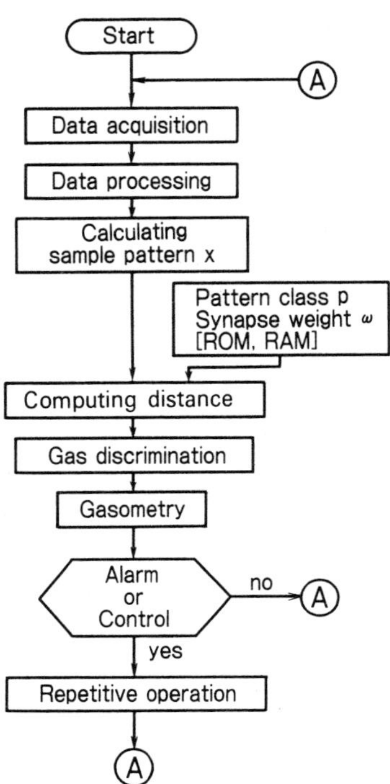

Figure 12. Pattern recognition algorithm

(a) Identification: For calculations, we measured the conductivities of the six sensor elements that were exposed to seven different gases with arbitrarily selected concentrations. The detection patterns were arithmetically determined. As we can see in Table 3, there was a great coincidence between the actual gases and identification results as shown by the starred values.

Figure 13 illustrates the results of smell identification obtained when the detector was exposed to anisole, cyclopentane, octane, trimethylpentane, hydrogen sulfide, *l*-menthol, and ammonia. The numbers along the horizontal axis refer to the standard pattern class of each smell material, i.e., 1=anisole, 2=cyclopentane, etc. As seen in the figure, the minimal value of distance for a smell was observed when the smell was compared with the corresponding standard pattern class. For anisole, the number one was the lowest value; for cyclopentane, the number two was the lowest, etc. The results indicated that this identification method was highly precise.

Table 3
Typical example of calculated distance

Sample Gas	Gas Conc. [ppm]	d^1	d^2	d^3	d^4	d^5	d^6	d^7	Identification result
Anisole	2.0	*0.001	0.202	0.121	0.283	0.111	0.165	0.152	Anisole
Cyclopentane	1.0	0.194	*0.002	0.111	0.283	0.036	0.129	0.248	Cyclopentane
Octane	2.0	0.130	0.118	*0.001	0.063	0.053	0.107	0.036	Octane
Trimethyl pentane	5.0	0.271	0.276	0.040	*0.002	0.187	0.190	0.036	Trimethyl pentane
Hydrogen sulfide	2.0	0.116	0.054	0.054	0.204	*0.001	0.040	0.144	Hydrogen sulfide
l-Menthol	5.0	0.167	0.123	0.086	0.208	0.029	*0.002	0.179	*l*-Menthol
Ammonia	0.5	0.147	0.233	0.022	0.053	0.164	0.164	*0.002	Ammonia

d^1,d^2,d^3,d^4,d^5,d^6 and d^7 mean distance to Anisole, Cyclopentane, Octane, Trimethyl pentane, Hydrogen sulfide, *l*-Menthol, and Ammonia, respectively

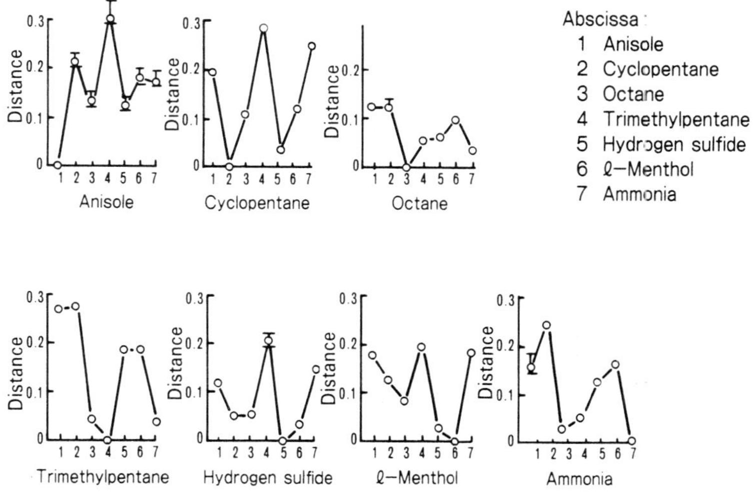

Figure 13. Pattern distance of smell material for each standard pattern class

(b) Quantification: Once the smell material was identified, the concentrations of each were derived by analyzing the detection signal output from the sensor element which exhibited the highest relative sensitivity to the detected scent. The calculation was done using numerical data catalogued in a large-scale integrated (LSI) memory and linear interpolation.

Figure 14 shows the specific pattern for ammonia obtained using the integrated sensor and the sensor element with the highest output. In this case, the element fabricated with SnO_2 was used because of its high relative sensitivity to ammonia.

Figure 15 illustrates the experimental results obtained by measuring the concentration of seven smell materials after identification. The results indicate the probability of highly accurate quantitative analysis for each sample.

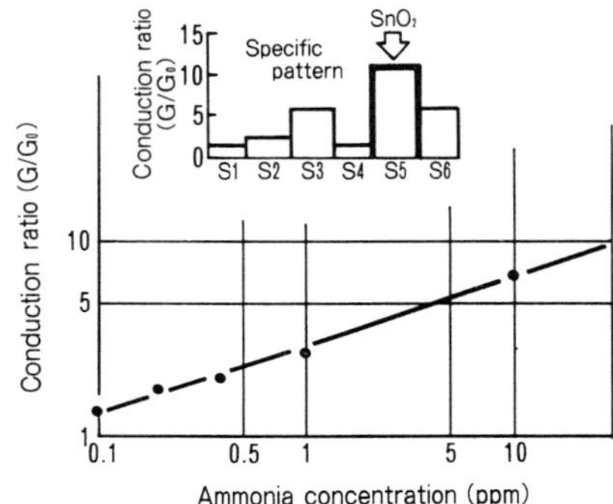

Figure 14. Specific pattern and sensitivity of SnO_2 element to ammonia

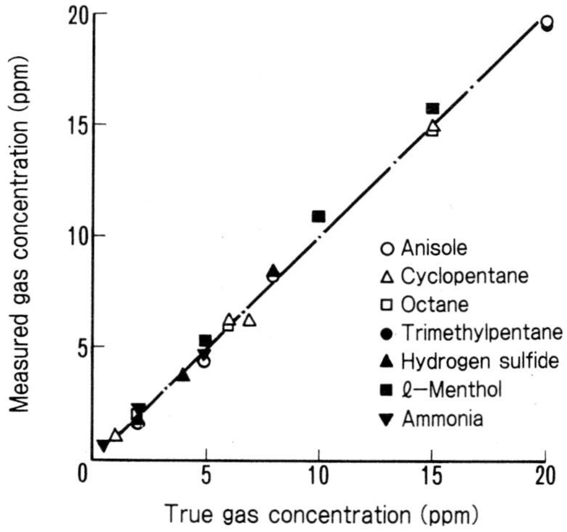

Figure 15. Experimental results of gasometry

5. SUMMARY

This chapter describes the identification and quantification of smells using an electronic system. Through this research, the following points have been proved.

1) An integrated sensor can be used to develop a specific pattern corresponding to a specific scent.
2) Smell recognition is possible by evaluating the distance of the specific pattern to standard pattern classes.
3) Using an integrated sensor and a microcomputer, identification and measurement of scents can be done successfully.

Using the sensing system and arithmetical processing method that we have described, any given smell or organic solvent vapor can be qualitatively and quantitatively identified with high reliability and high accuracy. We believe that as long as the standard patterns of gases to be detected differ from each other, a great number of gases can be identified and quantified.

REFERENCES

1. A. Ikegami et al., Thick Film Sensors and Their Integration, in Proc. European Hybrid Microelectronics Conf. (1982) pp.211-218.
2. M. Kaneyasu et al., A Method to Identify Smell by Integrated Sensor, in Proc. Soc. of Instruments and Control Engineers of Japan (1983) pp.271-272.
3. A. Ikegami and M. Kaneyasu, Olfactory Detection Using Integrated Sensor, in Digest of Technical Papers, Transducers '85 (1985) pp.136-139.
4. M. Kaneyasu et al., Smell Identification Using a Thick-Film Hybrid Gas Sensor, IEEE Trans. on Components, Hybrids, and Manufacturing Technology, Vol. CHMT-10, No.2 (1987) pp.267-273.
5. M.Kaneyasu et al., A Gas Detecting Apparatus, US Patent No. 4638443 (1985)
6. D.G.Moulton, Electrical Activity of Olfactory System of Rabbits With Indwelling Electrodes, Olfactory and Taste, Oxford: Pergamon (1963) pp. 71-84
7. S. Takagi, Kyukaku No Hanashi (in Japanese;Discussion of Olfactory Sense), Iwanamishinsho (1974)

Intelligent Sensors
H. Yamasaki (Editor)

Echo location systems

W. Mitsuhashi

Department of Communications and Systems Engineering,
University of Electro-Communications, Chofu, Tokyo 182 Japan

Abstract

A computational model of bionic sonar is proposed for some species of bats that emit only frequency modulated(FM) sounds. From an engineering point of view, this study assumes that the FM bats could detect phase information of their high-frequency sounds to perceive the range and speed of moving objects. Since the phase information contained in echo waveforms is an important clue to the perception of speed, a mock architecture is designed in this study so as to simulate a coherent correlation receiver. On the basis of the frequency-tuned mechanism in the bats' auditory periphery, this architecture comprises a bank of constant-Q filters uniformly distributed along the log-frequency axis. Numerical experiments show that an equivalent function to the correlation receiver is realized and the range and speed of a simulated object can be measured precisely by the proposed model.

1. INTRODUCTION

Nocturnal bats explore their environments by emitting ultrasonic sounds and detect flying insects by finding a slight difference in time-frequency structures between the emitted sound and returned echoes[1]. This behavior is called *echo location*. In accordance with the sound structures in time and frequency, echo locating bats are classified into two categories: CF-FM bats which emit constant frequency(CF) sounds followed by frequency modulated(FM) components, and FM bats emitting only FM sounds. The validity of these sounds for the purpose of echo location depends on how the information on moving objects is processed in the bats' auditory system.

In order to find prey and then pursue it, a bat must be able to perceive not only range but relative speed. It should be noted that a bat's flying speed generally reaches approximately 1% of the sound propagation speed. Thus, echoes returned from moving objects may be distorted excessively in time and frequency structures. It is well known that the CF-FM bats can perceive the object speed by detecting the frequency difference of CF components between the emitted sound and received echoes. Compared with the auditory systems of other mammals, the auditory periphery and the auditory cortex of the CF-FM bats have peculiarly evolved for the purpose of detecting the frequency shift[2, 3], because the perception of relative speed plays an important role for insectivorous bats that pursue flying prey.

On the other hand, FM sounds contribute effectively to improving the range resolution, and enable the bats to detect weak echoes buried in ambient noise[4]. However, these useful features of FM sounds do not perform well unless a correlation receiver is provided in the bats' auditory system.

There have been various arguments as to whether a correlation receiver exists in the bat's auditory system [5, 6, 7, 8, 9, 10]. Needless to say, the correlation operation may not be executed in a mathematical sense. However, it is generally assumed that a certain kind of biological analog of the correlation operation would exist in the bat's auditory system [5, 6]. For example, Suga has proposed a neural model for processing range information by cross-correlation analysis[11]. In his model the processing elements for speed information are composed of frequency sensitive neurons, and, hence, the mechanism for perceiving relative speed is constructed in a different manner from that for range perception.

In a recent study in an animal behavioral experiment Menne *et al.* have suggested that there may exist phase sensitive mechanisms in the auditory system of the FM bats[12]. On the basis of their experimental results, we will propose a phase sensitive model for the FM bats to perceive both the range and speed of a moving object.

2. ESTIMATION OF RANGE AND SPEED OF MOVING OBJECT

If an emitted sound $u(t)$ has a relatively long duration, the position of the reflection from a moving object is different for each temporal component of the sound. The distance of travel of each of these components thus differs individually. In consideration of this condition Kelly *et al.* have derived the following equation for a generalized representation of the echo reflected from an object moving with constant speed:

$$e(t) = \sqrt{s_o} \cdot u(s_o \cdot [t - \tau_o]), \tag{1}$$

where the parameters s_o and τ_o depend on the speed and the range of the moving object. In fact s_o denotes the dilation of the time dimension due to the Doppler effect caused by the object's motion, and τ_o indicates the delay time, i.e. time taken for the sound component $u(0)$ to propagate to the moving object and reflect back to the emitter[13]. In the above expression $\sqrt{s_o}$ is necessary in order to account for the fact that the sound energy does not change under the Doppler distortion. However, we will disregard the term $\sqrt{s_o}$ throughout this article because of its slight influence on the estimation of the object's motion.

In Eq.(1) it is assumed that $u(0)$ is transmitted at $t = 0$ and is reflected by the object just at $t = \tau_o/2$. This assumption does not necessarily mean that $u(t)$ must have an energy at $t = 0$. If the object range at the time of the reflection is r_o, the following relationships hold:

$$\tau_o = \frac{2r_o}{c} \quad \text{and} \quad s_o = \frac{c - v_o}{c + v_o} \simeq 1 - \frac{2v_o}{c}, \tag{2}$$

where c is the propagation speed of sound and v_o is the speed of motion of the object. The measurement of both range and speed is thus reduced to the problem of estimating the following two parameters: a propagation delay (τ_o) and a time scale factor (s_o).

2.1. The need of the FM bats for motion perception

The dilation of the time dimension due to the object's motion will change the sound waveform to cause a phase rotation. For a CF sound of frequency ω, the phase rotation is written as follows:

$$\omega t \longrightarrow \omega s_o \cdot (t - \tau_o) = \omega t - \frac{2v_o\omega}{c}t - \omega s_o \tau_o, \tag{3}$$

where the second term of the right-hand side is the Doppler frequency shift proportional to the speed of motion of the object. It has been demonstrated that the CF-FM bats utilize the Doppler shift as a clue to the perception of an object's motion[14, 11].

The phase offset $\omega s_o \tau_o$ in Eq.(3) is induced by the propagation delay. This term can not be utilized, however, as a clue to delay perception, because the value of this term exceeds in general the detectable limit of $\pm\pi$. Instead of detecting the phase offset, the CF-FM bats have adopted an alternative way of sensing the propagation delay of the FM location sounds. That is, this class of bat species can alter the function of their location sound in accordance with the motion parameters to be perceived.

On the contrary, the FM bats have to perceive both range and speed with only the FM sound. Of course, in the same manner as the CF-FM bats, the FM bats are likely to be able to perceive an object's range by detecting the propagation delay of the FM sound. However, the frequency of the FM sound varies in time, and the Doppler frequency shift is not fixed accordingly. Hence the FM bats probably do not utilize the Doppler shift as a clue to the perception of an object's motion.

The FM bats can estimate the range of objects whenever they emit the location sounds, and, consequently, they are able to perceive the range differences between sound emissions. For this reason there is possibly no necessity for the FM bats to be sensitive to the speed of motion of an object. However, in the wild state, unwanted echoes from undesirable objects could be observed in addition to those echoes reflected from prey insects. In such a noisy environment it must be more profitable for the FM bats to be able to discriminate the difference in relative speed between the insects and other objects. In the discussion that follows we will consider which type of processing mechanism we must design for the perception of the range and speed of motion of an object.

2.2. Detection of time dilation by the Mellin transform

Since the motion of an object affects the time scale of the returned echo, the perception of motion is equivalent to an estimation of the change in the time scale. Echo location is thus closely analogous to the pattern recognition problem that classifies various objects with respect to their scale and translation[15]. It is well known that the Mellin transform is suited to the recognition of a pattern whose size is magnified or reduced[16].

The Mellin transform of the emitted sound $u(t)$ is defined by

$$U_M(\Omega) = \int_0^\infty u(t) \cdot t^{-j\Omega-1} dt, \quad t > 0, \tag{4}$$

where $u(t)$ is assumed to have energy for $t > 0$. Converting the variable t into an exponential function $\exp x$, we can rewrite the above equation in a more convenient form:

$$U_M(\Omega) = \int_{-\infty}^{+\infty} u(\exp x) \exp(-j\Omega x) dx = \int_{-\infty}^{+\infty} \tilde{u}(x) \exp(-j\Omega x) dx. \tag{5}$$

This equation indicates that the Mellin transform is equivalent to the Fourier transform after the logarithmic conversion of the time variable t.

Let us assume that the propagation delay of the returned echo is canceled, i.e., $e(t) = u(s_o t)$. In this case the Mellin transform of $e(t)$ can be written

$$\begin{aligned} E_M(\Omega) &= \int_0^\infty u(s_o t) t^{-j\Omega-1} dt = \int_{-\infty}^{+\infty} \tilde{u}(x) \exp\{-j\Omega(x - \ln s_o)\} dx \\ &= \exp(j\Omega \ln s_o) \cdot U_M(\Omega). \end{aligned} \tag{6}$$

The time dilation due to the motion of the object is thus replaced by a time shift as a result of the logarithmic transform, and, consequently, a phase difference results from the Fourier transform. If the speed of motion of the object is much slower than the propagation speed of the sound, then we obtain from Eq.(2)

$$\ln s_o \sim -2v_o/c. \tag{7}$$

The Mellin transform of the delay-canceled echo therefore becomes

$$E_M(\Omega) = \exp(-j2\Omega v_o/c) \cdot U_M(\Omega). \tag{8}$$

Eq.(8) states that we can estimate the time dilation due to the object's motion by comparing the phase of the returned echo with that of the emitted sound on the basis of the Mellin transform. However, in order for this technique to function effectively, the time of origin of both sounds must precisely coincide. It is therefore impossible to apply the Mellin transform directly to the estimation of time dilation, because the translation of the echo waveform along the time axis is still unknown. In other words, we must measure the object range before estimating the speed.

2.3. Sound waveform suitable for estimating the motion of an object

A measurement of speed on the basis of the phase shift described in Eq.(8) will require a large amplitude of the Mellin transform at a specific value of Ω against noise or other spurious signals. To maximize the amplitude of the Mellin transform at $\Omega = \alpha$, for example, we must arrange for the sound $\tilde{u}(x)$ in Eq.(5) to be a sinusoidal function

$$\tilde{u}(x) = \tilde{a}(x) \cdot \exp\{j(\alpha x + \beta)\}, \quad x = \ln t, \quad t > 0, \tag{9}$$

where α and β are arbitrary real variables and $\tilde{a}(x)$ denotes an envelope function. If $\tilde{a}(x)$ is designed to be a Gaussian function, the sound energy can be localized both in the Ω domain and in the x domain. Expressing the sound $\tilde{u}(x)$ in terms of time t, we obtain a sound function of the following form:

$$u(t) = a(t) \cdot \exp\{j(\alpha \ln t + \beta)\}, \quad t > 0, \tag{10}$$

where the envelope $a(t)$ is represented by a log-normal function. Substituting Eq.(10) into Eq.(1), we obtain the returned echo as follows:

$$e(t) = \sqrt{s_o} a(s_o \cdot [t - \tau_o]) \cdot u(s_o \cdot [t - \tau_o]) \simeq \exp(-j2v_o\alpha/c) \cdot u(t - \tau_o), \tag{11}$$

where the effect on the envelope $a(t)$ of the change in time scale is assumed to be negligible.

From Eqs(11) and (8) we find that the time dilation caused by the object's motion yields a constant phase shift proportional to the speed of the object. One should note that the constant phase shift in both Eqs(11) and (8) corresponds to a constant frequency shift for the case of the CF sound given in Eq.(3). If the FM bats emit the sound defined in Eq.(10) and are sensitive to phase information, then they will be able to perceive the motion of an object by sensing the phase difference between the emitted sound and a returned echo. However, the problem of how to estimate the range, or the delay τ_o, still remains unsolved. The upper part of Fig.1 shows a typical example of the sound defined in Eq.(10), illustrated by a solid line, and its echo, given by Eq. (11) when $\tau_o = 0$, illustrated by a dashed line. Their transformed versions on the abscissa are also shown in the lower part of Fig.1. We find from these figures that the echo location sounds defined in Eq.(10) are transformed to sinusoidal waves by the logarithmic conversion of the time axis, and that the time dilation accordingly becomes the time shift.

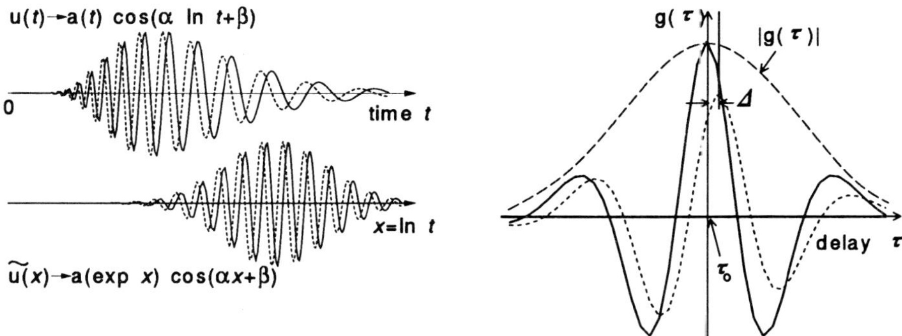

Figure 1: Upper plot: Examples of emitted sound(solid line) and returned echo(dashed line). Lower plot:Their transformed versions on the abscissa.

Figure 2: Cross-correlation of a simulated echo $e(t)$ with the sound $u(t)$.

3. LINEAR–PERIOD MODULATED SOUND

The instantaneous frequency of the sound defined in Eq.(10) is given by the time derivative of its phase as follows:

$$\omega = \frac{d(\alpha \ln t + \beta)}{dt} = \frac{\alpha}{t}. \tag{12}$$

Clearly the instantaneous period of this sound increases linearly with time, and accordingly the sound is called the *Linear–Period Modulated*(LPM) sound[17, 18]. It has been pointed out that the LPM sound has a typical feature of the FM location sounds produced by the little brown bats *Myotis lucifugus*[19].

Correlating the echo of Eq.(11) with the emitted sound $u(t)$, we obtain the following result:

$$g(\tau) = \int_{-\infty}^{+\infty} e(t) \cdot u^{\star}(t-\tau)dt = \exp(-j2v_o\alpha/c) \cdot \phi_{uu}(\tau - \tau_o), \tag{13}$$

where the symbol $\star$ represents complex conjugation and $\phi_{uu}(\tau)$ denotes the autocorrelation function of $u(t)$. Eq.(13) shows that the magnitude of the correlation has a maximum amplitude at $\tau = \tau_o$ independent of the object's motion. We also find that the phase value at $\tau = \tau_o$ is proportional to the speed of the object.

Numerical examples of the cross-correlation of $e(t)$ with $u(t)$ are illustrated in Fig. 2. In this figure the solid line represents the cross-correlation in the case where the object is stationary, while the cross-correlation of the echo reflected from an object receding with arbitrary speed is illustrated by the dotted line. However, the absolute value, or amplitude, of the correlation function shown by the dashed line does not change with the object's motion.

Note that the maximum amplitude of $g(\tau)$ arises from the fact that the individual frequency components of $u(t)$ are aligned so as to be in phase at the time τ_o as a result of the correlation operation. Thus we can consider that, when all frequency components are in phase, the time and the phase give important clues to the estimation of the range and speed of motion of an object.

3.1. Linearization of non-linear spectral phase

The energy of a frequency-modulated sound is dispersed in time, and the energy dispersion is characterized by the negative derivative of the spectral phase $\Psi(\omega)$ with respect to frequency ω:

$$d(\omega) = -\frac{\partial \Psi(\omega)}{\partial \omega}, \tag{14}$$

i.e., sound energy with instantaneous frequency ω_k appears at time $d(\omega_k)$[20]. Thus the elimination of non-linear terms in the spectral phase will yield the maximum amplitude of the resultant output. This process can be realized by delaying each of the frequency components to linearize the spectral phase of the returned echo.

The frequency spectrum of the LPM sound $u(t)$ is given by its Fourier transform

$$U(\omega) = A(\omega) \cdot \exp\{-j(\alpha \ln \omega + \theta_o)\}, \tag{15}$$

where $A(\omega)$ is the amplitude spectrum and θ_o indicates a residual phase independent of frequency ω[15]. The spectrum of the echo described in Eq.(11) is then written as follows:

$$\begin{aligned} E(\omega) &= \exp(-j2v_o\alpha/c)\exp(-j\omega\tau_o)U(\omega) \\ &= A(\omega) \cdot \exp\{-j(\alpha \ln \omega + \omega\tau_o + 2v_o\alpha/c + \theta_o)\}. \end{aligned} \tag{16}$$

In order to linearize the spectral phase of Eq.(16), we will use the mechanism shown in Fig.3 based on a model of the function of the auditory periphery. In this model we

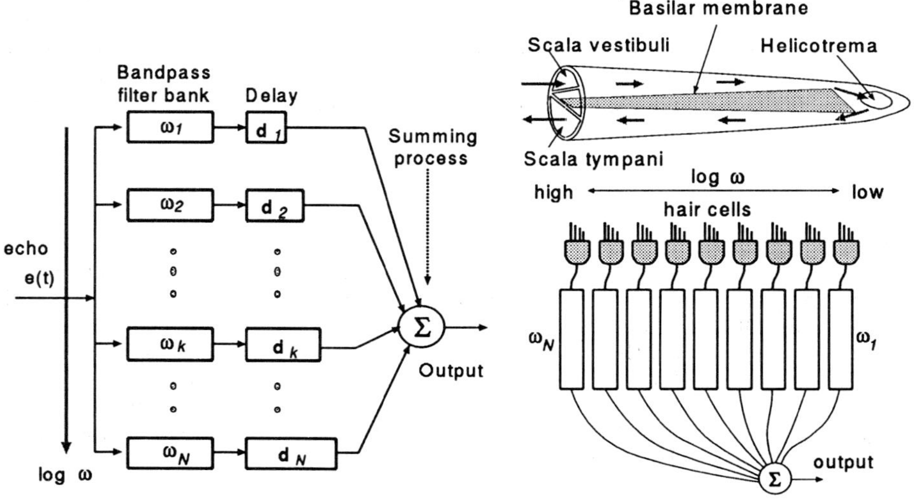

Figure 3: A functional model for the auditory periphery.

assume that the basilar membrane and hair cells function as bandpass filters. Each of the filters is designed to have a second-order transfer function according to observation of the frequency tuning characteristics of the auditory periphery[21]. Although the frequency selectivity of a second-order filter is rather inferior, the overall selectivity of the model can be improved by a mechanism such as lateral inhibitory connections between the filters.

The phase characteristics of a second-order transfer function in the neighborhood of its tuned frequency ω_k have a gradient of approximately $-2Q_k/\omega_k$, where Q_k denotes the Q value of the filter. Thus the phase characteristics of the bandpass filter with its center frequency ω_k can be expressed by means of a linear equation: $-2Q_k(\omega - \omega_k)/\omega_k$. Consequently, the phase spectrum of the output signal from the k-th filter shown in Fig.3 is given from Eq.(16) as follows:

$$\Psi_k(\omega) = -\{\alpha \ln \omega + 2Q_k(\omega - \omega_k)/\omega_k + \omega\tau_o + 2v_o\alpha/c + \theta_o\}. \tag{17}$$

In the following discussion we will consider how to linearize $\Psi_k(\omega)$ given in Eq.(17).

3.2. Design of delay devices

After being delayed for d_k through a delay line, Eq.(17) becomes

$$\Psi_k(\omega) = -\{\alpha \ln \omega + 2Q_k(\omega - \omega_k)/\omega_k + \omega d_k + \omega\tau_o + 2v_o\alpha/c + \theta_o\}. \tag{18}$$

Since the logarithm is a monotonic function with respect to its variable, the Taylor series expansion of $\ln \omega$ around the center frequency ω_k is given by

$$\ln \omega = \ln \omega_k + \frac{\omega - \omega_k}{\omega_k}. \tag{19}$$

Substituting Eq.(19) into Eq.(18), we obtain the spectral phase near the frequency ω_k:

$$\Psi_k(\omega) = -\{\alpha \ln \omega_k + \omega([\alpha + 2Q_k]/\omega_k + d_k + \tau_o) - \alpha - 2Q_k + 2v_o\alpha/c + \theta_o\}. \tag{20}$$

By differentiating the phase of Eq.(15) with respect to ω, we can confirm that the term α/ω_k corresponds to the group delay of the LPM sound at the frequency ω_k. Consequently, the group delay characteristics of the echo can be eliminated by designing the delay device so that

$$d_k = d_o - \frac{\alpha + 2Q_k}{\omega_k}, \tag{21}$$

where $2Q_k/\omega_k$ is the group delay characteristic of the k-th filter, and d_o is required to satisfy the causality, because d_k must have a positive value. The resultant spectral phase then becomes

$$\Psi_k(\omega) = -\{\alpha \ln \omega_k + \omega(d_o + \tau_o) - \alpha - 2Q_k + 2v_o\alpha/c + \theta_o\}. \tag{22}$$

3.3. Arrangement of bandpass filters

The first term in Eq.(22), $\alpha \ln \omega_k$, depends on how the center frequencies are determined. When we design the arrangement of filters so as to cancel the influence of the center-frequency distribution on the final result, the non-linear phase term is eliminated in Eq.(22). If the center frequencies are determined to have the following relation with arbitrary integer m:

$$\alpha \ln \omega_{k+1} = \alpha \ln \omega_k + 2\pi m, \tag{23}$$

then Eq.(22) can be rewritten as

$$\Psi_k(\omega) = -\{\alpha \ln \omega_1 + 2\pi(k-1)m + \omega(d_o + \tau_o) - \alpha - 2Q_k + 2v_o\alpha/c + \theta_o\}; \tag{24}$$

and finally we obtain

$$\Psi_k(\omega) = -\{\omega(d_o + \tau_o) + 2v_o\alpha/c + \Theta_o\}, \qquad \Theta_o = \theta_o + \alpha \ln \omega_1 - \alpha - 2Q_k, \tag{25}$$

where the term $2\pi(k-1)m$ is omitted because k and m are integer values.

If we design each of the band-pass filters to have the same Q value, then the spectral phase given in Eq.(25) is common to all the filters. Thus the summation of the delayed outputs of all the filters has a maximum amplitude at the time $d_o + \tau_o$ with the phase $-(2v_o\alpha/c + \Theta_o)$.

Furthermore, Eq.(23) indicates that the band-pass filters distribute uniformly over the logarithmic frequency: a bank of constant-Q filters is thus constructed. As a result, if we emit an LPM sound, we can estimate the range and speed of a moving object by investigating the phase distribution of the bank of constant-Q filters.

4. NUMERICAL EXPERIMENT AND THE RESULTS

An LPM sound was designed on the basis of the echo location sound of the FM bat *Eptesicus fuscus* and was generated digitally with a sampling interval of $2\mu s$. The instantaneous frequency of the sound was swept down from 55kHz to 25kHz through an appropriate design of its envelope $a(t)$. The LPM parameters, α and β, are determined as follows:

$$\alpha = c\pi/2, \qquad \beta = 0. \tag{26}$$

Under this condition, the phase rotation $2v_o\alpha/c$ reaches π at the speed of $v_o = 1$m/s: this is the estimated limit of the object's speed. Harmonic structures usually observed in the sounds of *Eptesicus* were not produced in the numerical experiments. Each constant–Q filter was equipped with a quadratic pair of second order *IIR* digital filters, and the Q value was set to 5π throughout the experiments.

The mean frequency f_c and the effective *rms* bandwidth B of the sound were 40.3kHz and 10.6kHz,respectively, being calculated from the following definitions:

$$f_c = \frac{m_1}{2\pi} \text{ and } B = \frac{1}{2\pi}\sqrt{m_2 - m_1{}^2}. \tag{27}$$

The variables m_1 and m_2 are

$$m_1 = \frac{(1/2\pi)\int_0^\infty \omega |U(\omega)|^2 d\omega}{E_u}, \qquad m_2 = \frac{(1/2\pi)\int_0^\infty \omega^2 |U(\omega)|^2 d\omega}{E_u}, \tag{28}$$

and

$$E_u = (1/2\pi)\int_0^\infty |U(\omega)|^2 d\omega, \tag{29}$$

where $U(\omega)$ denotes the one-sided spectrum of the LPM sound.

Using these definitions, *Cramér – Rao* lower bounds for the variance in the range and speed estimations are given by [18]

$$\sigma_\tau{}^2 \geq \frac{1}{(E_u/N_o)\cdot(m_2 - m_1{}^2)} \text{ and } \sigma_v{}^2 \geq \frac{m_2}{(E_u/N_o)\cdot(m_2 - m_1{}^2)\cdot(\pi)^2}. \tag{30}$$

4.1. Function of the proposed model as a correlation receiver

To examine whether the proposed model works as a correlation receiver we assume that an object moves in a noisy environment with relatively high velocities. Figs 4 and 5 illustrate the simulated echoes compared with the envelopes of summed outputs of the bank of constant-Q filters. Fig.4 shows that the energy centroids of the echoes translate in the direction of the time axis according to the change in time scale caused by the motions of the object. However, all the maximum amplitudes of the output envelopes are obtained at τ_o regardless of the object's speed. The Doppler tolerant property[19] of the proposed model is consequently well confirmed. In Fig.5, echoes reflected from three stationary objects are observed with additive noise. Regardless of the noise level, we can clearly discriminate individual objects against the serious noise.

4.2. Motion parameter estimation

Fig.6 illustrates synthesized outputs of the quadratic pair of constant-Q filter banks. The phase characteristics of each of the filters are also illustrated in the lower part of the figure, being superimposed on each other. These phase characteristics always intersect at the time $d_o + \tau_o$ with the phase value $-2v_o\alpha/c - \Theta_o$. In other words, we can estimate the range and speed by detecting the phase value possesed by all the filters simultaneously. An example of speed estimation is shown in Fig.7. Although biased estimates are observed in this figure, we can eliminate the biased amount by a proper choice of the parameter Θ_o.

The accuracy in parameter estimation is generally evaluated in terms of the square root of the variance in the estimation. Fig.8 illustrates the accuracy of motion parameter estimation by the proposed method, compared with *Cramér – Rao* lower bounds. The

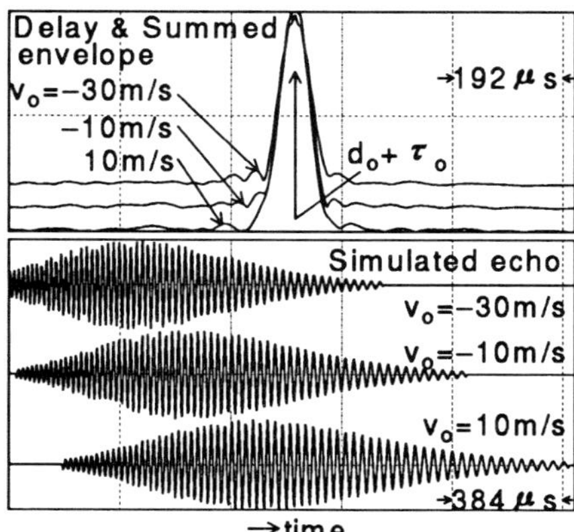

Figure 4: Doppler tolerant property. Regardless of the speed, output envelopes have their maximum values at the same time.

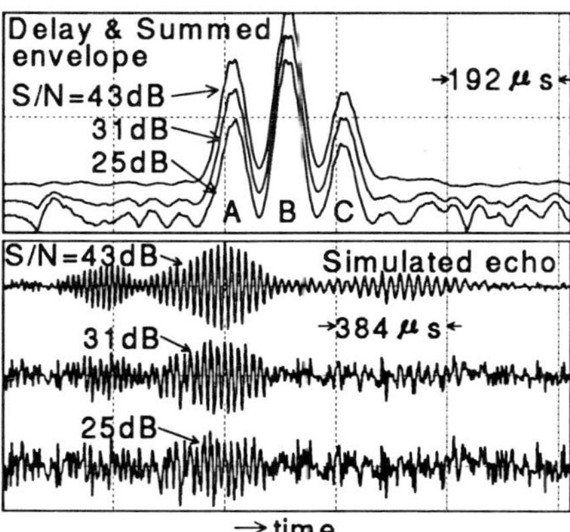

Figure 5: Simulated echoes reflected from three stationary objects contaminated with additive noise. The objects A, B and C are assumed to have the reflectances 0.75, 1 and 0.5, respectively. The summed outputs clearly indicate these objects.

accuracy based on envelope peak detection with the correlation receiver is also indicated in this figure. As a result, it is found that in motion parameter estimation the proposed method performs similarly to the correlation-detection method.

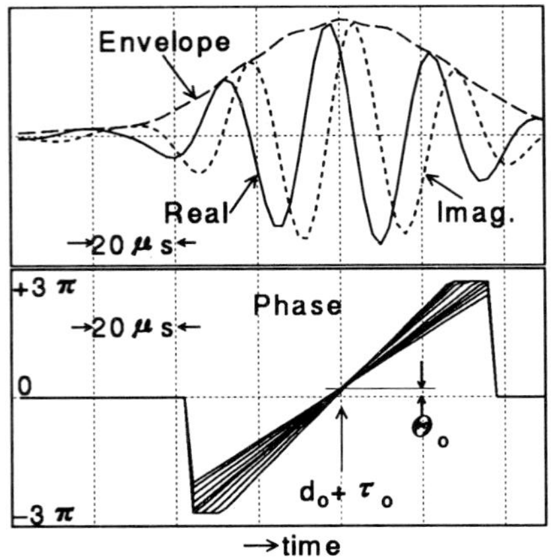

Figure 6: Synthesized output of the quadratic pair of filter banks and the phase characteristics of each of the filters, when $v_o = 0$[m/s].

Figure 7: Speed estimates by means of the proposed model. Since the measurable extent of the speed is designed to be in the limit of ± 1[m/s], estimated values beyond the limit are converted to the modulus value 1[m/s].

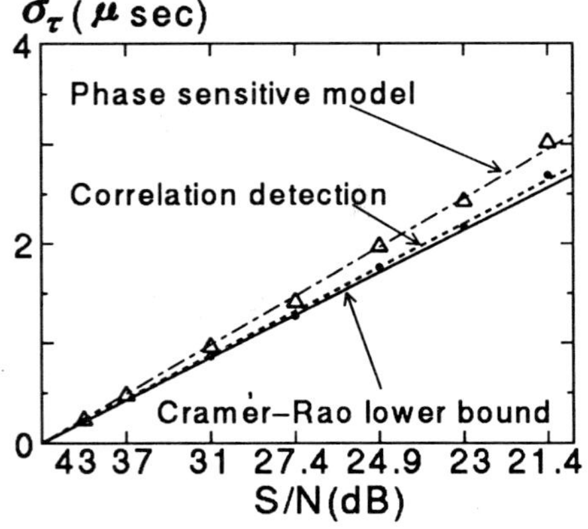

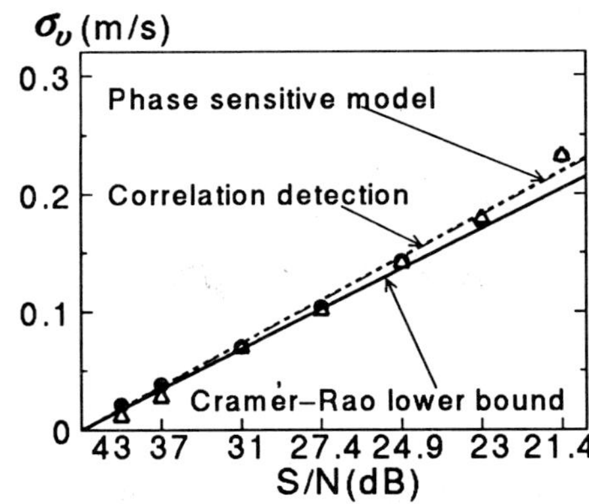

Figure 8: Dependence of estimating accuracy, σ_τ and σ_v, on the signal to noise ratio E_u/N_o.

5. CONCLUSION

Numerical experiments show that the bank of constant–Q filters can work effectively as an equivalent of a correlation detector. The phase sensitive model proposed in this study enables us to estimate the range and speed of a moving object. However, there is no evidence that the FM bats can sense an object's speed by utilizing the phase information of the received echo. This study has proposed a computational model of the echo-location system, and further investigation is necessary into how the FM bats actually perceive the speed of an object.

REFERENCES

1 Donald R. Griffin: *Listening in the Dark: The acoustic orientation of bats and men.* Yale University Press(1958)

2 G. Neuweiler: "Auditory processing of echos : peripheral processing," in *Animal sonar system*, R. G. Busnel and J. F. Fish, editors, pp.519–548, Plenum press(1980)

3 Nobuo Suga and W. E. O'Neill: "Auditory processing of echoes : representation of acoustic information from the environment in the bat cerebral cortex," in *Animal sonar system*, R. G. Busnel and J. F. Fish, editors, pp.589–611, Plenum press(1980)

4 C. E. Cook and B. Bernfeld: *Radar Signal.* Academic Press(1967)

5 J. A. Simmons: "The Resolution of Target Range by Echolocating Bats," *J. Acoust. Soc. Am.*, **Vol. 54**, pp.157–173(1973)

6 J. A. Simmons: "Perception of echo phase information in bat sonar," *Science*, **Vol. 204**, pp.1336–1338(1979)

7 G. Neuweiler: "How bats detect flying insects," *Physics Today*, pp.34–40(August 1980)

8 D. Menne and H. Hackbarth: "Accuracy of distance measurement in the bat *Eptesicus fuscus*: theoretical aspects and computer simulations," *J. Acoust. Soc. Am.*, **Vol. 79**, pp.386–397(1986)

9 B. Møhl: "Detection by a Pipistrelle Bat of Normal and Reversed Replica of its Sonar Pulse," *Acustica*, **Vol. 61**, pp.75–82(1986)

10 H. Hackbarth: "Phase Evaluation in Hypothetical Receivers Simulating Ranging in Bats," *Biol. Cybern.*, **Vol. 54**, pp.281–287(1986)

11 Nobuo Suga: "Cortical computational maps for auditory imaging," *Neural Networks*, **Vol. 3**, pp.3–21(1990)

12 D. Menne et al: "Range estimation by echolocation in the bat *Eptesicus fuscus*: trading of phase versus time cues," *J. Acoust. Soc. Am.*, **Vol. 85**, pp.2642–2650(1989)

13 E.J. Kelly and D.P. Wishner: "Matched filter theory for high-velocity, accelerating targets, " *IEEE Trans. Military Electron.*, **Vol. MIL-9**, pp.56–69(1965)

14 O. W. Henson, Jr. et al.: "The constant frequency component of the biosonar signals of the bats, *Pteronotus parnelli parnelli*," in *Animal sonar system*, R. G. Busnel and J. F. Fish, editors, pp.913–916, Plenum press(1980)

15 R.A. Altes: "The Fourier-Mellin transform and mammalian hearing," *J. Acoust. Soc. Am.*, **Vol. 63**, pp.174–183(1978)

16 P.E. Zwicke et al: "A new inplementation of the Mellin transform and its application to radar classification of ships," *IEEE Trans. Pattern Anal. Machine Intell.*, **Vol. PAMI–5**, pp.191–199(1983)

17 J.J. Kroszczyński: "Pulse compression by means of linear-period modulation," *Proc. IEEE*, **Vol. 57**, pp.1260–1266(1969)

18 R.A. Altes and D.P. Skinner: "Sonar-velocity resolution with a linear-period modulated pulse," *J. Acoust. Soc. Am.*, **Vol. 61**, pp.1019–1030(1977)

19 R.A. Altes and E.L. Titlebaum: "Bat signals as optimally Doppler tolerant waveforms," *J. Acoust. Soc. Am.*, **Vol. 48**, pp.1014–1020(1970)

20 Athanasios Papoulis: *The Fourier integral and its applications*, McGraw-Hill Book Company(1962)

21 B. M. Johnstone et al.: "Basilar membrane measurement and the travelling wave," *Hear. Res.*, **Vol. 22**, pp.147–153(1986)

Intelligent Sensors
H. Yamasaki (Editor)

High-Precision Micromachining Technique and High Integration

Isemi Igarashi

Toyota Physical and Chemical Research Institute
Nagakute, Aichi, 480-11 Japan

Abstract

Smaller sensors are more convenient to use in most cases. To collect a variety of information using them, it is important to establish the technology for making them multifunctional and giving them a composite form. For this purpose, various kinds of technologies developed for manufacturing semiconductor IC and LSI are directly applicable. That is, the development of microsensors formed on silicon substrates, coincides with the advance in semiconductor fabrication techniques. Among them, the high-precision micromachining technique is particularly important.
The difference between sensor and IC lies in whether it comes in contact with the open atmosphere or is kept sealed when working. No doubt the former faces severer conditions. For this reason, the packaging is the key parameter for a practical sensor.
This paper describes in detail the recent development of semiconductor pressure sensors which are most commonly used for various practical purposes today. It also relates how the diaphragm was developed and describes some working examples of composite function sensors, multiple ion sensors, and micro gas sensors.

1 Introduction

A device capable of converting a quantity of physical, chemical, biological, mechanical, or medical parameters to some other signal which can be processed is called a “sensor”. In many cases, the conversion to an electrical signal is the most convenient. Therefore, the output from all the sensors described below has also been converted to electrical signals. In recent years, the micromachining technique for fabricating semiconductors has made rapid progress, making it possible to popularize the application of IC and LSI in every industry. In other words, the development of this micromachining technique has made a great contribution to the advance of science and technology. Its effect on the sensor technology-related field is also not insignificant. In fact, the wide-spread effects on this field in the past several years have been far more than expected and can be described as follows.

① Both shape and dimension of sensors have been incomparably miniaturized As a result, their reliability, mass productivity, and manufacturing costs have been significantly reduced.

② Due to the unification of the body and signal-processing circuit of the sensor, its increasing integration as well as its added multifunction handling capability, the development of the intelligent sensor is becoming more and more realistic.

In the sections below, the details of these matters will be described using some of our developed sensors as examples.

2 Properties of silicon and strain gauge

Silicon has many excellent properties besides the fact that it is a good material for manufacturing ICs. Because it is one of the inexhaustible natural resources, making up almost 25% of the substances forming the surface layer of the earth, and is also a safe material, no special care is needed in its handling. Silicon is only the material that can be mass-produced artificially in the form of a pure and perfect crystal. Demand for polycrystalline silicon has been increasing. For example, the application of amorphous silicon thin film for solar cells or light→electricity conversion sensors has been drawing much attention recently.

A though the properties of silicon as an electronic material are relatively well-known, its mechanical and physical properties are not. They are shown in Tables 1 and 2.[1, 2]

[Table 1] Physical properties of silicon crystal for sensors

Physical properties	Effect used for signal conversion
Light・radiation	Optical electromotive force effect, photo electron effect, photoconductive effect, photomagnetic electron effect
Stress	Piezoresistive effect
Heat・temperature	Seebeck? effect, thermal resistance effect, p-n junction, Nernst effect
Magnetism	Hall effect, magnetic resistance effect
Ion	Ion-sensitive electric field effect

[Table 2] Physical properties of silicon crystal

Density	2.33 g/cc
Modulus of direct elasticity	1.9×10^{12} dyn/cm ^{2}
Bulk modulus	7.7×10^{11} dyn/cm ^{2}
Melting point	1412℃
Specific heat	0.18 cal/g・℃ (18 - 100℃)
Coefficient of thermal expansion	2.33×10^{-6}/ ℃ (at room temp.)
Thermal conductivity	0.3 cal/s・cm・℃ (20℃)

[Table 3] Elastic constants of semiconductors (at room temp.)

Crystal	Modulus of elasticity ($\times 10^{12}$ dyn/cm ^{2})			Compressibility ($\times 10^{-12}$ cm^2 /dyn)	Young's modulus [100] ($\times 10^{12}$ dyn/cm^2)	Young's modulus [111] ($\times 10^{12}$ dyn/cm^2)
	C_{11}	C_{12}	C_{44}			
Si	1.657	0.639	0.796	1.023	1.30	1.87
Ge	1.289	0.483	0.671	1.330	1.02	1.55
InSb	0.647	0.327	0.307	2.310		0.745
GaAs	1.188	0.538	0.594	1.325		1.42
GaSb	0.885	0.404	0.433	1.773		1.033

Among the signal conversion effects listed in Table 1, the most important and most commonly utilized ones are the piezoresistive effect, the Hall effect and the Seebeck effect. Table 2 shows the main physical properties of silicon, which decides the characteristics of sensors.
Table 3 shows the elastic moduli of various semiconductor materials.[3)] When they are utilized for some dynamic or pressure sensors, their properties as a spring material become important. As is apparent form the table, even though the magnitude of Young's silicon modulus changes with its crystalline direction, its maximum value is almost equal to that of steel, showing a good spring property. Figure 1 shows the temperature dependence of Young's silicon modulus. It is apparent that silicon can stand a strain of nearly 1% without causing plastic deformation at room temperature. Here it must be noted, however, that the experiment shown in Fig. use 1 was carried out using a silicon whisker crystal with almost no defect. In the case of its bulk crystal, the maximum fracture strain may be about 0.3%. According to some experiments conducted by the author, when strain gauges were bonded to both surfaces of the holding edge of a cantilever beam, the fracture of the tensile-side gauge took place around 0.3% while no fracture took place on the compression-side gauge, even after a strain several times higher than the former was added.[4)]
Here, it should be noted that these results were obtained when the surface processing layer of the bulk type gauge was removed by etching and its resulting surface, with almost no defects was bonded with epoxy resin. Thus, if the surface is converted to a no-defect condition to avoid any localized stress concentrarion, silicon can be used as an excellent spring material. In fact, the mechanical properties of materials used for sensors and actuators are very important.
In the section below, the semiconductor strain gauge based on the piezoresistive effect will be described.[5)] The gauge section of this strain gauge is formed on a certain area of silicon single-crystal plate using the same method as in the fabrication of integrated circuits. With this fabrication method, both a miniaturization and an increase in sensitivity of the gauge are quite easily achieved and many ideas for its application

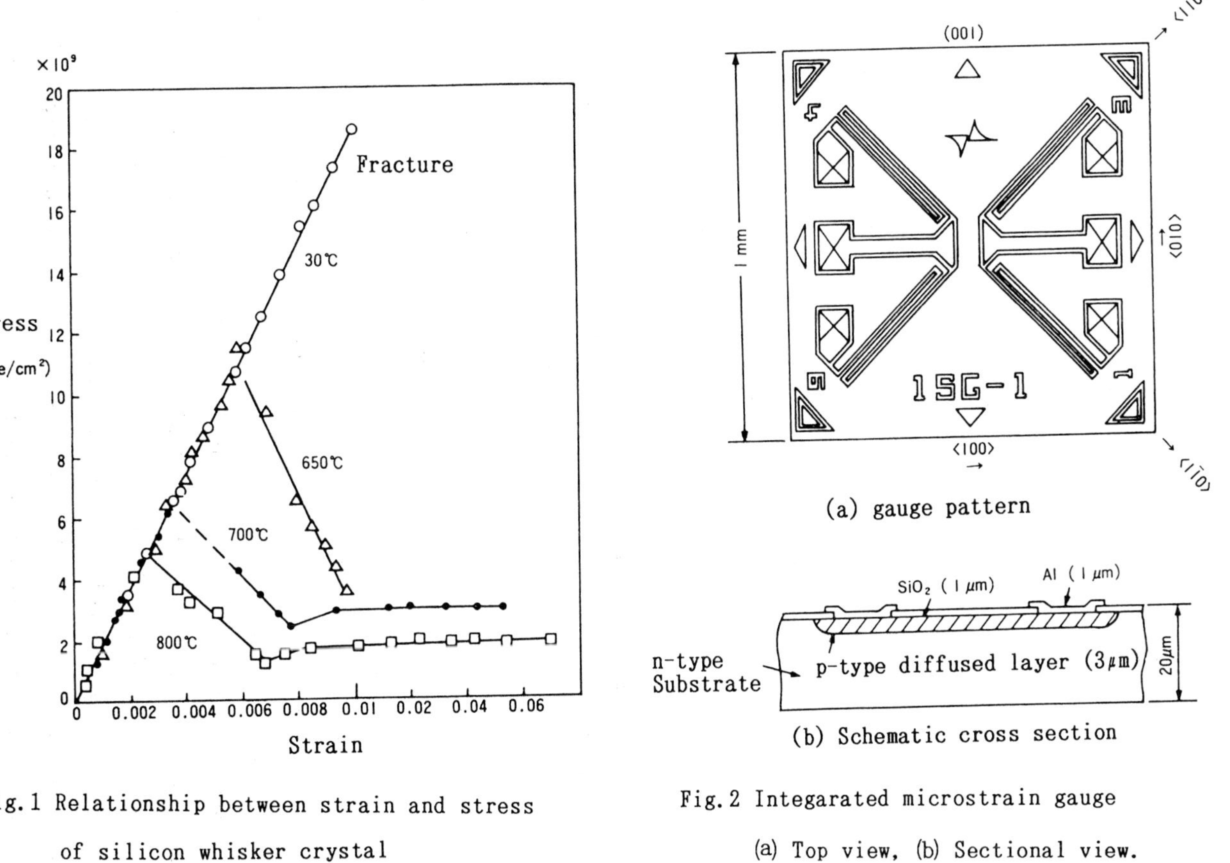

Fig. 1 Relationship between strain and stress of silicon whisker crystal

(a) gauge pattern

(b) Schematic cross section

Fig. 2 Integarated microstrain gauge

(a) Top view, (b) Sectional view.

in special load cells have been proposed.
The four gauges shown in Fig. use 2 which form an X-shaped pattern are integrated on the substrate. That is, the two sets of gauges right-angled to each other were formed on a thin silicon plate with a thickness of 20μm using the diffusion process. As shown in Fig. use 2 a, lead wires were taken from the points marked ☒ on both sides of the gauge. The sizes of the actually prepared gauges were 1 × 1mm² and 2 × 2mm² and their diffused layer was 3μm thick. The substrate and gauge were n- and p-types, respectively, and they were electrically insulated from each other, although unified together structurally. Fig. use 2 a and b show the top and sectional views of the gauge, respectively. These may be some disadvantages in handling because of the small size, but its suitability for mass production, its high sensitivity, and its miniature size are generally the great advantages over such handling inconvenience. In fact, its possible application for load cells appears to be promising.

3 Micronization and integrated pressure sensor

In recent years, there have been a number of published reports on microsensors for measuring the mechanical quantities prepared as a result of combining IC fabrication technology with the anisotropic etching of silicon and various thin film-forming techniques based on the excellent mechanical properties of silicon. This combination is known as the maicromachining technique today.[6, 7]
The manufacturing technology of the silicon pressure sensor is fundamentally based on IC fabrication technology.[6, 9] For this reason, the degree of its miniaturization is considered to have increased with the improvement in micromachining techniques for ICs. For example, Fig. use 3 a shows the miniaturization trend of silicon pressure sensor size, where the diaphragm size of the silicon pressure sensor built for practical use (ϕ for a circulay type and □ for a rectangular type) is used as an indicator. The down-sizing trend of the minimum size of an IC memory chip (DRAM) is also included in the figure for reference. As can be seen in the figure, the size of the diffusion type silicon pressure sensor was 15mm in

diameter when it was first produced, and was reduced to 1/20 in the following 15 years. It is predicted that its size will be reduced to less than 100μm sometime in 1993s. However, there are a number of technical problems to overcome in order to achieve this target, one of them being the etching processing for the diaphragm.

Fig. use 3 c and d show examples of the smallest pressure sensors fabricated using the conventional techniqe.[10] Because of its extremely small size, these pressure sensors can be used as a catheter tip-type pressure gauge and as a body-implanting type tonometer that can be inserted directly into a blood vessel. The sizes of the daiphragms are 0.5 X 1mm and 0.2 X 0.4mm, while the outher dimensions of the sensor chips are 3 X 1 X 0.2mm (L X W X H) and 1.5 X 0.5 X 0.15mm (L X W X H), respectively. Using a [100]-plane silicon substrate, diffusion type strain gauge were formed along the [110] direction. The diaphragm was applied by anisotropic etching on its back to form a 5-μm layer ovre its surface. To process this diaphragm see processing method A in Fig.3, double-sided lithography was used and strain gauges were formed on its surface with the application of deep silicon etching on its back side. Because of the relatively low processing accuracy of double-sided lithography, the practical minimum size of the diaphragm formed by applying this deep etching treatment is considered to be around 100μm. Therefore, to built a micro pressure sensor it is more desirable to process a chip using single-sided lithography, as shown in processing method B in Fig.3. Using method B, a higher alignment accuracy of lithography can be maintained, making it possible to process a diaphragm with a thickness less than 100μm. Therefore, application the cost reduction in manufacturing sensors can be expected.

At the present, however, more sensor chips belonging to the A type or double-side lithography type are circulating on the market. For instance, more than 10 million of these type A sensors see Fig.3 b are manufactured annually. Most of these sensor chips are used as intake depression sensors for automobiles. In Fig. use 3 d and f, a pressure sensor can be seen in the central section of the chip, while its signal processing circuit is fabricated outside the central section. The simplest type (f) is attracting attention as promising type of pressure sensor.

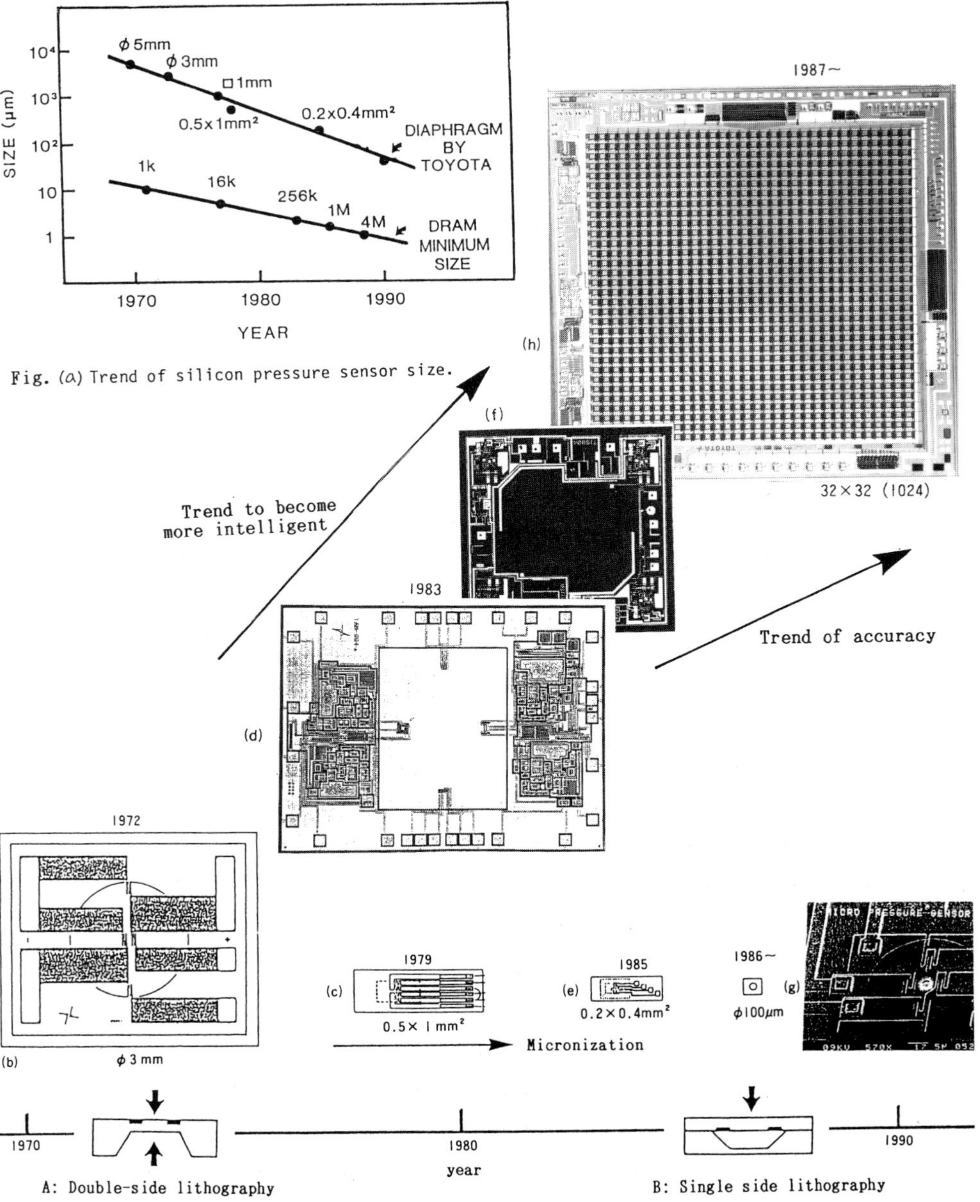

Fig. 3 Micronization & integration trend of silicon pressure sensors

In detailed explanation (legend) of Fig.3

Micronization and integration trends of a silicon pressure sensor. (a) the micronization trend of the sensing section of the silicon pressure sensor when compared with the integration trend of DRAMs. (b)→(c)→(d)→(e) the micronaization trend of the silicon pressure sensor with regard to time. (d)→(f)→(h) the intelligence trend of the integrated pressure sensor in recent years. The double-sided lithography, as shown in A, is the most common processing method for diaphragms, however, the application of single-sided lithography is becoming more common today for processing diaphragms less than 100μm thick.

Fig. use 3 g and h show a sensor unit and a pressure sensor array, respectively. In them, a microdiaphragm with a size of 100μm × 100μm and its CMOS signal processing circuit are formed on the same substrate.

Fig. use 3 g shows an enlarged photograph of the sensor unit, which is one of a number of sensor units forming the two-dimensional array. Each sensor unit consists of an analog switch, a power switch, and a logic circuit for address selection. The pressure sensor array is comprised of a matrix of 32 × 32 sensor units, with the distance between them being 250μm. Around the array are CMOS signal processing circuits. They include timing circuits, 10-bit counters, X- and Y- decoders, trigger circuits and amplifieres. Fig. use 3 h shows the external appearance of the chip, with a size of 10mm × 10mm. The signals from 1024 sensing units are output as the serial time-division analog waveform, while synchronizing with some externally supplied scan pulses. The rated clock frequency is 4 MHz and the time required for reading out one frame is less than 16 ms. The development of a pressure sensor array with a high two-dimensional resolution such as this expected to significantly contribute to the development of practical touch-perception imagers used in precision-work robots and in establishing the pressure-distribution measurement technique for micro-zones formed around injection nozzles.

Fig. use 4 outlines the manufacturing process of a microdiaphragm pressure sensor. Using [100]-plane silicon as a substrate, both the diaphragm and the cavity for the reference chamber are formed using the undercut-etching process. The section below will briefly explain the succeeding processes seen in Figure 4 a-e.

(a) Formation of the first silicon-nitride layer on silicon substrate, followed by the opening of a rectangular window on its central section.
(b) Formation of etching channels with polysilicon to cover the rectangular window.
(c) After the formation of the second silicon-nitride layer, polysilicon strain gauges are formed at the location where the diaphrahm is to be positioned. Then, etching holes, passing through the second and third silicon-nitride layers and reaching the polysilicon etching channels, must be opened.
(d) Undercut etching is then applied to the chip using an anisotropic etching solution, such as KOH aqueous solution. The etching proceeds throuh the etching holes while removing the polysilicon etching removing channels, leaving a cavity inside the silicon substrate.
(e) When all the etching channels are removed and a pyramidal cavity surrounded by [111]-planes is formed, the anisotropic etching operation stops automatically. After the formation of electrodes by aluminum deposition, a silicon-nitride layer is formed on the substrate surface using the plasma CVD technique in order to seal the etching holes.

4 Silicon monolithic miniature pressure-flow sensor

Pressure and flow are the important physical parameters in fluid cotrol. With thier simultaneous measurement, it becomes possible to obtain some new kinds of information, such as the power of the fluid and the impedance of the flow path. If, therefore, two types of sensors for measuring these parameters are combined, there will be various advantages, for example sensor size reduction, space-saving for installation, and or measurement cost reduction.

A monolithic miniature pressure-flow sensor, in which a pressure sensor and a thermal flow sensor with a heat-insulating structure were incorporated, was fabricated for experimental purposes. An explanation is given

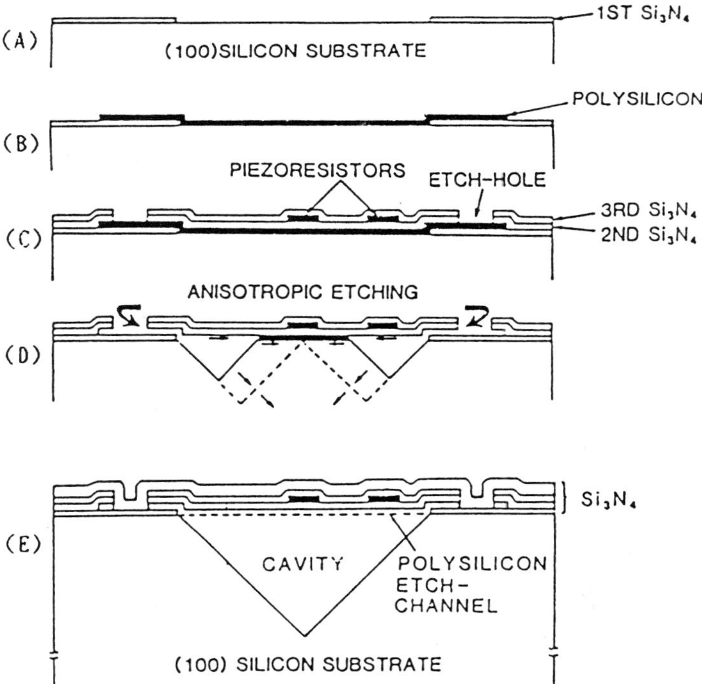

Fig. 4 Fabrication sequence of a microdiaphragm pressure sensor

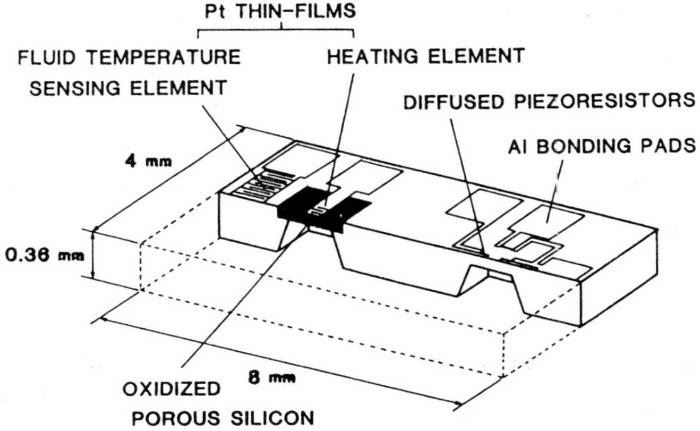

Fig. 5 Structure of the monolithic pressure-flow sensor chip.

below.

Fig. use 5 shows the basic structure of this monolithic miniature pressure-flow sensor. Photo 1 shows the phtographs of the chip (size 1mm X 5mm X 0.15mm). The chip has two diaphragms with a thickness of 10μm formed by etching on its back which corresponds to the pressure-sensing part (right side) and the flow-sensing part (left side). The diaphragm for the pressure-sensing part is made of single-crystal silicon with a size of 0.4 mm X 1mm and has four diffused gauges on its surface. The diaphragm for the flow-sensing part is made of oxidized porous silicon with a size of 0.4mm X 0.4mm, formed by the anode- forming treatment in hydrofluoric acid. A nickel thin-film resistive heating element is formed on the diaphragm. Flow can be determined by measuring the electric power consumed to maintain the temperature of the heating element always higher than the fluid temperature by a given degree (ΔT), since the electric power consumed is directly proportional to the amount of flow. Fluid temperature is determined by measuring the resistance change of the nickel thin-film resistive heating element formed on the tip surface of the chip.
With the built-in feedback circuit consisting of a bridge and operational amplifier circuits, the temperature difference (ΔT) between the thin-film resistive heating element for flow sensing and the fluid is always maintained at a given level. Some physical properties of flowing water were measured in one test using this chip installed in a catheter tube. As a result, it was found that the chip could produce a pressure sensitivity value of 6.2mV (Kgf/cm²) V for the pressure range 0 - 0.4Kgf/cm², while the flow-sensing accuracy was ±10%/FS for the flow range 0 - 50cm/s. To improve the flow-sensing accuracy of this sensor further, it seems necessary to introduce the concept of fluid dynamics to the design of its shape in the future.

5 Multiple ion sensor

If a miniature sensor capable of measuring many different kinds of ions simultaneously is developed, the concentration of those ions in urine or blood samples can be determined easily, providing doctors with important information for the patients' home health management as well as his clinical management. The pH value and the Na^+ and K^+ concentrations are all very important parameters physiologically. Accordingly, measurement of changes in their concentration is crucial when medical treatment must be given.

In an attempt to add multifunctional and composite features to ion sensors using the existing IC technology, utilization of the ISFET (ion-sensitive field-effect transistor) type of sensor appears to be promising. In this amplifier-type device, a change in current between its source and drain occurs because of a change in the electric charge of the ion-sensing film on the FET's insulating gate. Because of this working principle, increases

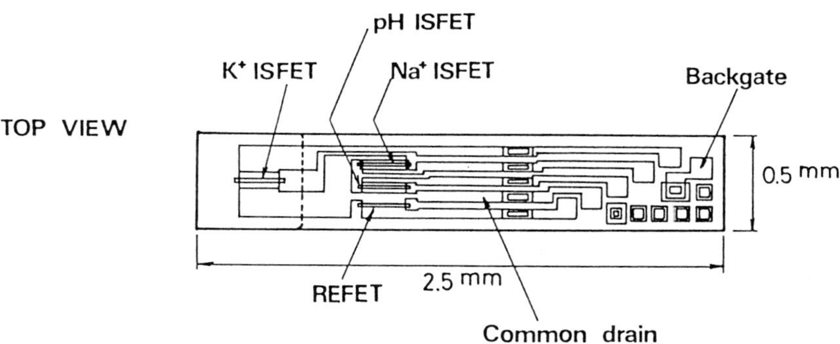

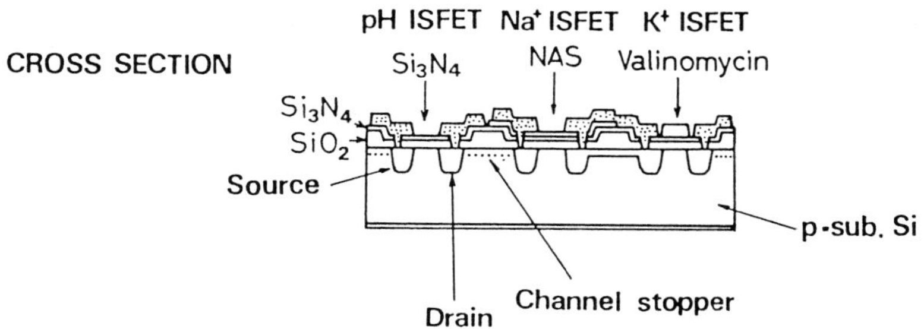

Fig.6 Structure of multiple ion sensor. (a) the top view of the sensor, (b) the sectional view of the same sensor. Here, the sensing films for pH, Na^+, and K^+ are made of silicon nitride, NAS glass, and Valinomycin, respectively.

in the degree of its integration and sensitivity can be achieved, if an appropriate ion-sensing film is selected.

Fig. use 6 shows an experimentally fabricated sensor.[12] The size of the sensor was 0.5mm in width, 2.5mm in length, and 0.15 in height and three ISFETs for measuring H^+, Na^+, and K^+ concentrations were incorporated in it. Each ISFET was fabricated using the n-channel FET process, where the p-type [100]-plane silicon substrate is treated with two types of films (silicon oxide SiO_2, thickness 100 nm and silicon nitride Si_3N_4, thickness 200 nm) in order to form the double-layer gate-insulating layer. The silicon nitride film formed here can also work as a hydrogen ion-sensing film. For sensing sodium and potassium ions, however, NAS (sodium aluminosilicate) glass and polyvinyl chloride containing Valinomycin were used.

Two technical modifications was made to the shape of the sensor to improve its durability, namely, the formation of sodium ion-sensing film using the ion implantation method and the use of polyester with a molecular weight of 1,500 - 8,000 for the plasticizer of the potassium ion-sensing film. As a result of these modifications, the life of ISFET for sodium ions kept in water was prolonged to about two months. The ISFET for potassium ion resulted in a better response than when the conventional dioctyl phthalate was used as plasticizer and its life was prolonged to about two months, which is 10 times longer than that of the conventional type.

6 Micro-oxygen sensor[13]

About 15 years ago the first practical oxygen sensor based on zirconia solid electrolyte was produced in order to cope with the problem of exhaust-gas purification. Since then, this device has significantly contributed to the protection of the global environment. In addition, it was successfully miniaturized to save its driving power and the amount of material used for making it. The trend in both micronization and sensitivity increase for oxygen sensors has advanced further in recent years. Today these sensors are widely used not only for onboard purposes, but also for an accurate, real-time measurement of oxygen consumption in various sporting activities as well as for the measurement of oxygen concentration in medical treatment, foods, or when a gas burner is used.

Fig. use 7 shows the structure of a thin-film limiting current type oxygen sensor.[14, 15] When current is supplied to the zirconia solid electrolyte, oxygen flows into the cathode and is then discharged from the anode. To limit the amount of oxygen arriving at the cathode, a porous alumina substrate is used. Using the sputtering technique, the films of platinum cathode, zirconia solid electrolyte, and platinum anode are deposited one after another to form the sensing part of this device. The platinum heater is formed on the back of the alumina substrate. The size of this device is 1.7 X 1.8 X 0.3mm³ and its response is 0.2 seconds. Its power consumption is less than 1 watt. Because of its small size and easy use, its application in a variety of fields can be expected.

Fig.7 Structure of thin-film limiting current type oxygen sensor. Zirconia solid electrolyte equipped with electrodes at its upper and lower surfaces is positioned above the porous alumina substrate, while the thin film platinum heater is positioned under the substrate.

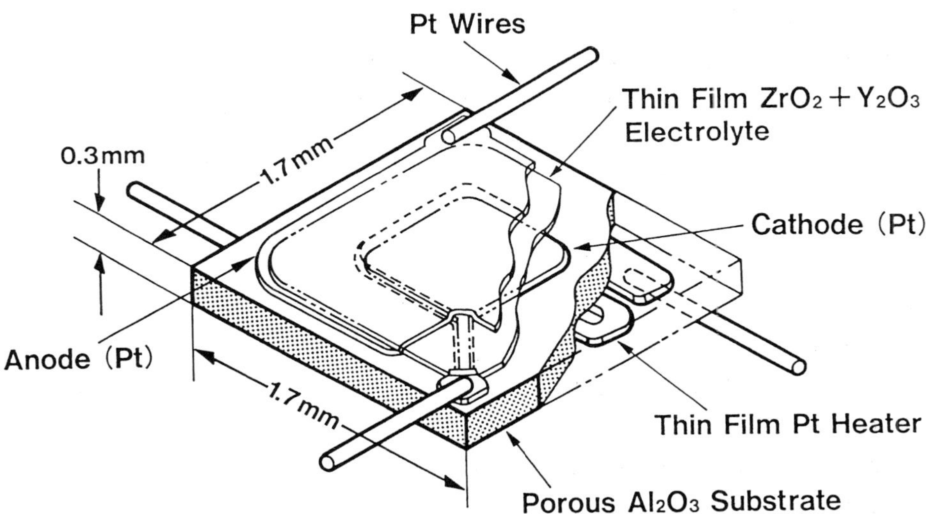

Fig. 7 Micro-oxygen sensor

7 Conclusion

More is expected of the development of microsensors because the progress and popularization of semiconductor fabrication techniques have helped spread this technique to many other fields, already providing various positive effects. No problems are expected even thouh the size of sensor's sensing part is close to that of a period. In the case of some fluids, for example, dynamic measurement will be possible if the information of hydrostatic pressure at one point of the fluid is available together with the information of time-dependent changes at the other points. In this case, the measurement can be made even without causing any disturbance to the fluid. Sensors suitable for micro-machines must be very small in volume and their three-dimensional integration for multiple functioning and combining will be the key subject to be solved.

Silicon is the most suitable material for micromachining. Many kinds of technology related to this material are already available. However, silicon has some unfavorable properties with regard to its mechanical strength, brittle fracture, and cleavage. Therefore, it is necessary to pursue basic studies on reinforcing methods for silicon, the possibility of developing other materials, and the use of metallic compounds or micromachining methods for oxidized film.

These items are equally applicable to all kinds of micromachines, including actuators and packaging as well as sensors. Since all are expected to have many difficult problems, it will be more enjoyable when they are solved one after another.

References

1) I.Igarashi: Rep. Toyota Phys. Chem. Res. Inst., 30th Anniv. Issue p.1 (1970)
2) C.Kittel: Solid Stage Phys., 91, p.85-102 (1956)
3) I.Igarashi: BME, 3-5, p.1 (1989)
4) I.Igarashi: Appl. Phys., 29, p.73 (1960)

5) K.Shimaoka, A.Tsukada and I.Igarashi: Non-destr. Insp., 30-9, p.756 (1981)

6) K.E. Petersen: Silicon as a Mechanical Material, Proc. IEEE, 70-5, p.420 (1982)

7) Proc. Transducers'87, Tokyo (1987), '89, Montoreux (1989), '91, San Francisco (1991)

8) S.Sugiyama, T.Suzuki, K.Kawahata, K. Shimaoka, M.Takigawa and I.Igarashi: Micro-Diaphragm Pressure Sensor, IEEE IEDM Tech. Digest, p.184 (1986)

9) S.Sugiyama et al.: Micro-Diaphragm Pressure Sensor, Inst. Electr. Commuc. Eng. JPN. Res. Mater., CPM86-115 (1987)

10) A.Nakajima, H.Inagaki and I.Igarashi and K.Kitano: (1st Rep.) Structure and performance in ME & BE, Spec. Issue 23, p.377 (1985)

11) O.Tabata, H.Inagaki, I.Igarashi: Silicon monolithic miniature pressure-flow sensor, Sensor Tech. Material, IEEE Trans. on Electron Devices, ED34, No.12, p.2456 (1987)

12) T.Ito, O.Tabata, T.Taguchi, H.Inagaki and I.Igarashi: ME & BE, Spec. Issue 26, (1988)

13) K.Saji: J. Electrochemical Soc., 134, 10, p.2430 (1987)

14) H.Takahashi, K.Saji, H.Kondo, K.Hayakawa and T.Takeuchi: Thin film limiting currenet type oxygen sensor, 100 Selected R & D Winner (1988)

15) H.Takahashi, K.Saji, H.Kondo, T.Takeuchi and I.Igarashi: Inst. Electr. Comme. Eng. JPN., ED86-163 (1987)

Intelligent Sensors
H. Yamasaki (Editor)

Integrated Magnetic Sensors

K.Maenaka
Department of Electronics, Himeji Institute of Technology
Shosha 2167, Himeji 671-22, Japan

Abstract

This chapter is a review of integrated magnetic sensors in which one or more magnetic sensors and peripheral circuits are efficiently combined in one silicon chip. Simple integration of two or three elements without an electric circuit and complex integration containing a signal processing circuit are described. Although GaAs Hall cells and MR elements can be integrated with a circuit, they are omitted here and only silicon sensors are featured.

1. INTRODUCTION

Since the late 1970's, integrated sensors (or intelligent sensors) have became of major interest. Especially in the field of magnetic sensors, many silicon magnetic sensors which can be fabricated using the standard integrated circuit process have been reported, and some review-type papers for such sensors have been published[1-3]. Because of the process compatibility of the sensors and circuits, these sensors can easily be integrated with peripheral circuits, such as an amplifier, a switching circuit and a calculation circuit. Some integrated sensors, for example Hall IC's which include a Hall cell and an amplifier on one chip, are now commercially available.

In this chapter, various integrated magnetic sensors including commercial products and research examples will be illustrated, and the methods and the effects of such integration will be discussed.

2. INTEGRATION OF TWO OR THREE ELEMENTS

There are some sensors in which two or three elements are effectively integrated in one silicon chip. Prior to a description of integration with circuitry, these types of sensors will be described.

2.1 DAMS(Differential Amplification Magnetic Sensor)

Figure 1 (a) and (b) show the structure and the equivalent circuit of a DAMS[4,5]. The DAMS has a Hall cell with a differential amplifier in one chip. The Hall cell is formed in the base region of the differential transistor pair and is inseparable from the transistors. In this sensor the current difference between the collectors C_1 and C_2 is the amplified signal of the Hall voltage. The external load resistors, R_L, convert the current difference between the two collectors into a voltage signal. Thus the absolute sensitivity can be large, and it reaches 120 V/T with a load resistance of 100 kΩ. Figure 2 illustrates the magnetic characteristic of the DAMS. In the figure, output saturation is observed owing to the limit of the power supply voltage. This saturation can be avoided by decreasing the load resistance (of course, a decrease in load resistance leads to a decrease in the absolute sensitivity). The weak point of this sensor is an incompatibility with the standard IC process.

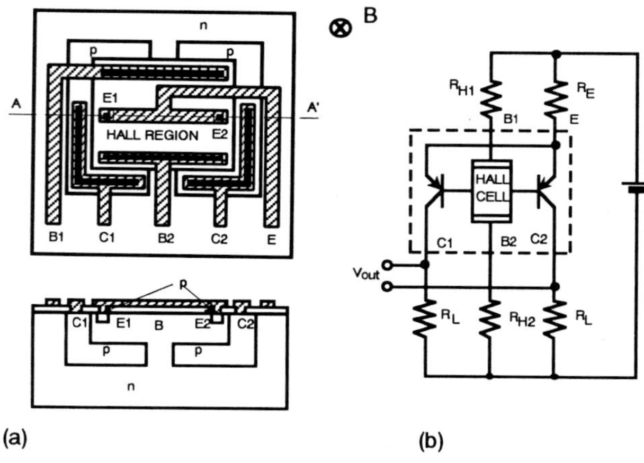

Figure 1. Structure (a) and equivalent circuit (b) of DAMS.

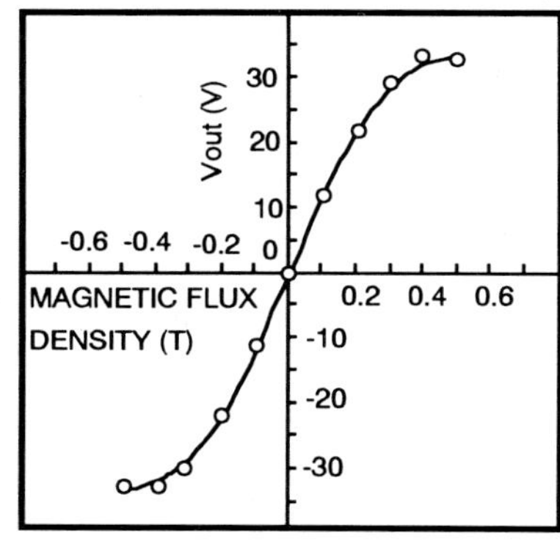

Figure 2. Characteristic of DAMS.

2.2 Complementary MOS magnetotransistors with two drains

Magnetic sensors with a transistor structure are called magnetotransistors. For example, there is an MOS magnetotransistor with two drains, in which the two drain currents are differentially changed with the applied magnetic field[6,7]. Using such magnetotransistors, a compound sensor can be realized. Figure 3 (a) and (b) show the structure and the equivalent circuit of the complementary MOS configuration using n- and p-type MOS magnetotransistors[8]. The n-type

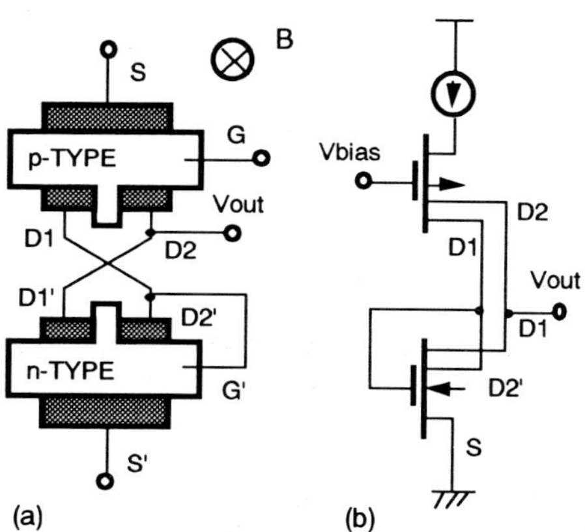

Figure 3. Top view and equivalent circuit of complementary MOS magnetotransistor.

MOS magnetotransistor acts as an active load of the p-type MOS magnetotransistor. By this configuration, when the magnetic field is applied to both of the magnetotransistors as shown in the figure, one of the drain currents, I_{D1}, of the p-type magnetotransistor, increases and the drain current of the n-type magnetotransistor, I_{D2}', decreases. Simultaneously, the opposite situation occurs for the set of I_{D2} and I_{D1}'. Thus, when both transistors are operating in the saturation condition, the small current change in the drains makes a large voltage change on the drains. The magnetic characteristic is shown in Figure 4, and the absolute sensitivity of 1.2 V/T is obtained independently of the bias current.

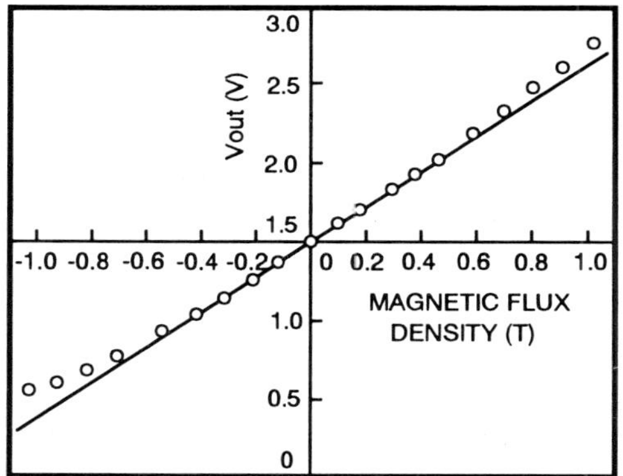

Figure 4. Characteristic of complementary MOS magnetotransistor.

2.3 Multidimensional magnetic sensor

The vertical Hall cells (VHC) that detect the magnetic field parallel to the chip surface have been presented[9,10]. Using two VHC's rectangularly located to each other, a two-dimensional magnetic measurement can be realized. Moreover, when the lateral Hall cell (LHC) that detects the magnetic field perpendicular to the chip surface is added to the two-dimensional magnetic sensors, three-dimensional magnetic sensors can be achieved. Since both the VHC and LHC are fabricated using a standard bipolar IC process, the realization of three-dimensional integrated magnetic sensors including bipolar signal processing circuitry is easy. This type of multidimensional magnetic sensor has a relatively high spatial resolution (~100μm) and excellent inde-

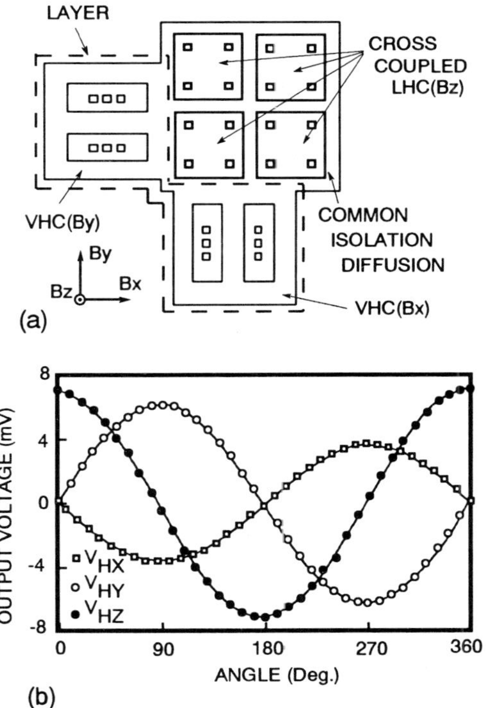

Figure 5. Top view(a) and characteristic(b) of three-dimensional magnetic sensor.

pendence of the measuring axes. Since the sensitivities of the VHC and LHC are different (the VHC and LHC have product sensitivities of about 50 V/AT and 300 V/AT, respectively) because of the difference between their structures, the adjustment of the sensitivities is required for practical three-dimensional applications.

Figure 5 (a) and (b) show the top view and the magnetic characteristics of the three-dimensional magnetic sensor[11], which includes two VHC's and a cross coupled quad pair of LHC's for reducing the offset voltage and its drift. The magnetic characteristic was measured under the turning magnetic field of 0.1T on the x axis around the three-dimensional magnetic sensor with an inclined angle of 120 degrees to the z axis. The sensitivities of the axes are equalized by adjusting the supply currents of the individual Hall cells. In the figure solid lines indicate the theoretical value, and a measurement with a maximum error of 3 % is achieved.

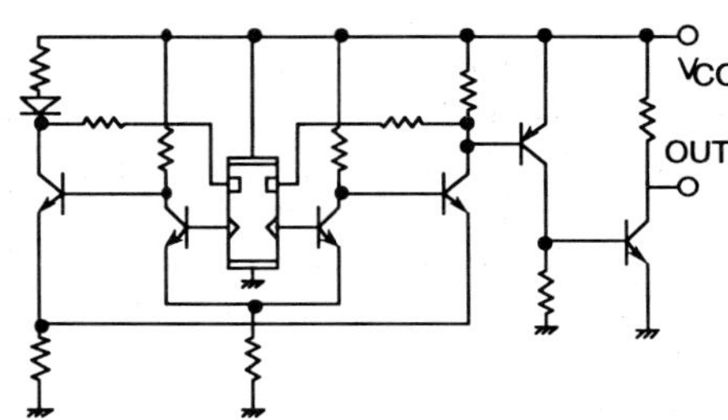

Figure 6. Switching-type Hall IC

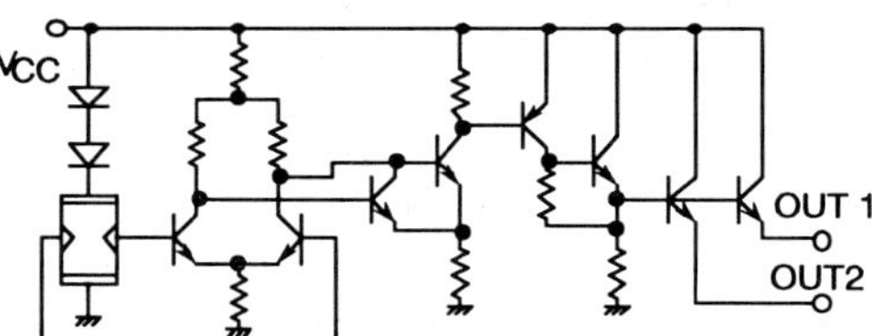

Figure 7. Switching-type Hall IC

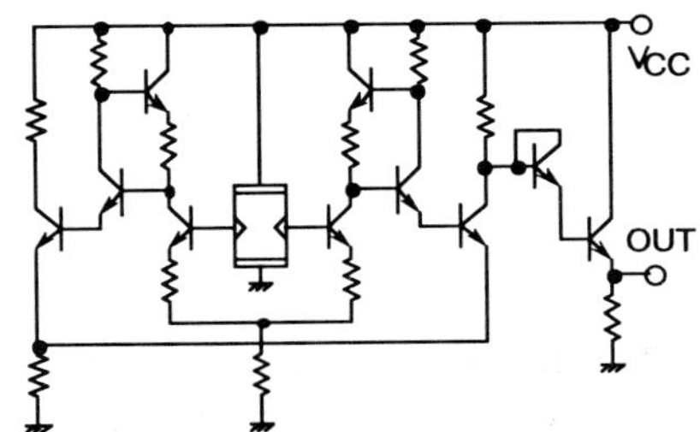

Figure 8. Analog-type Hall

3. INTEGRATION OF MAGNETIC SENSORS AND PERIPHERAL CIRCUITRY

3.1 Hall cells + amplifier or switching circuitry

Almost all of the Hall IC's on the market can be classified into this category. Since the bipolar technology is suitable for designing the peripheral circuits such as an amplifier and a switching circuit, the Hall cell in the standard bipolar structure, especially LHC, is commonly used for a detector. However, magneto-resistive (MR) elements and MOS Hall cells are also adopted for a few kinds of Hall IC's[12,13]. The circuit to be integrated with the Hall cell is a differential amplifier with an impedance converter and/or a switching circuit with hysteresis characteristics. In some cases, a driving circuit for the Hall cell (a constant-current drive circuit with temperature compensation, etc.) is combined. The practical circuits are shown in Figures 6[14], 7[15], 8[16] as examples. The circuit shown in Figure 6 is the switching Hall IC in which the Hall cell has control electrodes. The control electrodes act as a positive-feedback element, and they realize the hysteresis and switching characteristic for the applied magnetic field. Figure 7 shows another type of switching Hall IC. This Hall IC has two output electrodes of the emitter-follower in order to be suitable for the key matrix network of key boards. Two diodes connected to the Hall cell are for the temperature compensation of the sensitivity. The circuit shown in Figure 8 is for an analog output.

3.2 Hall cell + power driver

For a brush-less motor control, a magnetic sensor can be used for the detection of the rotation of the motor. In such an application, simple Hall cells or Hall IC's as shown in Figures 6, 7 and 8 are generally used for the detection of the rotation, and the external circuitry, including operational amplifiers and power transistors, are used for the actual control and drive of the motor. However, there are some reports that describe an integrated power Hall IC which contains a Hall cell, various motor control circuits, and bipolar power transistors in one chip, as shown in Figure 9[17]. Since the power consumption is large in the power Hall IC, it is necessary to design carefully for the sensitivity change and the offset drift caused by the high temperature.

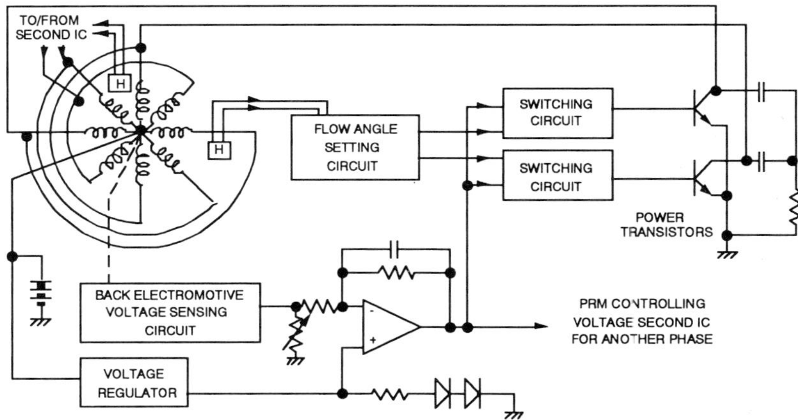

Figure 9. Power Hall IC (Motor controller).

3.3 CDM(etc.) + temperature compensation

Some magnetic sensors, such as the CDM (Carrier Domain Magnetometer) and SSIMT (Suppressed Sidewall Injection MagnetoTransistor), have relatively large temperature dependence with respect to the sensitivity, etc. For these sensors, a temperature compensation circuit is often incorporated as a part of the integrated magnetic sensor. In on-chip temperature compensation the measured temperature exactly matches the temperature of the sensing element because they can be closely located on the same chip, and rapid and exact temperature compensation is possible. This is one of the merits of integrated sensors. Examples can be found in Ref.[18] and [19] for CDM and SSIMT.

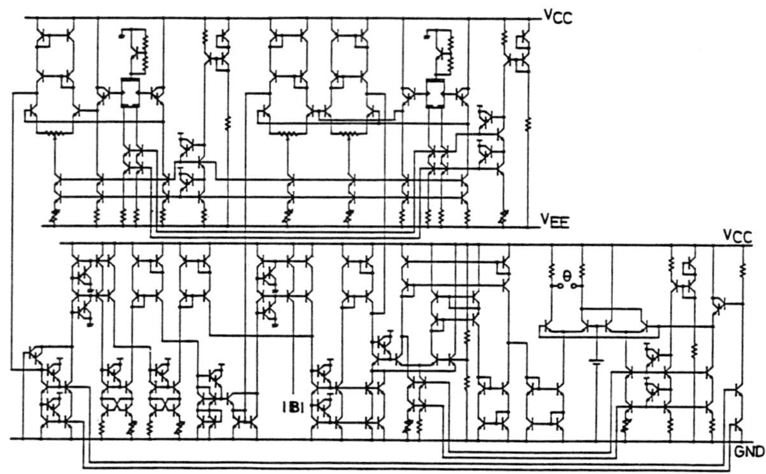

Figure 10. Circuit diagram of two-dimensional integrated magnetic sensor.

3.4 Two-dimensional Hall cell + calculation circuit(1)

The two-dimensional magnetic sensor as mentioned in **2.3** has been incorporated with a calculation circuit for solving the direction of an applied magnetic field[20]. The direction, θ, of the applied magnetic field on an x-y plane(chip surface) can be calculated by the arc-tangent equation, $\theta=\tan^{-1}(B_y/B_x)$, where B_x and B_y are the measured magnetic components for the x and y directions. In the actual integration of Ref. [20], the arc-tangent is approximated by

$$\theta=\tan^{-1}(B_y/B_x)\approx 1.57B_y/(0.63B_x+(0.88B_x^2+B_y^2)^{-1/2})\,. \quad (1)$$

This approximation is fairly good, and the maximum error is 0.24 % for $B_y/B_x < 0$ and $B_y < 0$. The design of the circuit is based upon the translinear circuit technique[21], which has many advantages, such as current outputs/inputs, high precision, wide dynamic range and insensitivity to temperature variations. Figure 10 shows the circuit diagram of the integrated magnetic sensor. In the experimental device, the system was separated into two IC chips (one included the two-dimensional magnetic sensor and amplifiers, and the other included the calculation cir-

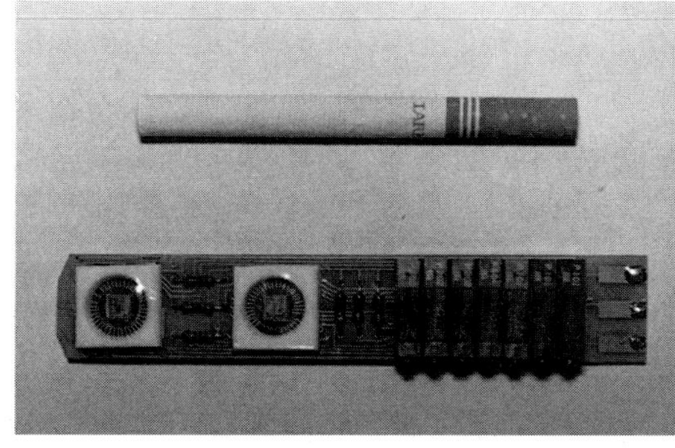

Figure 11. Photgraph of a two-dimensional magnetic sensor.

cuit); they were mounted on a glass epoxy printed circuit board as shown in Figure 11. The measured characteristic is shown in Figure 12, where the horizontal axis represents the direction of the applied magnetic field and the vertical axis the output voltage. The maximum error of ± 2 %/FS is obtained under the applied magnetic flux density of 0.1 T.

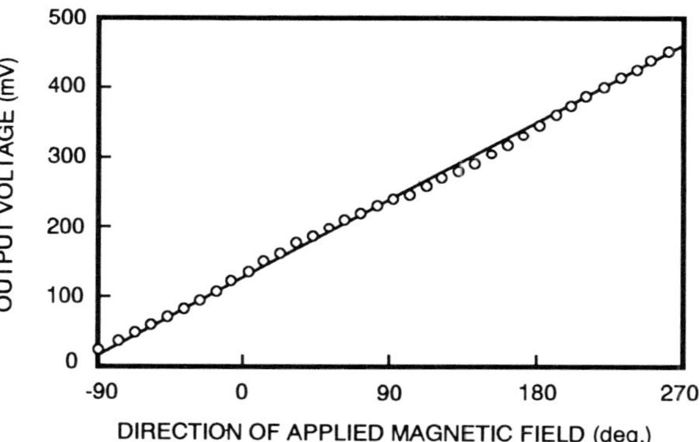

Figure12. Characteristic of two-dimensional integrated magnetic sensor.

This type of integrated magnetic sensor can be used for a magnetic compass, a tension arm controller of a magnetic tape driver, a precise motor control, etc. Moreover, it can be extended to measurements in threedimensions using a three-dimensional magnetic sensor.

3.5 Three-dimensional Hall cell + calculation circuit

In the conventional measurement using a Hall cell, etc., which is a unidirectional measurement, it is very difficult to detect a magnetic field whose direction is unknown and whose direction changes with time. However, the omnidirectional measurement can be observed by the integration of a three-dimensional magnetic sensor and calculation circuit[22]. In this integrated sensor, the three-dimensional magnetic sensor provides three components of the magnetic field, B_x, B_y and B_z, and the calculation circuit achieves the absolute operation, $(B_x^2+B_y^2+B_z^2)^{-1/2}$.

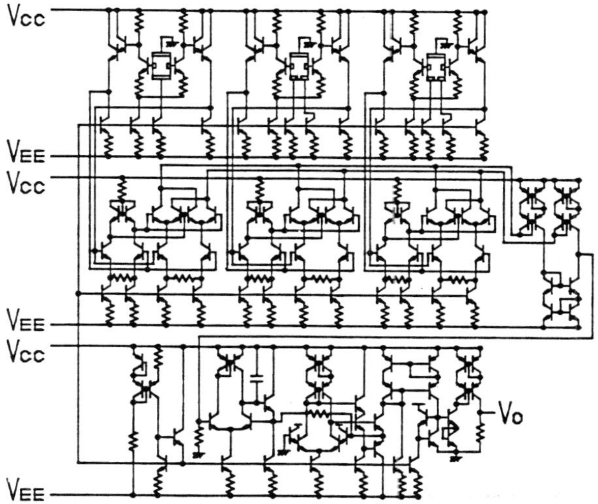

Figure 13. Three-dimensional integrated magnetic sensor.

Figure 13 shows the circuit diagram of the integrated magnetic sensor that shows the omni-directional measurement. The three-dimensional magnetic sensor as mentioned in section **2.3** is used as a detector, and the bipolar analog circuit is adopted for the calculation. The circuit is composed of differential amplifiers, square circuits using the Gilbert multiplier[23], and a square-root circuit based on the translinear circuit technique. The size of the experimental chip is 4.8 x 4.8 mm^2. As an example of the measurement, Figure 14 shows the output voltage plotted on a polar graph for various angles of the applied magnetic field of 100 mT with the incline angle of the sensor as a parameter. A maximum error of ±4 % for the dynamic range of 34 dB is obtained.

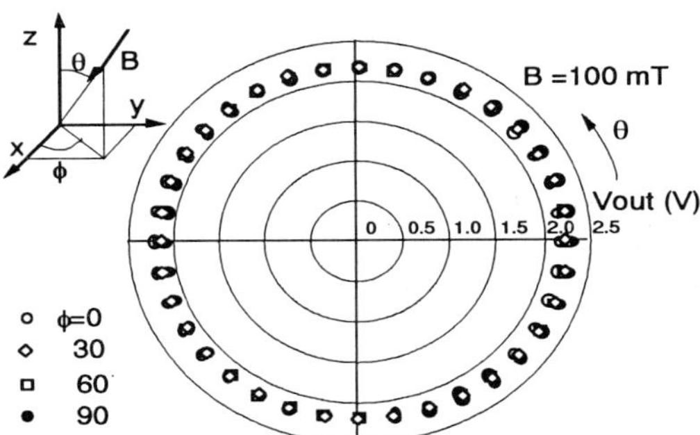

Figure 14. Characteristic of three-dimensional magnetic sensor.

3.6 Hall cell / Magneto- transistor + modified operational amplifier

Figure 15 shows the conceptual diagram of the universal magnetic sensor named as MOP (Magneto OPerational amplifier)[24,25]. The MOP has a modified operational amplifier circuit which has two sets of the input differential amplifier stage, one for the electrical signal similar to a normal operational amplifier and another for the magnetic signal. The magnetic signal is added to the electrical signal and amplified by the large gain amplifier stage. The electrical input is used for the negative or positive feedback, and various magnetic operations can be achieved by changing the feedback characteristic. Similarly to normal operational amplifiers, many types of magnetic operations (e.g., simple magnetic amplification, magnetic switching with hysteresis, magnetic integration and differentiation, magnetic controlled oscillation, log and anti-log con-

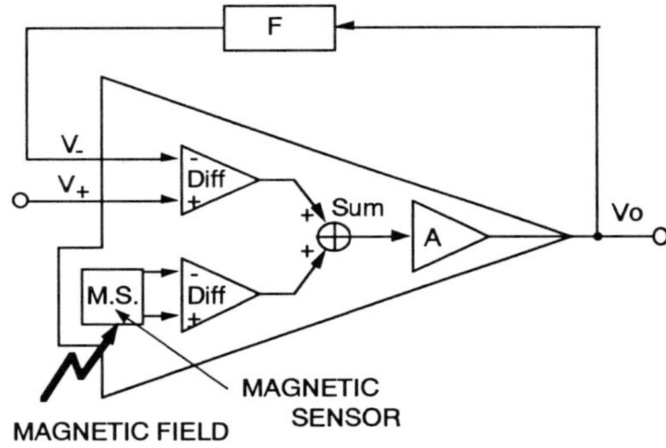

Figure 15. Conceptual diagram of MOP.

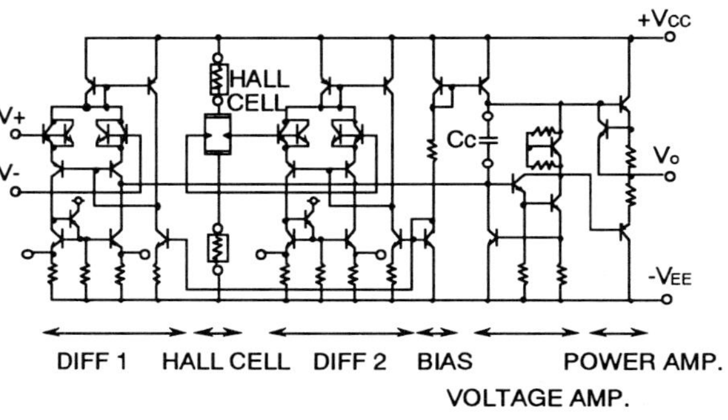

Figure 16. Circuit diagram of MOP.

versions, current output, etc.) can be achieved according to the feedback elements employed.

Two examples of the MOP have been reported, one using SSIMT as a magnetic detector and another using a Hall cell. Figure 16 is the circuit diagram of the latter example. Although many applications have been demonstrated using the MOP, one application is shown in Figure 17. This is the magnetic controlled oscillator whose frequency of oscillation is controlled by the applied magnetic field. From the figure, a sensitivity of around 10 kHz/T or 250 %/T is obtained.

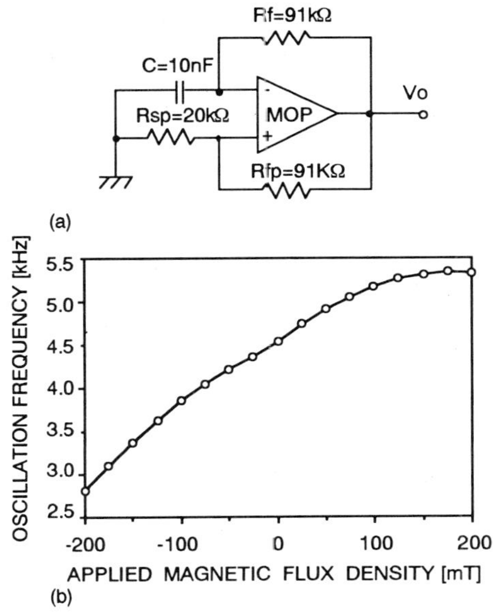

Figure 17. Circuit(a) and measured characterictic(b) of magnetic controlled oscillator.

4. CONCLUSIONS

Various types of integrated magnetic sensors have been discussed. Integrated magnetic sensors are classified and listed in Table 1 according to the type of magnetic detector and the type of function. The numbers in the table refer to the ref-

Table 1
Classification of integrated magnetic sensors

<table>
<tr><th>Function</th><th>Hall cell</th><th>Magneto-transistor</th><th>CDM</th><th>Others</th></tr>
<tr><td>Amplification/ Switching</td><td>4,5,13,14,15,16, 26(GaAs)</td><td>8,27</td><td></td><td>12(MR)</td></tr>
<tr><td>Power amplification</td><td>17,28</td><td></td><td></td><td></td></tr>
<tr><td>Temperature compensation</td><td></td><td>19</td><td>18</td><td></td></tr>
<tr><td rowspan="2">Calculation</td><td></td><td rowspan="2"></td><td rowspan="2"></td><td rowspan="2"></td></tr>
<tr><td rowspan="2">20,22</td></tr>
<tr><td rowspan="2">Multi-dimension</td><td rowspan="2"></td><td rowspan="2"></td><td rowspan="2"></td></tr>
<tr><td>11</td></tr>
<tr><td>Universalization</td><td>24,25</td><td></td><td></td><td></td></tr>
</table>

erence numbers listed below. In the table, Hall ICs (area enclosed by the thick rectangle line) are commercially available, but most of the other types are still being studied and are not as yet commercially available. However, it is true that the magnetic sensor is the most suitable sensor for integration and workers in all fields expect the realization of a practical integrated sensor. We are looking forword to the appearance of novel and epoch-making integrated magnetic sensors.

REFERENCES

1 S.Kordic, Sensors and Actuators, 10(1986)347.
2 H.P.Baltes and R.S.Popovic, Proc. IEE,74 (1986)1107.
3 CH.S.Roumenin, Sensors and Actuators A, 24(1990)83.
4 S.Takamiya and K.Fujikawa, IEEE Trans. Electron Devices, ED-19 (1972)1085.
5 R.Huang, F.Teh and R.Huang, IEEE Trans. Electron Devices, ED-31 (1984)1001.
6 P.W.Fry and S.J.Hoey, IEEE Trans. Electron Devices, ED-16 (1969)35.
7 A.Nathan, A.M.J.Huiser, H.P.Baltes and H.G.Schmidt-Wernmar, Can. J. Phys., 63 (1985)695.
8 R.S.Popovic, IEEE J. Solid State Circuits, SC-18 (1983)426.
9 K.Maenaka, T.Ohgusu, M.Ishida and T.Nakamura, Electronics Lett., 23 (1987)1104.
10 K.Maenaka, H.Fujita and T.Nakamura,Tech. Dig. of the 8th Sensor Sym.(1989)219.
11 K.Maenaka,T.Sawada and S.Maeda, Tech. Dig. of the 11th Sensor Sym., (1992)75.
12 T.Usuki, S.Sugiyama, M.Takeuchi, T.Takeuti and I.Igarashi, Proc. of the 2nd Sensor Sym. (1982)215.
13 S.Hirata and M.Suzuki, Proc. of the 1st Sensor Sym. (1981)305.
14 Telefonaktiebolaget L M Ericsson, U.K.Patent 1,441,009.
15 J.T.Maupin and M.L.Geske, C.L.Chien ed., The Hall effect and its applications, New York,Plenum Press (1980)421.
16 Matsushita Electronic Corp., Hall IC series catalog, A-017/2.
17 Y.Kanda, M.Migitaka, H.Yamamoto, H.Morozumi, T.Okabe and S.Okazaki, IEEE Trans Electron Device, ED-29 (1982)151.
18 S.Kirby, Sensors and Actuators, 4(1983)25.
19 K.Maenaka, H.Okada and T.Nakamura, Sensors and Actuators, A21-A23 (1990)807.
20 K.Maenaka, M.Tsukahara and T.Nakamura, Sensors and Actuators, A21-A23 (1990)747.
21 B. Gilbert, Electron. Lett. 11(1975)14.
22 K.Maenaka, T.Ohgusu, M.Ishida and T.Nakamura, Tech. Dig. of the 7th Sensor Sym. (1988)43.
23 P.R.Gray and R.G.Meyer, Analysis and design of analog integrated circuit, 2nd ed., Wiley, 1984.
24 K.Maenaka, H.Okada and T.Nakamura, Sensors and Actuators, A21-A23 (1990)807.

25 K.Maenaka, N.Higashitani and S.Maeda, Tech. Dig. of the 10th Sensor sym. (1991)161.

26 T.R.Lepkowski, G.Shade, S.P.Kwok, M.Feng, L.E.Dickens, D.L.Laude and B.Choendube, IEEE Electron Device Lett., EDL-7 (1986)222.

27 L.W.Davies and M.S.Wells, Proc. IREE Australia (1971)235.

28 P.R.Emerald, Tech. Dig. on Custom Integrated Circuit Conf. (1987)280.

Intelligent Sensors
H. Yamasaki (Editor)

Intelligent Sensors in Process Instrumentation

Kinji Harada Yokogawa Electric Corporation, Musashino-shi, Tokyo, 180 Japan

1. Introduction

In the second half of the 1970's microcomputers began to be widely used in the process control instruments installed in a control room. In the first half of the 1980's microcomputers were recognized to be highly reliable with respect to their actual results and over the past several years have spread to use in field instruments which must withstand severe environmental conditions. Not only has measurement accuracy been improved but various intelligent functions have been added by the application of microcomputers. In the other words so-called intelligent sensors have become a reality in the field instruments used in process instrumentation.

This section describes the general features of intelligent field instruments and explains several examples of them such as flowmeters, differential pressure transmitters and a control valve positioner. In addition the present status of the field bus is briefly described because of its close relationship with intelligent field instruments.

2. Purpose of making field instruments "intelligent"

The development of intelligent field instruments has made progress, with sophisticated signal processing functions and communication functions, by using microcomputers. Both measuring characteristics and communication capabilities have been greatly improved. For example, field instruments have become able to make calculations using the multiple sensor's signal, to pass accurate judgement on the operational conditions of the process and to acquire knowledge of the control process. The use of microcomputers has been an essential condition of the realization of intelligent field instruments.

There are many different reasons for making field instruments intelligent. They depend on what kinds of instruments they are and under what conditions they are used. Furthermore the sensing elements used in the instruments should have the potential to be made intelligent. The following are considered to be general reasons for the provision of intelligent field instruments.

(1) Improvements in measurement accuracy

a. Linearization of the relationship between input and output signals
b. Automatic zero-point calibration
c. Automatic compensation of errors caused by environmental disturbances such as changes in ambient temperature
d. Automatic compensation of errors caused by changes in the process condition such as fluid temperature and fluid pressure in flow measurement

(2) Improvements in operational capability and maintenance ability

a. Remote maintenance operation utilizing digital communication functions (zero-point and span adjustments, change of measurement range, etc.)
b. Integration of different range sensors by widening sensor's rangeability (improvement of the flexibility to user's specification changes, reduction of the inventory for maintenance service, etc.)
c. Storage and readout of sensor data and process control data (tag no., measurement range, historical data of maintenance, set-point value, etc.)
d. Self-check and self-learning functions
e. Emergency detection and alarm operation (out of range of measurement variable, unusual environmental conditions, etc.)

(3) Improvements in communication functions and reduction of system failure
a. Monitoring of important data (input signal, output signal, self-check signal,etc.)
b. Communication with upper system and surrounding systems
c. Fault detection and fault prediction of control system

Therefore the reasons for making intelligent sensors are in order to improve measurement ability, maintenance ability, immunity to the process and environmental disturbances and capability of judging conditions, and as a result to improve economy, to prevent system failure and to maintain high quality of the products.

3. Examples of intelligent field instruments

Some examples of widely used intelligent field sensors, including an intelligent valve positioner, are described below.

3.1 Electromagnetic flowmeter

The electromagnetic flowmeter is based on the principle of magnetic induction. Its basic principle is shown in Fig.1. The electromagnetic flowmeter is limited in its use in that the measured fluid must have electric conductivity. However, the flowmeter has such features as low pressure loss, since there is no obstacle in the measuring pipe, and a linear output signal with respect to flow rate can be obtained. Until now a four-wire method in which the signal and power lines are wired separately has been mostly adopted. Therefore compared with the differential pressure transmitter which adopts a two-wire method there wasn't a severe limitation of power consumption in the signal processing circuit design, and microcomputers could be introduced earlier on. Let's examine the history of its progress.

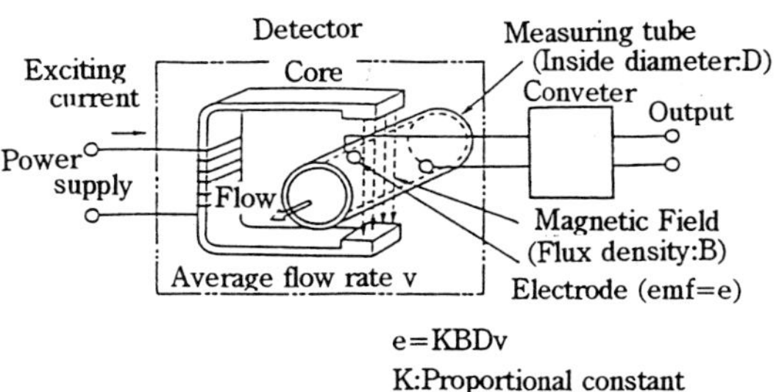

Figure 1. Principle of electromagnetic flowmeter

The method of exciting the electromagnetic flowmeter to produce a magnetic field has been changed as follows. At an early stage of its development the commercial frequency (50 or 60 Hz) was used for the generation of the magnetic field. This suffered from zero-point instability since the exciting current was always changing and derivative noise in the magnetic flux arose when the detector electrodes became soiled.

In the second stage the exciting current was changed so that it used a rectangular waveform with low frequency around 1/8 of the

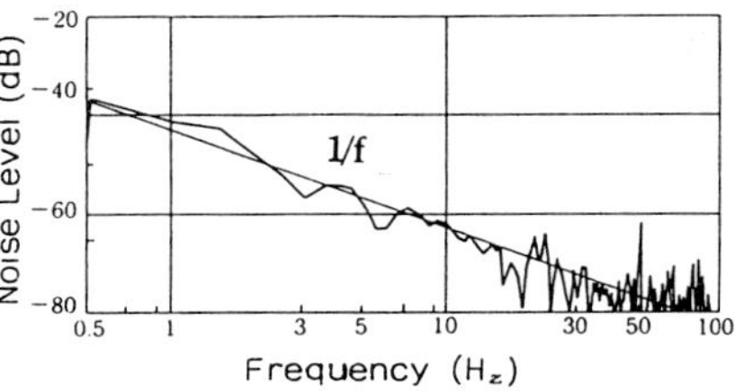

Figure 2. Slurry noise spectrum

commercial frequency. An emf signal was sampled just when the exciting current became stable. This method reduced the zero-point instability induced by the magnetic flux derivative noise remarkably. However output signal drift caused by slurry in the fluid became noticeable in some applications. The slurry noise is caused by collisions between solids in the fluid and the electrodes. An example of a noise spectrum caused by slurry is shown in Fig.2 [1].

In the third stage the slurry noise was removed and various intelligent functions were added by employing a microcomputer.

The principle and the circuit block diagram of the intelligent electromagnetic flowmeter are shown in Fig.3 and Fig.4 [1,2]. A high-frequency excitation (70 ~ 100Hz) is superimposed on the conventional low-frequency excitation (about 7Hz). Two emf signals are separately demodulated in synchronization with each exciting frequency, are passed by the low or high pass noise filter and are added together.

The slurry noise level in the low-frequency region is higher than that in the high-frequency region as shown in Fig.2. A high s/n ratio can be obtained with the high-frequency excitation. Though the noise level is higher for the low-frequency excitation, noises

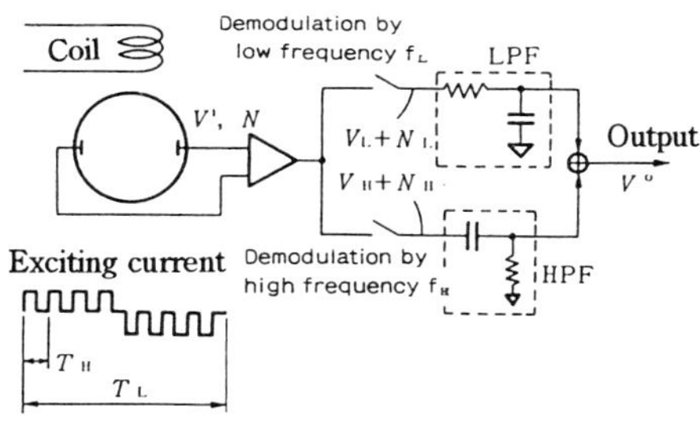

Figure 3. Principle of electromagnetic flowmeter using dual frequency excitation

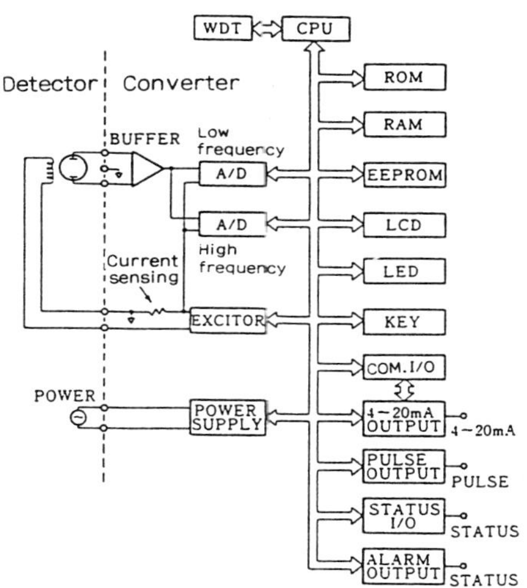

Figure 4. Circuit block diagram of electromagnetic flowmeter

are averaged by the low pass filter which has a long time constant, and a high s/n ratio is obtained also. Therefore the combined signal has a superior immunity to the slurry noise. The emf signal tends to be affected by the magnetic flux derivative noise with the high-frequency excitation, but it is eliminated by the high-pass filter since the noise frequency is low.

It became possible, by using a microcomputer, to generate the two kinds of sophisticated exciting current, to control the sampling timing of the flow rate signal and to construct flexible digital filters. Furthermore, in order to improve reliability and maintenance ability, many functions such as providing information on abnormal signals and circuit trouble, signal generation to hold the control process in a safe condition, the apability of taking measurements in either direction of flow, batch control application, etc., zero–point adjustment, change of measurement range and setting of damping constant can be carried out using a hand–held terminal or a terminal in the control room, as shown in Fig.5. The digital signal for communication is superimposed on the standard analog signal (4–20mADC) and it doesn't affect the analog signal during a communication exchange.

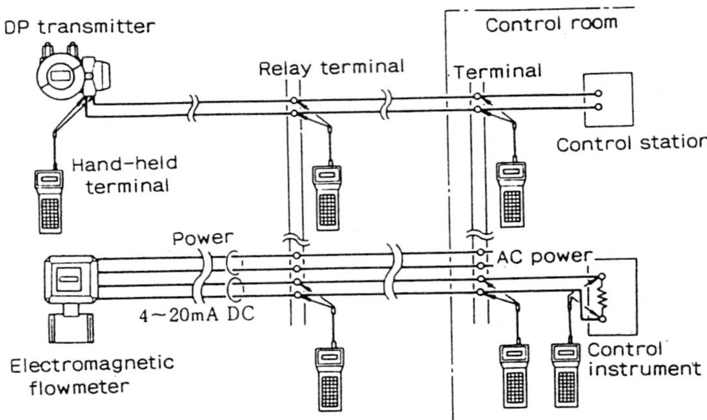

Figure 5. Communication using hand–held terminal

Development of the intelligent signal transmission system is now in a transition period from a conventional analog signal system to a full digital signal system using a field bus. At present the protocols used for the communication are not standardized. There is no interchangeability between the systems supplied by different manufacturers though the communication methods are very similar.

(2) Vortex flowmeter

A vortex flowmeter utilizes Karman vortices generated in turbulent flow. Boundary layer separation occurs on both sides of the vortex shedder (the vortex producing body) placed in the flow and the vortices rotating in opposite directions are alternately shed on each side as shown in Fig.6. The following relationship holds between vortex frequency(f) and flow rate(v).

$$f = St \times v / d$$

Where d is the width of shedder, St is a constant of proportionality called the Strouhal number.

The vortex frequency proportional to flow rate is measured by detecting the alternating force in the vortex shedder.

Fig.7 shows an example of a vortex flowmeter using piezoelectric sensing

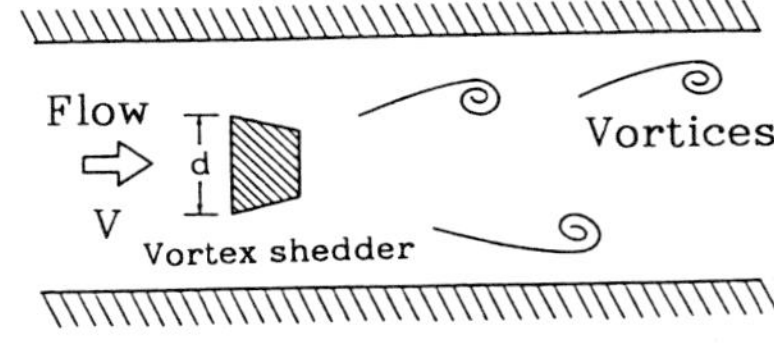

Figure 6. Generation of vortices

elements for the vortex force detector [3]. Two sensing elements are arranged so as only to detect the bending moments (S_1 and S_2) produced by the vortices and to cancel noise components (N_1 and N_2) induced by the mechanical vibration. The vortex frequency signal is sent to a signal processing circuit and converted to a 4–20mA output signal. Change of measurement range and setting of various parameters can be carried out from a remote terminal and these data are stored in EPROM. In case of power failure the integrated flow rate just before can be stored in EPROM.

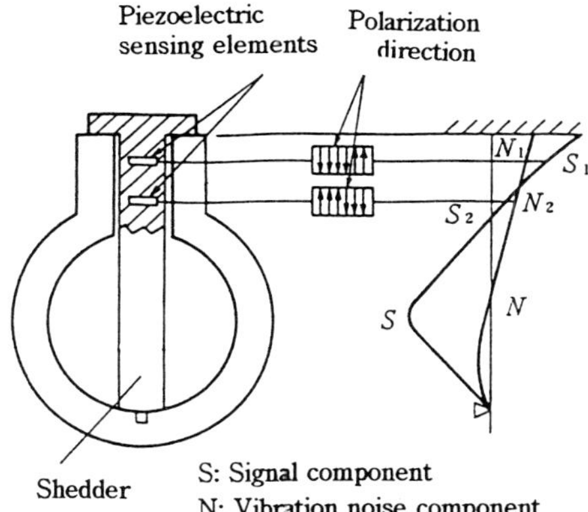

Figure 7. Cross section of vortex flowmeter using piezoelectric sensing elements

3.3 Differential pressure transmitter

A differential pressure (DP) transmitter is commonly used for process flow measurement utilizing the relationship that the square of the differential pressure caused by an orifice is proportional to the flow rate. It is also used for the measurement of liquid level and density.

A pneumatic force–balance type transmitter was widely used in the 1950's and in the early 1960's. In the 1960's electronic transmitters using a similar force–balance principle were developed and were used widely. Subsequently the force–balance type electronic transmitters have been replaced by the deflection type ones which use variable electrostatic capacitances or piezo–resistors for the sensing elements. Much effort has been made in the fabrication technology of the sensing elements and the characteristics have been steadily improved to realize intelligent sensors.

Recently one of the most common DP transmitters developed is a resistive type. Piezo–resistors diffused on a Si (Silicon) diaphragm are used for DP detection. Si is a superior elastic material with no creep. High quality sensing elements can be realized utilizing semiconductor fabrication technology.

The DP transmitter uses a different signal transmission method from the electromagnetic flowmeter. The latter uses a four–wire method as mentioned before. The former uses a two–wire method where signal and power lines are commonly used. The power consumption of the electronic circuit of the transmitter must be within 4mA (corresponds to 0% of the output signal) even when the transmitter is equipped with a microcomputer. It was one of the important conditions for realization of intelligent DP transmitters that the microcomputers and circuitry should be operated with low power.

An example of an intelligent DP transmitter which uses diffused Si resistors for the DP sensor is shown in Fig.8 [4]. Generally a resistive strain gauge has large temperature coefficients for resistivity and sensitivity. The gauge is also affected by the stress caused by static pressure change. These effects must be removed to measure the DP accurately. In this transmitter a temperature sensor and a static pressure sensor are fabricated on the same Si substrate in addition to the DP sensor, as shown in Fig.9. The effects caused by temperature and static pressure changes are fully examined in the factory beforehand. These data are stored in ROM and a signal from the DP sensor is compensated by using signals from the temperature sensor and the static pressure sensor.

Fig.10 shows another type of intelligent DP transmitter [5,6,7]. It uses resonant vibrators for the DP sensing element. Two vibrators are arranged on a Si diaphragm as shown Fig.11. They oscillate at their natural frequencies using self–oscillation circuits including the vibrators. When a DP is applied to both sides of the diaphragm, the diaphragm is deformed to cause axial forces on the vibrators. One is stretched to increase its frequency, and the other is compressed to decrease its frequency. An output

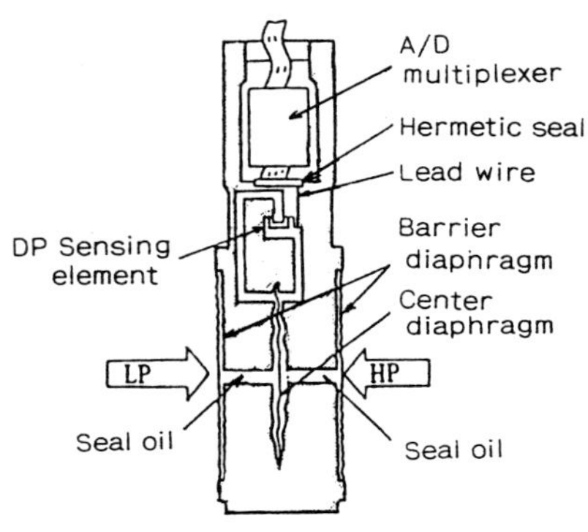

Figure 8. Construction of DP transmitter using piezo–resistive strain gauge

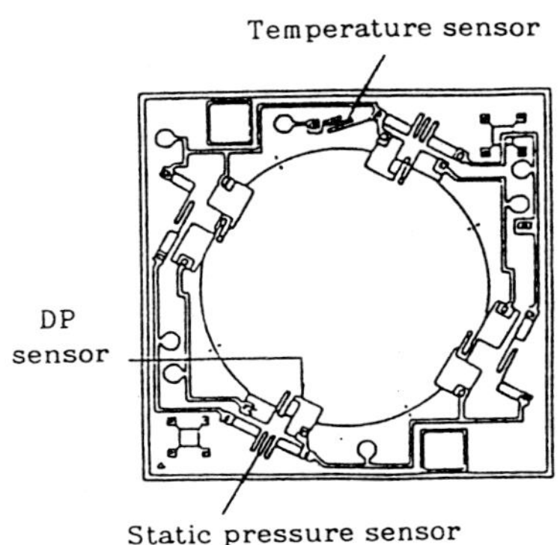

Figure 9. Piezo–resistive composite sensor for DP transmitter

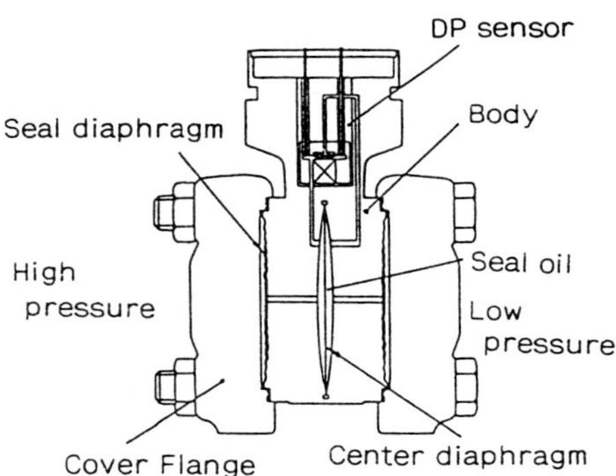

Figure 10. Construction of DP transmitter using resonant vibrators

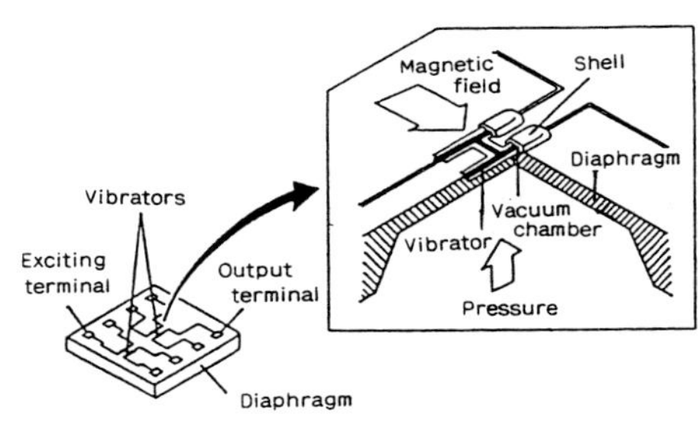

Figure 11. Construction of resonant vibrators for DP transmitter

signal is extracted from the beat frequency ($f_1 - f_2$) of the two vibrators' frequencies (f_1 and f_2). The effects of ambient temperature and long term drift are reduced by such a differential construction.

A resonant sensor essentially has high resolution and excellent repeatability. Compared with a diffusion piezo-resistive sensor, which is a type of physical sensor, the resonant sensor is a structural type sensor, and the characteristics mainly depend on the elastic constants of the material and the mechanical dimensions. The sensor shown in Fig.11 is made of a single crystal of Si which has superior elastic properties and is precisely fabricated using micromachining technologies. Repeatability, hysteresis error and long term stability were improved and miniaturization was realized compared with the conventional resonant sensors made of metal materials.

It is an important matter that intelligent DP transmitters should have a wide rangeability so that the measuring range can be changed easily from a remote terminal. In order to achieve this it is desirable that the input-output relationship is expressed accurately on a simple calibration curve. The transmitter shown in Fig.11 has a rangeability of 40 with an accuracy of ± 0.2% in each measurement range. Fig.12 shows a circuit configuration of the transmitter. Elevation and suppression also can be achieved remotely. The communication method is the same as in the electromagnetic flowmeter.

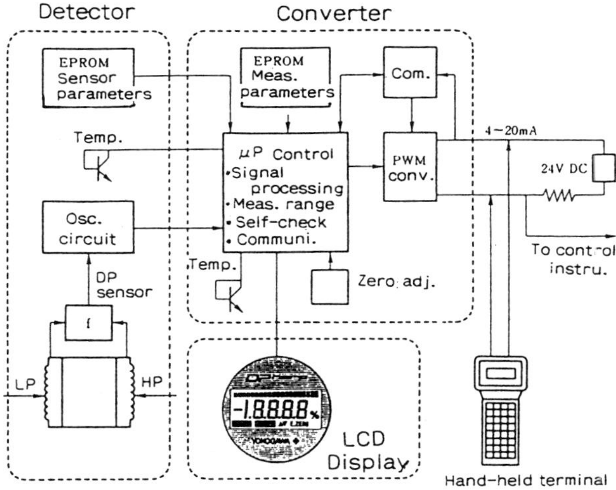

Figure 12. Circuit configuration of DP transmitter using resonant vibrators

3.4 Control valve positioner

A pneumatic control valve is most commonly used as an actuator for process flow control. A positioner is an instrument that receives an electric signal from a controller, converts it to a pneumatic pressure in order to control a valve position. Recently, intelligent positioners equipped with microcomputers have been proposed to improve the characteristics described below.

a. Improvement of response characteristics

An oscillation that sometimes occurs with a mismatch of system loop gain and load capacity can be suppressed by using appropriate control software. The response characteristics will be improved in changing from a small valve to a large one.

b. Improvement of flow rate characteristics

The relationship between valve position and flow rate can be chosen freely so as to meet the control process.

c. Improvement of communication function

It frequently occurs that operators want to know whether a control valve is being controlled correctly according to the controller output (so-called answer back). Such information can be obtained using a communication function. The control range and flow characteristics can be set from a remote terminal. Furthermore, an abnormal status of the pneumatic output signal and valve position and abrasion of valve grand packing can be detected. Fig.13 shows a block diagram of an example of an intelligent valve positioner [8].

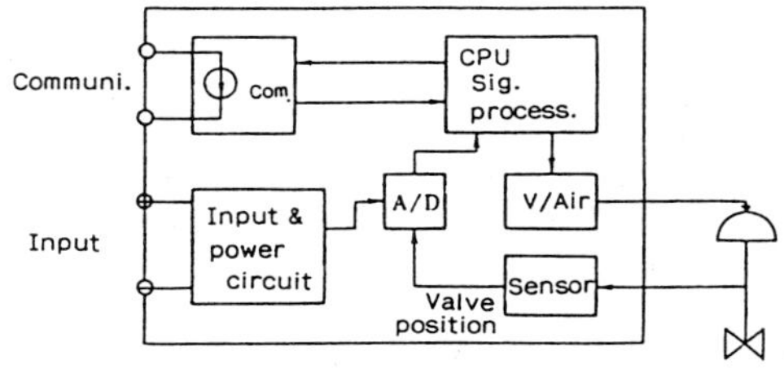

Figure 13. Block diagram of intelligent valve positioner

4. Intelligent sensors and field bus

As described above a 4–20mADC analog signal is still used as a transmission signal though the transmitters have come to be equipped with microcomputers and to have digital signal processing functions. Moreover there is no interchangeability between intelligent transmitters supplied by different manufacturers. International standardization of the field bus for digital communication systems is now expected.

The present analog signal system is an open system and possesses interchangeability between products from different manufacturers. When the field bus is promoted, the concept of the open system should be inherited. The promotion of the field bus has been investigated by the ISA SP50 committee (Standards and Practices group 50). The committee gives the following definition of the field bus.

「Field bus is a digital, serial, bidirectional communication link between intelligent sensors and actuators mounted in an industrial process area (the "field") and controlling, higher-level devices in a control area (the "control house").」

"Controlling, higher level devices in a control area" means DCS (digital control system) and PLC (programmable logic controller). Various forms such as construction of communication data, transmission speed of data and signal waveform should be defined as the standard protocols for digital communication exchange between instruments. Therefore the ISO (International Standard Organization) defines the whole system with respect to the seven-layer reference model as shown in Fig.14. The field bus actually consists of the 1st layer (physical layer), the 2nd layer (data link layer) and the 7th layer (application layer) among these seven layers. Standardization work done by the ISA SP-50 committee will be authorized as an international rule by the IEC committee. At present the standardization work has mainly progressed with the 1st layer.

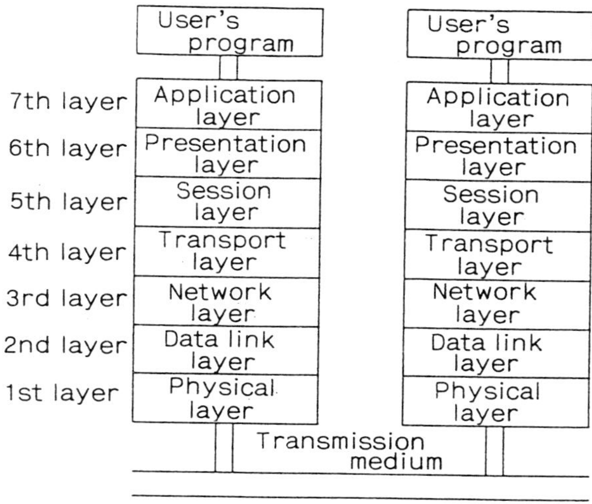

Figure 14. OSI seven- layer reference model

Merits are considered as follows when the field bus is introduced.

(1) Increase in the amount of and improvement of in the quality of field information

Much more field information can be obtained by a communication link and the quality of the information will be improved by its check function.

(2) Improvement of maintenance ability

It becomes easier to access the field instruments from a control room. Therefore check and adjustment work of the field instruments, wiring check and start- up work will become much easier.

(3) Reduction of installation costs

Plural field instruments can be connected using a single signal cable. Substantial reduction in wiring costs in the field area can be expected.

Intelligent sensors in process instrumentation are still in the process of development and rather far from the ideal intelligent sensors as yet. It is expected that more sophisticated intelligent functions such as self- check, automatic calibration, self- recovery from fault and check of environmental conditions will be steadily realized in accordance with the introduction of the field bus.

References

1 K.Kuromori, S.Gotoh, K.Nishiyama and T.Nishijima, Yokogawa Tech. Report, No.32- 3 (1988) 129

2 K.Nishiyama, H.Ohta, Y.Kuroki and N.Shikuya, Yokogawa Tech. Report, No.35- 2 (1991) 65

3 T.Saegusa and N.Kayama, Yokogawa Tech. Report (International Edition), No.4 (1987) 43

4 M.Fujiwara, Automation, No.35–9 (1990) 25
5 K.Harada and H.Kuwayama, Journal of Society of Instrument and Control Engineers, No.28–6 (1989) 509
6 T.Saegusa, S.Gotoh, H.Kuwayama and M.Yamagata, Yokogawa Tech. Report, No.36–1 (1992) 21
7 T.Saegusa and M.Yamagata, Journal of Society of Instrument and Control Engineers, No.31–6 (1992) 689
8 M.Ishizuka, Keiso, No.33–6 (1990) 29

Intelligent Sensors
H. Yamasaki (Editor)

Fault Detection Systems

Takahiko Inari

Faculty of Biology-Oriented Science and Technology, Kinki University
Iwade-Uchida, Wakayama, 649-64 Japan

1. Introduction

One of the features of fault detection is that the targets of detection or recognition are "states" of the devices, pieces of equipment, machines or systems. The fault detection systems consist of, therefore, some sub-systems each using different kinds of techniques, such as detection of physical signals from sensors, signal processing and information processing. In this sense fault detection may be considered as an example of a typical intelligent sensing system.

In fault detection the techniques of signal processing and information processing are, of course, very important, since the "state" is derived from the physical data by means of these techniques, based on the knowledge derived from human experience or from experimental results. However, the development of sensors for difficult targets to detect the physical data is also very important, because the current level of detectable information is limited, and the regions and reliability of detection should be improved and widened.

In this paper some examples of newly developed systems are described. One example is the optical sensor network system, which is characterized by the use of an optical-fiber network connecting optically the devices in the field with the central computer system. In this system, the detection operations themselves are performed directly by means of optical methods or the electric signals from the usual sensors are changed into optical signals in a sensor-interface processor. The other example is the group of new sensors for the detection of faults which previously had been difficult to detect practically. These sensors are based on various optical principles.

2. Features of Fault Detection Systems

The final aim of a fault detection system is to discriminate the abnormal "states" of a system or piece of equipment being monitored from the normal states; moreover, discrimination between types and degrees of abnormal states is required frequently.

The technical fields required in the development of fault detection systems are divided roughly into two groups as follows:

1. The abnormality and its type and degree are derived from the signals from the usual sensors, which are not specially developed. In this group the signal processing or information processing techniques are very important. Since the discrimination between states is based on human experience or experimental facts, processing techniques combined with knowledge engineering, such as application of the neural network, or expert systems for waveform analysis, have been recently developed.

2. On the other hand, the real difficulties arising in fault detection systems, in many cases, have been caused by the difficulties of access for the usual sensors because of physical hazards or narrow spaces, or by the inadequacies of the sensors themselves with respect to the detection of physical data from the faults. Therefore the development of new sensors has been required.

A combination of the techniques in the two groups described above is, of course, necessary for the development of new systems. The techniques in group 1 will be systematized relatively easily, but the techniques in group 2 will require difficult, individual and special developments.

In this paper some examples of the sensors or sensing systems in group 2 that have been developed are described in the following sections. Most of the sensors or sensing systems are based on optical techniques.

3. Instrumentation System Using Optical Fiber Network

Recent process-control or factory-automation systems and their monitoring or diagnosis systems have become more sophisticated, because of the requirements of a high level of reliability and high quality control. Therefore, highly reliable system structures are required for these sophisticated systems. One of the problems with such reliable systems is the effect of the electromagnetic noise arising in the environment of plants or factories, and another is the accidental danger of explosion caused by electric sparks. The instrumentation systems that using optical-fiber techniques have been expected to overcome these problems.

The basic concept[1)] of an instrumentation system using an optical-fiber network presented in this paper is to structure the system using as many optical fibers or optical devices as possible, with the result that a noise-free and intrinsically safe system can be achieved. The concept of the system is shown in Figure 1. Although the usual pieces of information are used, such as temperature, vibration, the sensors used are based on optical techniques. The sensors and transmitters in this system are divided into two types.

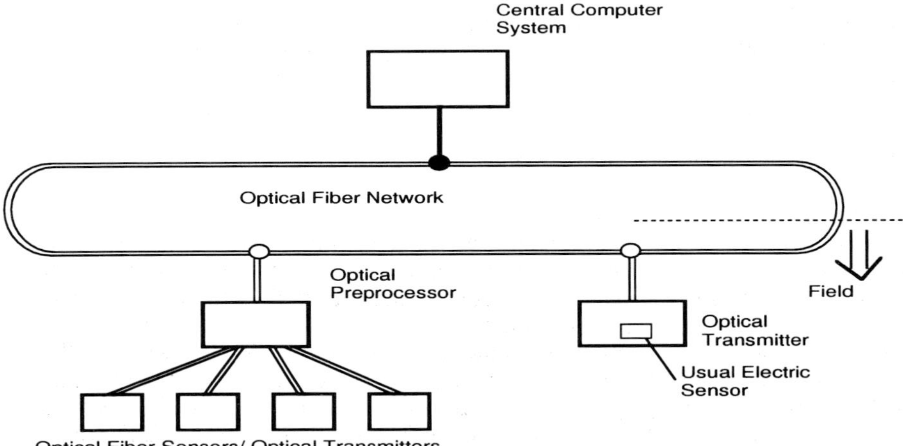

Figure 1. Concept of instrumentation system using optical fiber techniques

The first type is that using the usual electric sensor, and the electric signal is directly changed into an optical signal in the sensor-transmitter equipment placed in the field. This is called an optical transmitter in this paper. The block diagram of the optical transmitter is shown in Figure 2. The second type is that which takes advantage of the newly developed optical-fiber sensors, and the interface equipment is placed in the field. The interface equipment includes an optical emitting diode, a photo-detector, optical guiding systems, and their power sources. This is termed an optical preprocessor in this paper. The block diagram of the optical preprocessor is shown in Figure 3. In this example, four optical fiber sensors are connected with the preprocessor by means of optical switching devices. The double and the single lines in these figures represent the optical and the electric paths respectively. In the case of the optical transmitter shown in Figure 2, plural sensors may also be connected with the transmitter. In both cases minimum electric devices and power sources are designed to be placed in the field equipment.

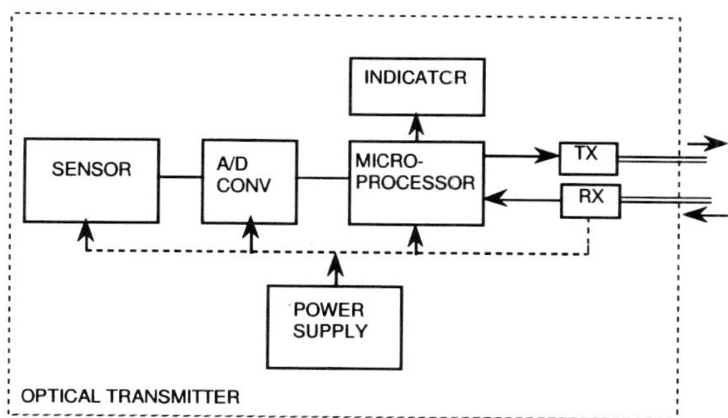

Figure 2. Block diagram of optical transmitter

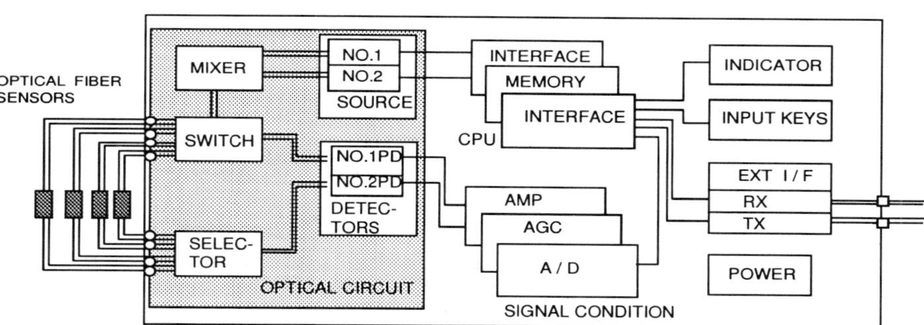

Figure 3. Block diagram of optical preprocessor

The microprocessors included in the optical transmitter or preprocessor are used not only for control of the equipment and signal processing for the individual sensors, but also for complex processing of the signals from the set of sensors selectively positioned on the machines being monitored to derive the data corresponding to the faults. In Figure 4, an example of a display from a fault detection or diagnosis system is shown. This picture is displayed on the CRT of the central monitoring computer of the instrumentation system. The current data from the acceleration and temperature optical-fiber sensors are displayed as a pattern, and the raw waveform of the vibration is shown on the right-hand side of the figure. This raw signal may be selected for display if necessary. The beginning of the fault, and its type and degree may be determined from the data group, the trends of and the relationships between the signals.

Advances in such optical equipment being closely connected optical-fiber networks may produce highly reliable and sophisticated control and monitoring system.

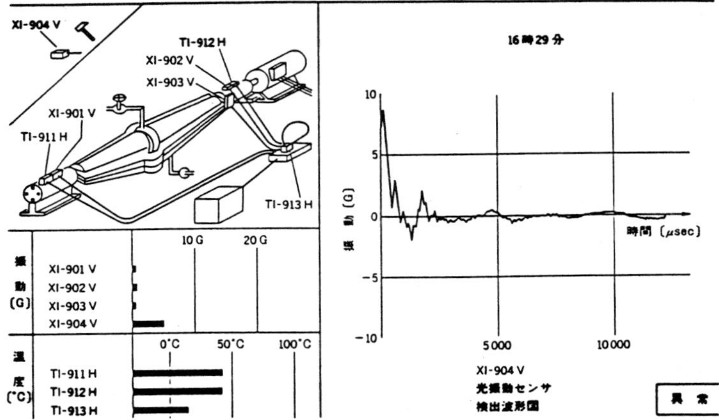

Figure 4. An example of a display derived from a fault detection system

4. New Sensors Based on Optical Techniques

Some examples of the new optical sensors developed specially for fault detection are described in this section.

4.1. Multifunctional sensor using laser light[2)]

In the situation where there are many points to be inspected in a plant in order to prevent accidents induced by faults in the machines, an inspection patrol system such as a mobile robot is required. For such a purpose, a compact, multifunctional and non-contact sensor to detect the faults is needed. The sensor described below can simultaneously detect the vibration of a machine, steam leaks from some equipment, and the distance between the sensor and the target by using only one laser beam. The configuration of the sensor is shown in Figure 5.

First, the Doppler effect is used to detect the vibration. The frequency of the light from the He-Ne laser source is modulated from f_1 to $f_1 + \Delta f_1$ by the A/O (Acousto-Optic) module, and the target is illuminated. The Doppler shift Δf_D is induced in the scattered light by the vibration of the target, and is detected by the detector which produces an output signal with a frequency of $\Delta f_1 + \Delta f_D$. The retaionship of directions of polarization should be considered in setting the optical interference system for vibration detection.

An electronic system for processing the Doppler signal with various noises and fluctuations is used to derive and track the component with the Doppler frequency Δf_D, and to detect the vibration frequencies and amplitudes.

Secondly, a steam leak can be detected by using the polarization properties of the light wave. Steam vapor consists of spherical particles with diameters from several μm to several tens of μm. Because the direction of polarization of the incident light wave is same after scattering by such particles, the polarization directions of the light waves scattered from the steam vapor and received by the receiving system are the same as that of the laser source. On the other hand, the light waves scattered from the background such as a wall, have various directions of polarizations as the result of scattering by the rough surfaces of the background. The detection system uses the difference in the polarization properties of the light scattered from the steam vapor and that scattered from the background.

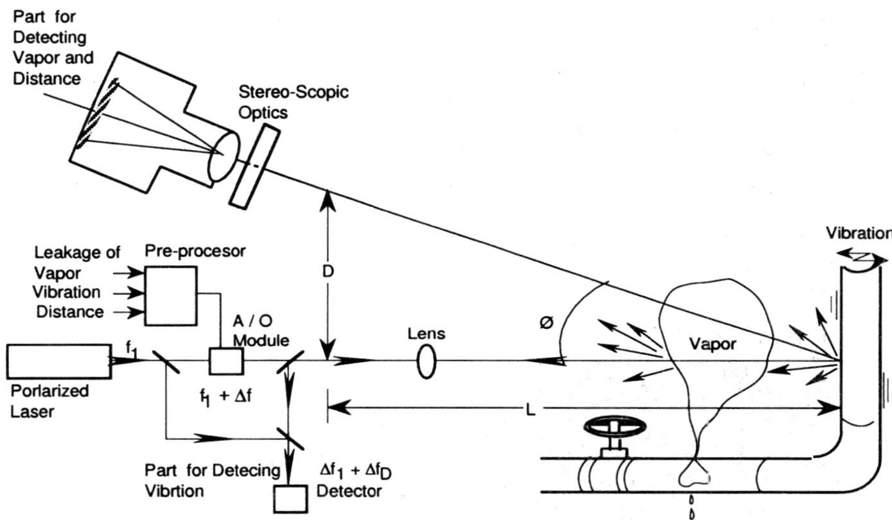

Figure 5. Optical configuration of multifunction sensor using one laser beam

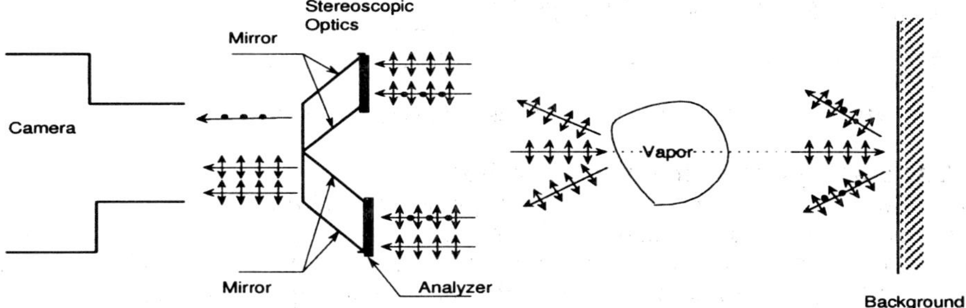

Figure 6. Principle of vapor detection using polarization properties

The receiving system for the scattered light comprise a TV camera and stereo scopic optics placed in front of it. The stereoscopic optics consist of two parts, and divide the range of vision into two ranges. A polarization analyzer is positioned in front of each of the two apertures of the stereoscopic optics system. One of the analyzers only passes light with the same direction of polarization as that of the laser source, and the other only transmits light polarized in a direction perpendicular to that of the laser source. As a result, the image on the TV camera formed by light waves passing through the former analyzer is due to the scattering from both the steam vapor and the background, and the image formed by light passing through the latter analyzer is only due to scattering from the background. The relative polarization directions are shown in Figure 6. Both images can be observed on the TV camera simultaneously as shown in Figure 7. In the figure, the spot images B are from the background wall, visible in both ranges, and the spot image A are from the steam vapor, which appears in only one range. Hence the steam leak can be detected separately from the disturbance due to the background.

Thirdly, the distance between the sensor and the target can be determined by use of the triangulation principle. In the construction of the sensor, as shown in the Figure 5, the direction of the receiving system including the TV camera is different from that of the incident light beam, so the distance between the sensor and the target can be calculated by the triangulatior ɔrinciple from the location data of the spot image on the TV camera.

Thus, it is clear that the sensor described above is multifunctional with respect to the detection of vibrations and steam leaks, and the measurement of distances. Possible specifications of the trial system are shown in Table.1.

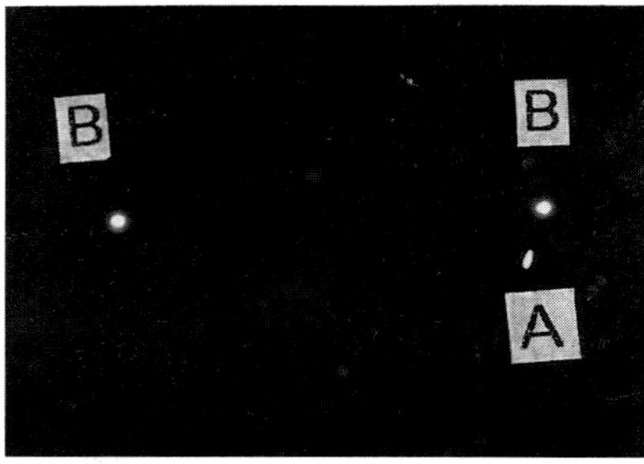

Figure 7. Spot images in the two image ranges:
A: from vapor, B: from background, appears in both ranges

Table 1. Posible specifications of tial system

Vapor leak detection	Density : > 7.7×10^4 particles / cm^3
Vibration detection	Frequency : 10 Hz ~ 1 kHz Amplitude : 10 ~ 0.1 mm
Distance measurement	Range : 500 mm ~ 5 m Accuracy ± 1.5 ~ ± 105 mm
Data transmittance	Capacity : 20 kb / s

4.2. Gas detection system using laser diode[3)]

It is very important for the purposes of plant inspection to detect the presence of gases such as methane. Especially for a mobile inspection system, the abilities of remote and selective detection for the specified objects in various environments are required. Though a gas detection method using the absorption spectra characteristic of a target gas has been used, realization of a gas sensing system for mobile inspection with the features described above has been difficult.

The sensing system described here also uses the detection of the absorption spectrum of the target gas, which is methane in this case, by means of the light beam from a laser diode with narrow bands in the emitting spectrum. The configuration of the sensing system is shown in Figure 8. The light beam from the laser diode incident on a background wall is scattered and received by the receiving system. If the wavelength of the emitting narrow band coincides with that of the absorption band of the gas, the intensity of the received light is reduced by the presence of methane between the system and the wall. The intensity of the received light is, however, also influenced by variations in the background. To compensate for such influences two emitting narrow bands are used, one of which coincides with that of the absorption band, and the other does not. By using the ratio of the intensities of the two bands, the influences due to the scattering by the background can be cancelled out.

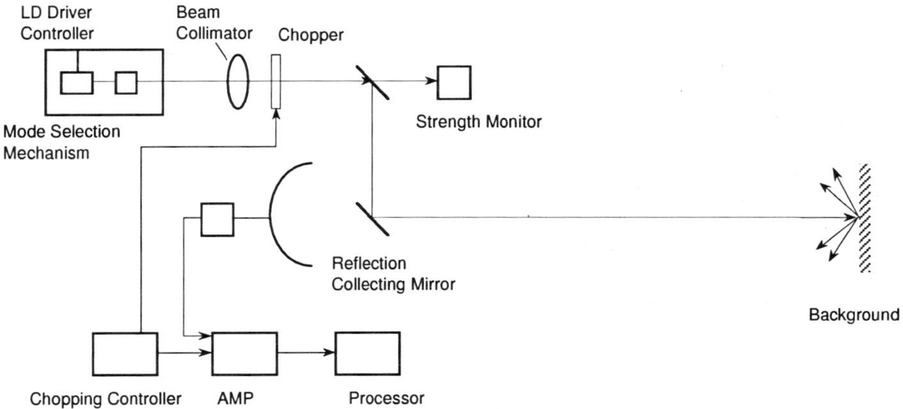

Figure 8. Configuration of gas detection system

The technical problem in the method is in forming the two emitting narrow bands from the laser diode, and in tuning one of the emitting bands to the absorption band of methane. In this system two longitudinal modes of the laser oscillation are used as the emitting bands. The wavelengths of the modes are tuned by the temperature control of the laser diode. As shown in Figure 9, the wavelength of the peak of the characteristic absorption band of methane is 1.3 μm, to which one of the modes is tuned. The wavelength of the other peak of the modes is out of tune with that of the absorption band of methane. The temperature control of the laser diode is achieved by means of a two-stage Peltier element. The accuracy of the tuning is within ±0.01nm. In the trial system the two emitting bands, one of which is for detection and the other for reference, are used interchangeably. The selection of the wavelengths and the change between the two bands are achieved by means of an etalon. An experiment using the trial system proved that this method can detect methane with a concentration of 100ppm without any interference from the environment.

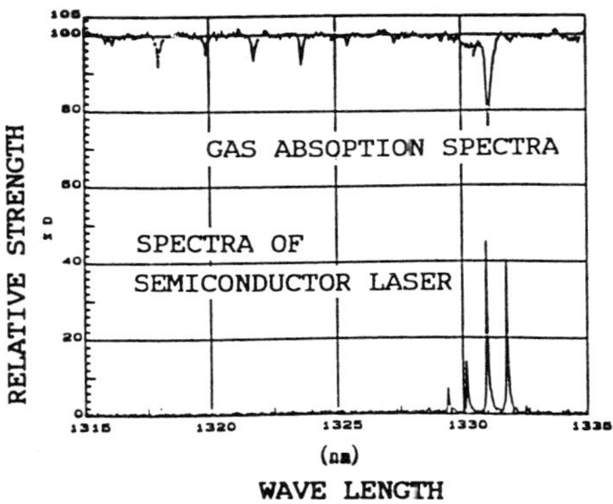

Figure 9. Relaion between the gas absorption and emission spectra

4.3. Inspection sensor for the inner wall of a pipe[4)]

Although the inner walls of pipes used for piping or heat exchangers of installations in some plants should be inspected periodically, few sensors capable of real inspection are available, especially for the shapes of the inner walls.

The construction of a newly developed sensor is shown in Figure 10. The light beam from the laser diode is so reflected by the conical mirror that a circular pattern of the light is projected onto the inner wall. The light reflected from the illuminated circumferential surface of the wall are collected by a lens that forms a circular image on the image sensor as shown in Figure 11. The circular image shows the shape of the inner wall, so, if there is some defect on the surface, the normal pattern as shown in Figure 11(a) is deformed as shown in Figure 11(b). The two-dimensional image sensor is used to detect the circular image. A high-speed image processor recognizes whether there are irregularities in the image or not, and derives quantitative data for the image shape. The dimensions of the inner wall are calculated from the circular-image data using the triangulation principle. The overall inner surface of the pipe can be inspected by passing the sensor along the axis of the pipe, so a special housing for the sensor and the driving mechanism supporting it should be used.

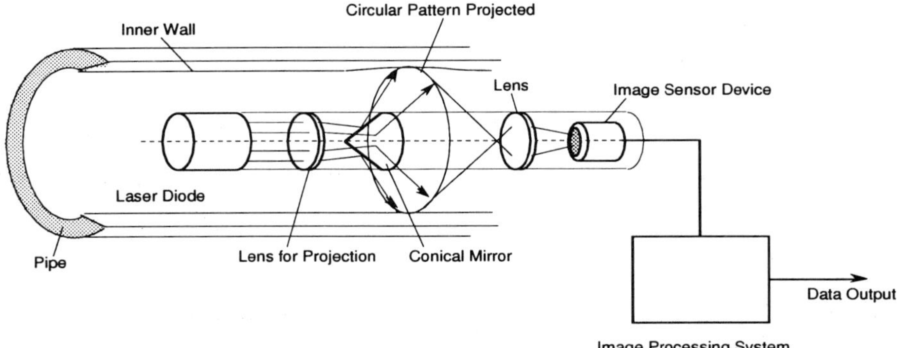

Figure 10. Construction of sensor for inspection of inner wall of a pipe

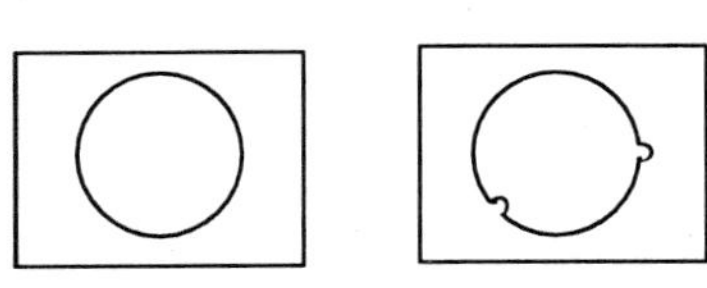

Figue 11. Circular images formed by reflection from inner wall (a) normal, (b) abnormal case

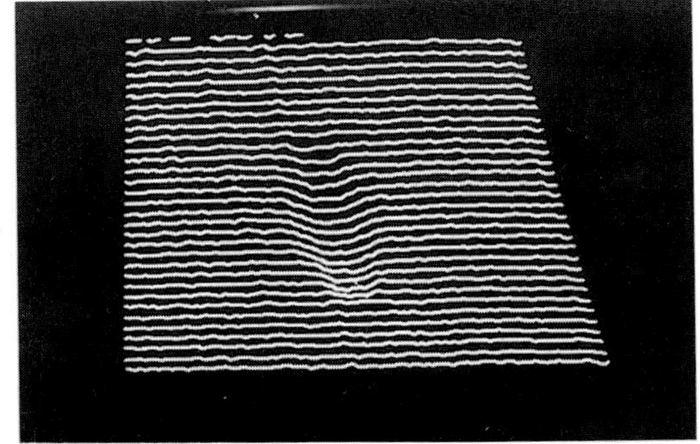

Figure 12. An example of displays showing inspection results

Figure 12 shows an example of the displays resulting from a practical inspection test. The experimental test proved that the abilities of the sensor are within a detection resolution of ±0.1mm at a transit speed of 30mm/second and an interval of inspection of 1mm along a pipe. Using this sensor, not only can defects like pits be detected but also corrosion and the accumulation of various deposits on the inner surface of pipes can be inspected in plants.

5 Conclution

Fault detection sensing systems have remarkable features. A typical feature is that the aim of a fault detection system is the "states" of the objects, which requires complex detection systems involving plural and various sensors. On the other hand, many inspection targets have been unattainable because of the lack of suitable sensors.

In this paper an optical-fiber network system and the optical apparatuses used in the network, which are useful for construction of intrinsically safe systems, and optical sensors developed specially for difficult inspection targets have been presented. The optical techniques may be useful for the further development of sensing systems in this field.

References

1. Y.Uezumi, K.Takashima, K.Kasahara, T.Inari, Investigation of Advanced Instrumentation System Using Fiber Optic Techniques: (Fiber Optic Network and Optical Field Equipments), Proc. Int. Workshop Industrial Automatic Systems, (1987)39-44
2. M.Kamei, T.Nakajima, T.Inari, Multifunctional Instrumentation System Using Various Characteristics of Light Wave, J. Soc. Instrument and Control, 31(1991) 1319-1326(in Japanese)
3. T.Nakajima, M.Kamei, A Gas Detection System Using Semiconductor Laser, Technical Digest of the 10th Sensor Symposium, (1991)203-206
4. T.Inari, K.Takashima, M.Watanabe, J.Fujimoto, Optical Inspection System for the Inner surface of a Pipe Using Detection of circular Images Projected by a Laser Source, Measurement, 13(1994) 99-106

Intelligent Sensors
H. Yamasaki (Editor)

Visual Inspection Systems

Kunihiko Edamatsu

Technical Skills Development Institute,
Fuji Electric Human Resources Development Center Co.,Ltd.,
1, Fuji-machi, Hino-shi, Tokyo 191, Japan

Abstract
This paper describes automated visual inspection systems for industry. The typical system consists of TV cameras with lighting, an image processing unit, and a mechanical unit. Several visual inspection systems in various industries are presented and the features are described. The first system deals with the inspection of pharmaceutical capsules and tablets. The second system deals with label inspection. The third system deals with the collation of stamped characters and marks. The last example describes a general-purpose inspection system.

1. INTRODUCTION (1) (2)

In recent years customers needs have tended to be various and the life-cycle of products has become shorter. In order to produce various products, flexible automation is necessary. NC machines and industrial robots have been introduced into production processes for the purpose of flexible automation. However, visual inspection in most manufacturing processes depends mainly on human inspectors. Product reliability is unstable, because the performance of human inspectors is generally inadequate and variable. Though the human faculty in pattern recognition is excellent, the accuracy of human visual inspection is low for dull, endlessly routine jobs. The inspection for visual appearance is the most difficult task for an artificial system. However, advances in pattern recognition and electronics technologies have resulted in better, and cheaper, machine vision systems, and machine vision systems have also contributed greatly to improvements in the reliability of electronics products. Advances in electronics technology could not be accomplished without the aid of machine vision systems.

2. FEATURES OF VISUAL INSPECTION SYSTEMS (3) (4)

2.1. Visual Inspection of Manufacturing Process

Product inspection is an important step in each production process of industrial manufacturing. Because product reliability is essential in mass-production facilities, 100 percent inspection of all parts, subassemblies and finished products is often carried out. The most difficult inspection task is that of inspecting for visual appearance.

Figure 1 shows the inspection at each step of a production process. Feedback data from inspection systems is used to improve each production process in order to improve product quality. To make the reliability of the data stable, the introduction of automated visual inspection systems is necessary. In general, material visual inspection is simple but the inspection speed is very fast. While the visual inspection of finished products is various and complex, the inspection speed is not so fast.

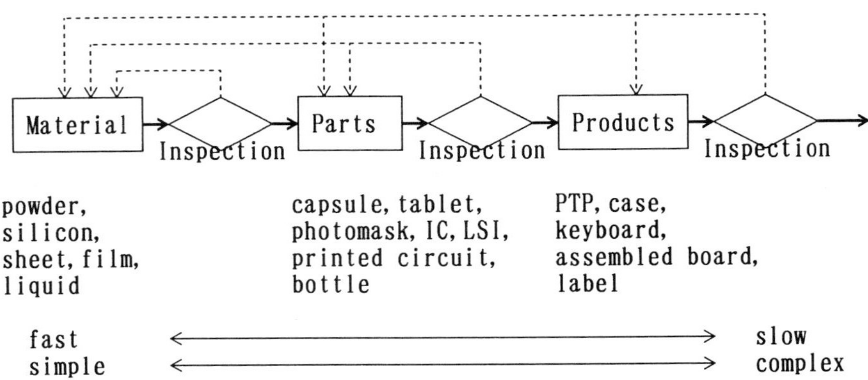

Fig. 1. Inspection in production processes

2.2. Features of Human Visual Inspection

The human visual system has the adaptability that enables it to perform in a world of variety and change. On the other hand, the visual inspection process requires workers to observe the same type of image repeatedly in the detection of anomalies.

Figure 2 shows a qualitative comparison between the human and the machine vision systems in the visual inspection process. The human faculty is flexible and superior to machine vision systems in respect of comparative and integrated judgement, but a human is subject to emotion and preconception. The faculty also differs among individuals. The accuracy of human visual inspection is low for dull, endlessly routine jobs. Slow, expensive or erratic inspection results. The inspection process is normally the largest single cost in manufacturing. Therefore an automated visual inspection system is obviously the alternative to the human inspector.

2.3. Advantages gained by Automating Visual Inspection

Many advantages arise from the automation of visual inspection. One obvious advantage is the elimination of human labor. Also the manufacturing environments in the semiconductor and pharmaceutical industries must be kept clean. Therefore, the elimination of human labor from the visual inspection process is necessary.

The advantages of automating visual inspection are as follows:

①Humanization of the workplace

・freeing humans from dull and routine labor.

②High-speed inspection
- matching high-speed inspection with high-speed production.

③Labor Saving
- saving human labor costs.

④Improvement of Product Quality
- stabilizing the inspection process.
- analyzing inspection data statistically.

⑤Clean Environment
- eliminating human inspectors.

⑥Unfavorable Environments
- performing inspection in unfavorable environments.

⑦Precise Assembly
- introducing a machine vision system employing a microscope.

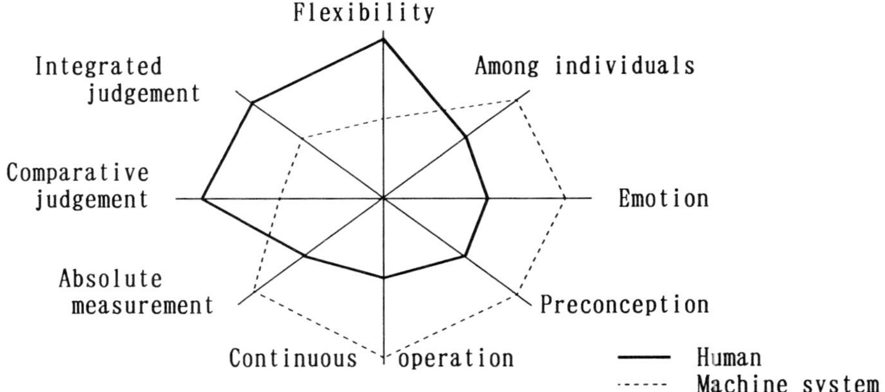

Fig. 2. Qualitative Comparison between human and machine system in the visual inspection process

3. A VISUAL INSPECTION SYSTEM

Figure 3 shows a simple diagram of an automated visual inspection system. The inspection system includes a TV camera with lighting, an image processor and a transport mechanism. The feeder/aligner arranges the objects so as to facilitate observation by the TV camera. The transport mechanism moves the objects to be inspected to the TV camera station. When the transport moves into the field of view of the TV camera a position sensor detects the objects. By means of a trigger input the TV camera observes the objects and sends a video signal to the image processor to be analyzed. After analysis, decisions are made and the image processor directs the sorter to reject the objects if they are defective.

3.1. Image Acquisition [5] [6]

The first step in achieving successful machine vision is to design a suitable image acquisition system. The lighting and the image capture techniques are important in solving a particular image processing problem.

Various lighting techniques are available. Many involve the application of everyday experience, experimentation and common sense, using a combination of cameras, illumination and optics. A robust image is required for further processing and the observed image must not be too susceptible to changes in ambient lighting and other relevant variations, nor require specialized and delicate illumination or optics.

One useful tool in machine-vision lighting applications is the xenon flash tube used as a strobed light source. The biggest advantage is obtained in the inspection of moving objects, owing to the fact that the duration of the pulse of light is very short: typically 5 to 200 microseconds. The apparent freezing of the image is caused by the fact that most TV cameras, whether solid state or otherwise, act as temporary storage devices. Synchronized with "firing" the strobe at an appropriate time, the image of the objects is stored during the short strobe interval and then read out by the scanning mechanism of the camera.

Another image freezing technique involves the use of some sort of shutter function. TV cameras using mechanical or electronic shutters have been developed. The image quality in this case is almost equivalent to that of the stroboscopic method. In addition, such a shutter system has the advantage of being able to employ lighting techniques that use conventional incandescent lamps. This results in a preferable flexibility of lighting technique.

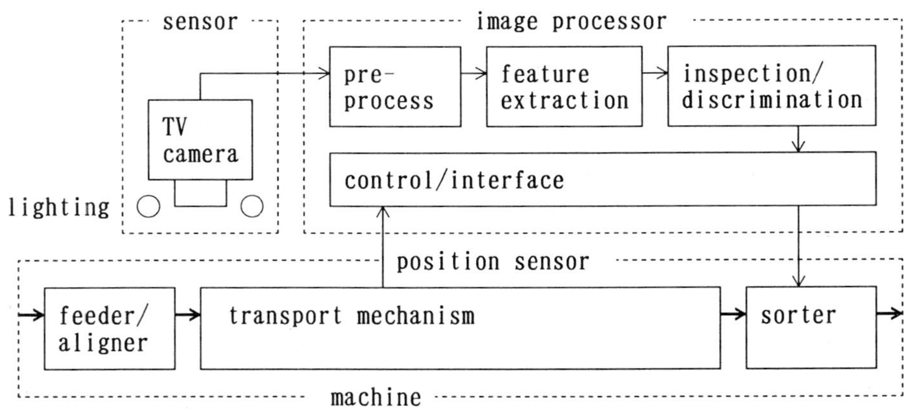

Fig. 3. Visual Inspection System

3.2. Image Processing (7) (8)

The image processor comprises pre-processor, feature extraction, inspection/discrimination and control/interface, as shown in Figure 3. The video signal from the TV camera is converted to a digital image in the pre-processor. The digital image is stored in a frame buffer memory or a run-length coding memory. The digital image is a binary image or a grey image. The binary image is a dual-level picture, that is a black and white image, and the grey image is a multi-level picture. Then the digital image is processed by using a combination of suitable algorithms such as filtering, expansion, contraction, sharpening, smoothing, enhancement, and so on. After preprocessing the digital image, the object features are extracted by means of algorithms for identification of the object. Features considered are "area, length, contour length, holes, corners, position, orientation, radial length, center of gravity", and so on. The techniques of inspection/discrimination are the methods of feature matching and pattern matching. In the feature-matching method the features extracted from the object and those defined for a standard object are compared. The pattern-matching method compares the recorded image with a standard image previously stored in memory.

3.3. Mechanical Arrangement

The mechanical arrangement consists of a feeder/aligner, a transport mechanism, a sorter and a position sensor. The mechanical apparatus is different in each application. The feeder/aligner must arrange each object in order to facilitate observation by the TV camera. The transport mechanism moves the objects to the TV camera station continuously or intermittently. The sorter ejects defective objects as a result of the decisions made by the image processor. The feeder/aligner and sorter must be designed carefully so as not to damage sound objects.

4. APPLICATIONS

4.1. Inspection for the Pharmaceutical Industry

The automation of the pharmaceutical industry is remarkable, and the spread of the G.M.P. (Good Manufacturing Practice) concept, the official guidance of the Ministry of Welfare, and improved quality control through mechanization are expected. However, it appears that visual inspection will remain as a part of unautomated production processes. A typical example of such visual inspection is the 100 percent inspection of all capsules by human inspectors. The typical defects are "crushed ends, bad joints, crushing, cracks", and so on, as shown in Table 1. They are generated when the capsules are filled. An automated capsule checker has been developed [(9)] and Figure 4 shows the configuration of the checker.

The checker is composed of two TV cameras and associated stroboscopic light sources for inspection of both sides of the capsules, logic circuits for two-dimensional parallel signal processing, two 8-bit microprocessors, and mechanisms to feed, transport, turn over and eject the capsules. Capsules fed through a hopper are automatically inspected on both sides and finally separated into either "good" or "no good" chutes at a rate of 20 capsules s^{-1}. To obtain a high-rate inspection capability, each TV camera simultaneously captures two capsules in its image field and outputs the video signals to analog circuits. Then the video signals are digitized into binary signals to be transformed to 320 × 240 matrix image elements by means of sampling circuits. These signals are serially transmitted to a two-dimensional local memory composed of high-speed shift registers and two-dimensional parallel processing circuits. The capsule-form features are extracted by this circuit and memorized directly in random access memories. Two 8-bit microprocessors discriminate the features at a processing rate of 40 patterns s^{-1}.

Table 1 Typical inspection items

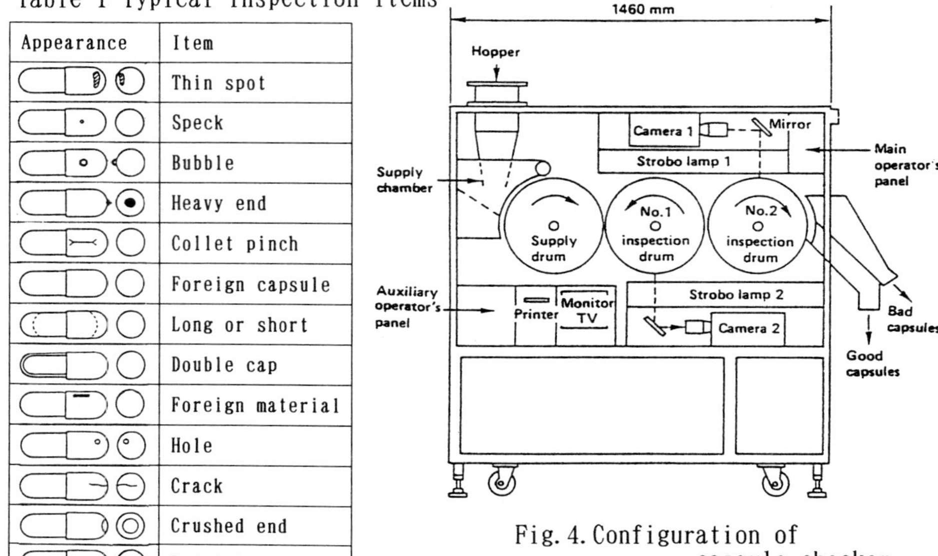

Appearance	Item
	Thin spot
	Speck
	Bubble
	Heavy end
	Collet pinch
	Foreign capsule
	Long or short
	Double cap
	Foreign material
	Hole
	Crack
	Crushed end
	Bad joint

Fig. 4. Configuration of capsule checker

This checker judges the exterior faults of a capsule from the four featuring quantities of the capsule's projection, i.e., area, axial length, straight length of joint, and circular shape, as shown in Figure 5. Also a capsule checker for printed capsules has been developed [10]. As shown in Figure 6, the inspection capability of these checkers is confirmed to be of the same or a higher level in comparison with human visual inspection.

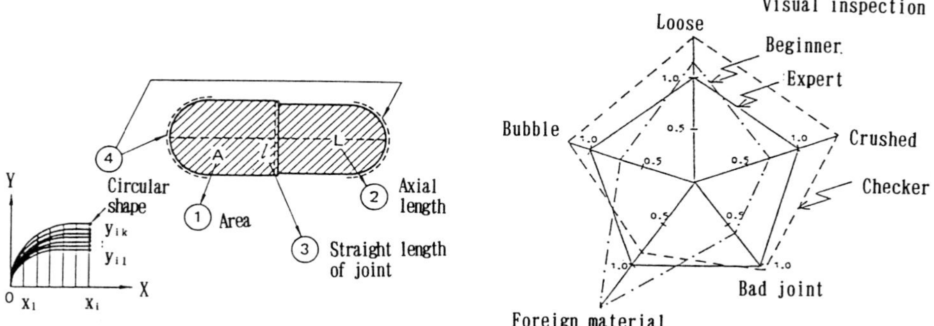

Fig. 5. Inspection algorithms

Fig. 6. Performance of the checker

A tablet checker [11] and an empty-capsule checker [12] have been developed.
The tablet checker is a sister product of the capsule checker and many of its parts, such as the strobe lighting-TV camera system, the 3-drum 2-

row conveying system and microcomputer judgement system, are identical to those of the capsule checker. The empty-capsule checker has photo-sensors instead of TV cameras and the inspection speed is 80,000 capsules per hour. The checker can be used for any color of opaque and transparent capsules.

4.2. Label Inspection [13]

Labels are stuck automatically onto a bottle by a labelling machine, but sometimes unacceptable labellings occur, which are classified as shown in Figure 7. This apparatus is designed to inspect labels for such defects as "shift, tilt, upside down, reversed, folded, stain" and so on. Moreover, the apparatus can inspect various kinds of labels. The inspection system in operation is shown in Figure 8.

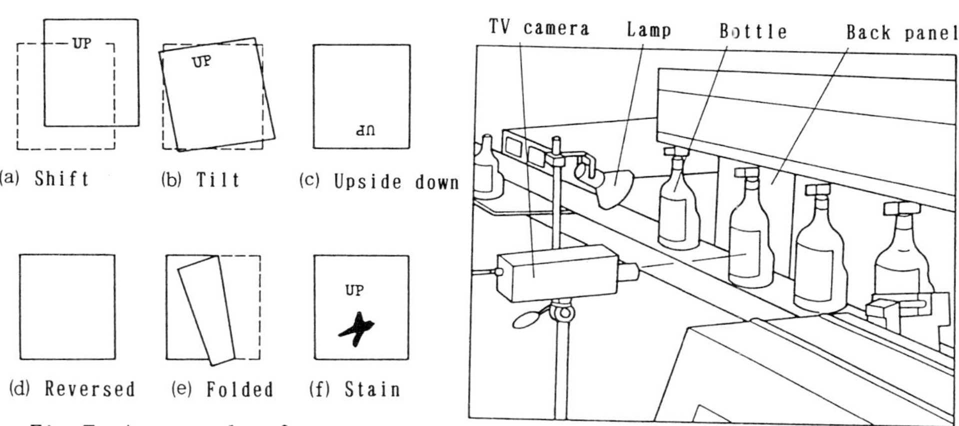

Fig.7. An example of unacceptable labelling

Fig.8. System in operation

The system is composed of a TV camera, illuminating lamps, a position sensor and an image processing system. Bottles are carried by a conveyor, and en route labels are stuck onto them. When a bottle arrives at the observation point the TV camera captures the bottle's image. The image processing system, which is composed of a microprocessor and image processor, receives the video signal and processes the image data to inspect the label on the bottle. When labels are judged to be unacceptable the result is transfered to the ejector which ejects the unacceptable bottle.

4.3. Character Recognition for Industry

In the automobile industry, a stamping machine marks a engine type on a cylinder block, a worker reads it and collates it with the scheduling slips of the production process. In order to introduce automatic inspection it is necessary to overcome the difficulties of noise elimination and character recognition. A stamped-character inspection apparatus has been developed[14]. The configuration of an automated inspection line is shown in Figure 9.

After stamping, a cylinder block is carried to an inspection position. A TV camera observes the stamped engine-type characters; the image has the appearance of a shadow under symmetrical illumination. The inspection apparatus recognizes the characters and collates with a preset engine type. At the same time it inspects for mis-stamp, legibility and defects of the stamped face. If the stamped characters are no good the cylinder block is removed. The video signal is converted into a binary image with an adaptive threshold, and the binary image is memorized in RAM1 as parameters of line segments in DMA mode. An 8-bit microcomputer processes the connectivity analysis and the recognition algorithm.

The engine type is composed of four characters, and there are thirteen kinds of characters: ten numerals and three letters. An example of measured distances and minimum distances is shown in Figure 10. In this case the character "4" can be distinguished from the others. The distance is a difference value in a pattern matching method.

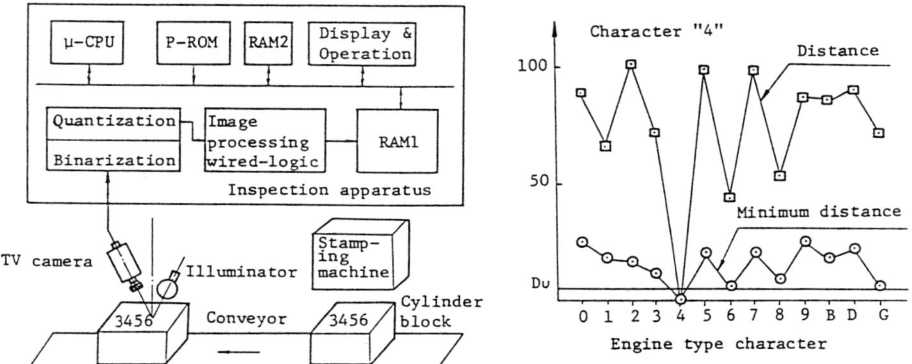

Fig. 9. Block diagram of the apparatus and configuration of the inspection line

Fig. 10. An example of measured distances

Moreover, a stamped-mark reader has been developed [15], which can recognize the stamped marks on compressors. To simplify automatic recognition, a new code mark "dot pattern code" is employed. These binary images of stamped characters and marks are affected by the following factors: (a) roughness of stamped face; (b) shape of stamped grooves; (c) contaminants such as oil and soot; (d) a coating of paint; (e) a curved surface. The surface should be cleaned and polished.

4.4. General-Purpose Visual Inspection System [16]

A general-purpose inspection system has been developed using an extended template matching method and a pattern matching method. Fairly high-level recognition can be permitted by setting an arbitrary form of windows. All of the parameters related to recognition and decision can be set by the user so that the same equipment is adaptable for use with various objects or to different purposes. The operation of the general-purpose inspection system consists of two phases: the training phase and the inspection phase.

In the training process a TV camera observes a standard object in a suitably illuminated state. Its video signal is converted into a binary image and the X, Y projection data, 1- and 2- dimensional moment data, and transition-point data are extracted. The normalization of the position and orientation is important. The position of the object is computed from the above data. The calibration of the orientation is possible by adding an optional function [17]. The inspection area can be set to the inspection point corresponding to the position of the object as a reference pattern. Besides setting by means of an approximate polygon of specified points, matching of the reference pattern to the object is also possible by displaying the images (by expanding/shrinking a binary image of the standard object as shown in Figure 11a). Upper and lower limit values are set for the area of the binary image of the object in the reference pattern. Setting of a matching reference pattern as an outside and inside area pair of the object is also possible by expanding or shrinking the displayed images as shown in Figure 11a. For the matching reference pattern, upper and lower limit values are set for the areas of the excessive and insufficient images against each reference pattern.

The training results are stored in the reference pattern memory.

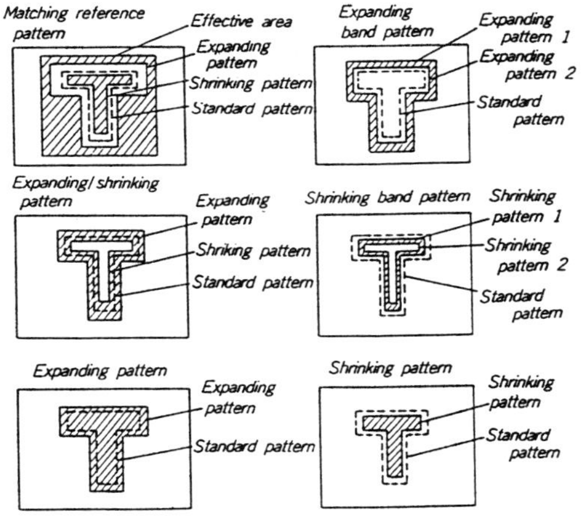

(a)Various reference patterns

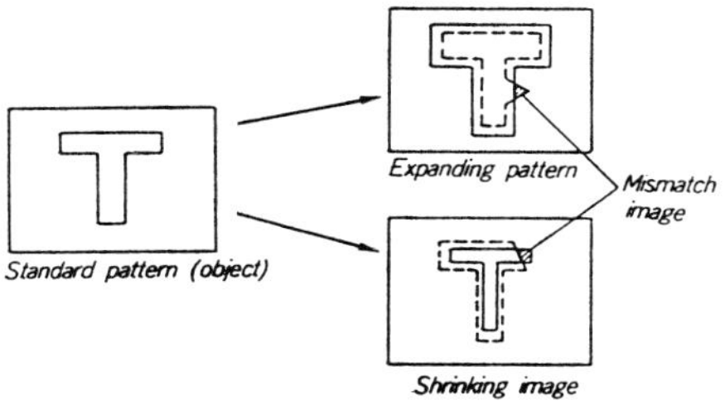

(b)Matching judgment image

Fig. 11. Pattern matching method

Next, when the image of the inspected object is input, the position of the image is computed. In order to normalize it to the reference position the input image is transformed to the reference position pattern (Affine transformation). Even if the object is rotated, the coordinate transformation to the reference direction is possible by calibrating the rotation. After conversion, the area of the image corresponding to the reference pattern is determined. As shown in Figure 11b, the area of the mismatching image is counted as a mismatching value. The count value is read by a microprocessor and is compared to the preset upper- and lower-limit values, a good/no-good judgement is performed and the result is output to the sorter. In addition there is a function which sets a specification area (aperture) at the input image. The function can control the threshold level by comparing the gray level in this area with the standard level and perform the binary conversion which is stable against illumination changes.

5. CONCLUSION

Automated inspection systems provide a powerful tool for quality control. Data from the inspection system can be used for trend analysis, quality assurance and production management. The successful development of inspection systems will depend strongly on further progress in artificial intelligence theory and VLSI technology.

REFERENCES

(1) R.T.Chin, C.A.Harlow: Automated Visual Inspection; A Survey, IEEE Trans. PAMI, Vol.PAMI-4, No.6, Nov., pp.557-573, (1982).
(2) B.Batchelor, P.W.Heywood: Automation in Manufacturing Intelligent Sensing ,NATO ASI Ser E (NLD), pp.199-220, (1987).
(3) V.Kempe: Vision in Industry, Rob.Comput.Integr.Manuf., Vol.3,No.2, pp.257-262, (1987).
(4) A.Buts, G.Smeyers: Some Examples of High Speed Industrial Vision Applications, Proc.2nd.Int.Conf.Mach.Intell., pp.205-211, (1985).
(5) A.Novini: Fundamentals of Machine Vision Lighting,Tech.Pap.Soc.Manuf.Eng.(USA), MS86-1034, pp.4'11-4'23, (1986).
(6) H.Yamamoto, T.Hara, K.Edamatsu: Autofocus Camera System for FA,SPIE Vol.850 ,Optics Illumination and Image Sensing for Machine VisionII,pp.28-32,(1987).
(7) Y.Nitta: Visual Identification and Sorting with TV Camera Applied to Automated Inspection Apparatus, 10th.ISIR, pp.141-152, (1980).
(8) K.Edamatsu, A.Komuro, Y.Nitta: Application of Random Pattern Recognition Technique to Quantitative Evaluation of Automatic Visual Inspection Algorithms ,Rec.1982 Workshop IEEE Conf. on Industrial Applications of Machine Vision, pp. 139-143, (1982).
(9) K.Edamatsu, Y.Nitta: Automated Capsule Inspection Method, Pattern Recognition, Vol.14, Nos.1-6, pp.365-374, (1981).
(10) K.Muneki, K.Edamatsu, A.Komuro: Automated Visual Inspection Method for Printed Capsules, SPIE, Vol.504, Application of Digital Image Processing VII, pp.56-63, (1984).
(11) K.Nakamura, K.Edamatsu, Y,Sano: Automated Pattern Inspection Based on "Boundary Length Comparison Method", Proc. 4th IJCPR, pp.955-957, (1978).
(12) N.Miyoshi, Y.Tachibana, T.Kawasaki, M.Kishi: Automatic Visual Inspection System for Empty Medicinal Capsules, IECON'84, pp.133-136, (1984).
(13) T.Yamamura: Automated Label Inspection Apparatus,'83 IFIP, pp.169-172,(1983).
(14) Y.Hongo, A.Komuro: Stamped Character Inspection Apparatus Based on the Bit Matrix Method, Proc. 6th ICPR, pp.448-450, (1982).
(15) A.Komuro, K.Edamatsu: Automatic Visual Sorting Method of Compressors with Stamped Marks, Proc. 5th ICPR, pp.245-247, (1980).
(16) M.Miyagawa, K.Ohki, N.Kumagai : Flexible Vision System 'Multi-Window' and this Application, '83 ICAR, pp.171-178, (1983).
(17) Y.Ishizaka, S.Shimomura: High-speed Pattern Recognition System based on a Template Matching, SPIE, Vol.849, Automated Inspection and High Speed Vision Architectures, pp.202-208, (1987).

Intelligent Sensors
H. Yamasaki (Editor)

Artificial Olfactory System Using Neural Network

Takamichi Nakamoto and Toyosaka Moriizumi

Faculty of Engineering, Tokyo Institute of Technology, Tokyo, 152, Japan

1 Introduction

In the fields of the food, drink and cosmetic industries, environmental testing and others, odors are sniffed by panelists trained especially in the discrimination of odors; this is known as a human sensory test. The result of a human sensory test, however, is inevitably affected by the inspector's state of health and mood, and therefore an objective and practical evaluation method such as an artificial odor sensing system is highly desirable in such fields. It is thought that mimicking the biological functions of a living body is an effective means of realizing an odor sensing system.

Although there are various explanations of the olfactory mechanism, there is still some ambiguity. Amoore proposed a stereochemical theory and asserted that an odor was composed of several fundamental smells [1]. According to his theory, there exist olfactory receptors that react specifically to the fundamental smells. However, his theory is not accepted generally today since there are a few hundred thousand kinds of odors and the theory cannot explain how such a wide variety of odors are discriminated.

Kurihara proposed an across-fiber pattern theory, and insists that the response profile in many olfactory cells with a membrane composed of lipid and protein is transformed into fiber patterns and the odor quality is recognized in the brain and that the olfactory cells do not specifically respond to the odors[2]. It seems that his theory can explain the discrimination between the wide variety of odors, while some people believe that protein contributes to specific odor reception [3].

In sensor technology, a gas sensor is in general sensitive to various kinds of odors, and Kurihara's model is useful for developing an artificial odor sensing system. The recognition of the output patterns from plural sensors with partially overlapping specificity by an artificial neural network is promising from a biomimetic viewpoint.

Persaud and Dodd observed the output pattern from plural semiconductor gas sensors, and showed that the discrimination between different kinds of gas was possible using the patterns [4]. Thereafter, many researchers reported that gas identification could be achieved from observation of the patterns of the outputs from plural quartz-resonator sensors [5], SAW (Surface Acoustic Wave) sensors [6] and electrochemical sensors [7] following conventional multivariate analysis.

The present authors have reported an identification system using a quartz-resonator sensor array and neural-network pattern recognition for whiskey [8] and flavors [9]. In order to raise the recognition capability, two methods, i.e. selection of the membranes and reduction of the data variation, were adopted and enabled the identification of closely similar whiskey aromas[10], [8]. Furthermore, it was successfully extended to perfume and flavor identification [11], [12]. In this paper, the authors review their developments of an odor sensing system and discuss the present and future strategies of the system.

2 Principle

The sensor used here is a quartz resonator coated with a sensing membrane [13], and is connected to an oscillation circuit. The oscillation frequency depends upon the mass loading caused by gas sorption; i.e., it decreases because of gas sorption and returns to the original value after desorption.

The sensor characteristics depend upon the coating membranes, and the output pattern from plural sensors with different membranes is specific to a particular kind of odor. The pattern is recognized by the artificial neural network shown in Fig.1, which has a three-layer structure made up of input, intermediate and output layers. First, the network is repeatedly trained until the output-layer neurons present the desired outputs specifically to certain odors. The learning rule used here was the back propagation algorithm [14]. The authors also developed the network called FLVQ (Fuzzy Learning Vector Quantization) [15], where fuzzy quantities were used as internal representations, and by which inputs from unknown categories could be discriminated from known ones.

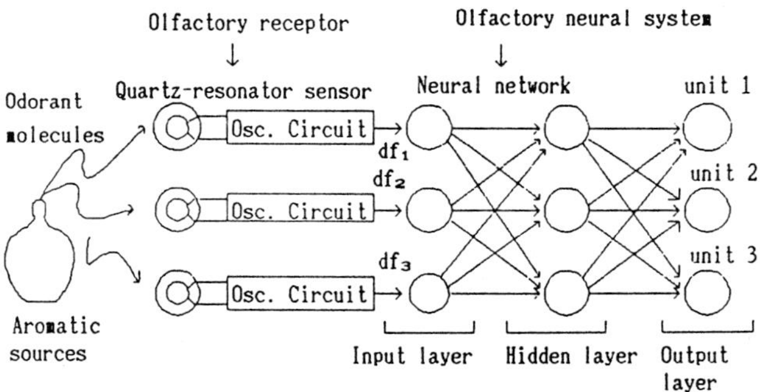

Fig.1 Schematic diagram of odor sensing system

3 Experiment on whiskey aroma identification

It was found from a preliminary experiment that the system could be used for odor identification of various kinds of liquors [16]. In order to examine the system capability, very similar aromas of Japanese whiskeys (Table 1) were tested for identification.

Table 1 Japanese whiskeys (Suntory Co. Ltd.) with very similar aromas

No	Samples
0	Old
1	Reserve
2	Royal
3	Ageing 15
4	Yamazaki

The experiment was performed as follows. First, each sample was measured ten times and the neural network, made up of eight input units, seven hidden ones and five output ones, was initially trained 20,000 times using the data. During the identification measurements, adaptive training was performed 500 times at every round of the measurement on the five samples to compensate for data drift [16].

The liquors are mainly composed of ethanol and water, and the difference in the other components (less than 1 wt.%) contributes to the identification. Therefore, one sample signal was used as a reference signal, and the reference signal was subtracted from the original data of the other samples, thereby enhancing the fine difference among the samples.

Table 2 List of membranes used in the sensor array

No.	Sensing membrane	Classification
1	Dioleyl phosphatidylserin	lipid
2	Sphingomyerin (egg)	lipid
3	Lecithin (egg)	lipid
4	Cholesterol	sterol
5	Perfluorinated bilayer	synthesized lipid
6	PEG (Polyethylene glycol) 20M	GC
7	Ethyl cellulose	cellulose
8	Acetyl cellulose	cellulose

Table 3 Identification result of five whiskey aromas in Table 1

Sample No.	Identification Result				
	0	1	2	3	4
0	10	0	0	0	0
1	0	9	1	0	0
2	0	1	9	0	0
3	0	0	0	10	0
4	0	0	0	1	9

At first it was difficult to identify these whiskey aromas and the recognition probability was less than 50%. Then, two approaches for raising identification capability were adopted. One was the appropriate selection of the sensing membranes suitable for the aroma identification by multivariate analysis [10], and the other was the suppression of the data variation during the measurement [8]. The membranes shown in Table 2 were chosen by discrimination and cluster analyses [10], [8], and the experiment on odor identification was performed, producing the result in Table 3. The columns indicate the identified categories and the rows are the sample numbers. The matrix number means the number of times the odor was identified as the corresponding odor; ten measurements were performed for every sample. The average recognition probability was improved to 94%. On the other hand, the human sensory test was performed by four hundred testees at a campus festival to compare the identification capability of the system with that of man. The recognition probability of man was below 50%, and it was concluded that the capability of the system exceeds that of a layman.

4 Experiment on flavor identification

The application of an odor sensing system is not restricted to the inspection of whiskey aromas, and perfume and flavor identification is possible. Typical perfumes (citrus, floral bouquet, cypre, modern bouquet and oriental) and flavors (orange, strawberry, apple, grape and peach) could be identified completely by the method described in the previous section[11].

An experiment on discriminating fine differences among samples was performed for the purpose of discussing the limit of the system discrimination capability. As it is important in the food and drink industry to extract any samples with slightly different aromas caused by the presence of small amount of foreign materials, the samples used were orange flavors with the foreign substance, 2-butylidene cyclohexanone, which has a smell different from that of the orange flavor. Five orange samples with different concentrations of 2-butylidene cyclohexanone were studied so that the detection limit of the foreign material concentration could be determined.

The measurement sequence was the same as that in the previous section, and the sensor-array output pattern was analyzed by principal component analysis, which is used to visualize multidimensional data by dimensional reduction[17], as given in Fig.2. The concentration of 2-butylidene cyclohexanone was varied from 0 to 1 vol.%. It is evident that the samples with a concentration above 0.5% could be completely discriminated from those with a concentration below 0.5%. The identification experiment by neural network was also performed and the same threshold was obtained.

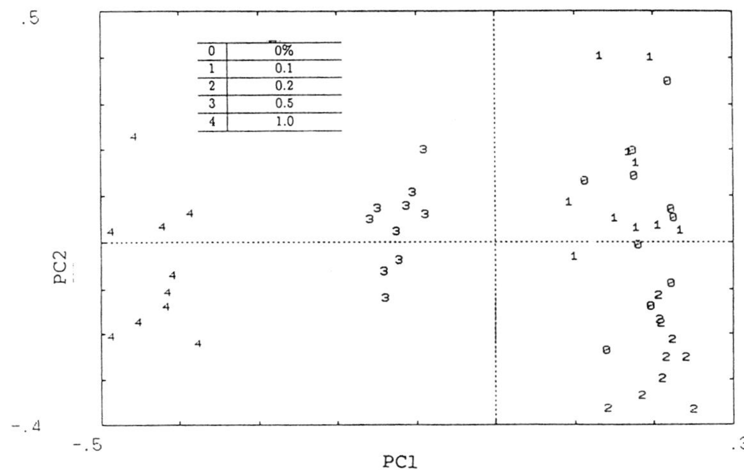

Fig.2 Scatter diagram sensor-array output pattern of the orange flavors with small amount of foreign substance (2-butylidene cyclohexanone)

The human sensory test for detecting 2-butylidene cyclohexanone in orange flavor was also performed. Twelve testees, who had little experience of the human sensory test, tried to pick up the samples with 2-butylidene cyclohexanone. The threshold was found to exist around 0.5%. The detection limit obtained by the test at the campus festival was less than 0.25%. It can be said from these tests that the detection limit of the fcreign substance was near to that of the human sense.

5 Extension to other flavors

The application of an odor sensing system was extended to other flavors. Nut flavor was tested since it has a large variation of arcma among different production lots and the present system was expected to be used to monitor this variation. Unfortunately, the outputs obtained from the sensors with the membranes listed in Table 2 were too small to identify the nut flavor. Hence, membrane selection was carried out for the identification of the nut flavor.

The method of selecting sensing membranes is briefly described here. The discriminatory ability of the sensing membranes can be evaluated by a statistical index called Wilks' Λ.[17],[18] A two-sensor case is illustrated in Fig.3. Wilks' Λ qualitatively expresses the ratio of the data variation within a group to that among the groups. When the data variation among the groups is large and the variation within a group is small, Wilks' Λ becomes small, and a good pattern separation exists.

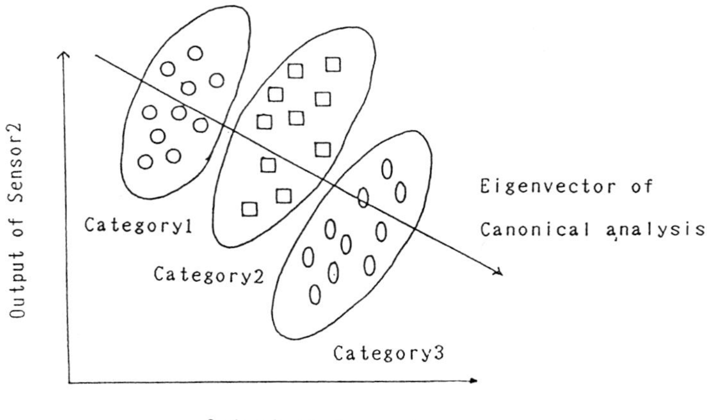

Fig.3 Qualitative explanation of Wilks' Λ.

The contribution of sensor j to the discrimination can be quantified using the partial Λ, which is the ratio of Λ with sensor j to that without sensor j. A partial Λ obeys the F distribution, and the partial F, which is the F value of a partial Λ, can be used for the sensor selection [12].

The investigated sensing membranes were cellulose substances, lipids and stationary-phase materials used in gas chromatography. Several candidates for GC membranes such as quadrol, triethanolamine and sucrose diacetate hexaisobutyrate were thought to be sensitive to an organic sulfide compound which is a major component in the nut flavor. Lipid materials are components of a biological cell, and the cell membranes are composed of mixtures of several lipids. Nomura and Kurihara [19] have reported that the properties of a liposome of mixed lipids drastically changed upon exposure to an odorant. Hence, mixed lipid membranes were also considered in the selection. The membranes were made as thick as possible to increase the sensor outputs. The Q values of the quartz resonators, however, were kept larger than 8,000 after the membrane coating.

The measurement was performed in the same way as that in section 2, the experimental parameters such as flow rate, temperature of the samples and sensors, and desorption time of sensors being modified for nut flavors [12].

The samples used were two types of nut flavor diluted with propylene glycol. The flavors were obtained from different lots and had slightly different smells. Although professional panelists could discriminate between the odors, laymen would have difficulty in identifying them. A more difficult sample consisting of a mixture of two flavors was also tested.

Table 4 Partial F of membrane materials.

Membranes	Classification	Partial F	Selected membranes
Acetylcellulose	CL	1251.6	○
Ethylcellulose	CL	37.2	
PEG1000	GC	< 0.1	
PEG4000	GC	303.6	○
PEG20M	GC	< 0.1	
Quadrol	GC	1214.8	○
Ucon 50HB 2000	GC	69.4	
Diethylene Glycol Succinate	GC	119.0	○
Siponate DS-10	GC	< 0.1	
FFAP	GC	87.8	
Triethanolamine	GC	< 0.1	
Tricresylphosphate	GC	1227.5	○
Sucrose Diacetate Hexaisobutyrate	GC	11.5	
Phosphatidylcholine	Lipid	< 0.1	
PS(50%)+L(50%)	Mixed Lipid	471.3	○
PS(50%)+SM(50%)	Mixed Lipid	82.6	
SM	Lipid	294.6	○
SM(67%)+CH(33%)	Mixed Lipid	16.3	
SM(50%)+CH(50%)	Mixed Lipid	2.7	
SM(33%)+CH(67%)	Mixed Lipid	581.7	○
L(50%)+SM(50%)	Mixed Lipid	67.3	
L(33%)+SM(67%)	Mixed Lipid	61.1	
L(67%)+CH(33%)	Mixed Lipid	76.1	
L(50%)+CH(50%)	Mixed Lipid	< 0.1	
L(33%)+CH(67%)	Mixed Lipid	47.3	
L	Lipid	2.8	

GC: Stationary-phase material for GC, CL: Cellulose material, PS: Phosphatidylserine, L: Lecithin (egg), SM: Sphingomyelin, CH: Cholesterol, PE: Phosphotidylethanolamine
PEG: polyethylene glycol; the PEG numbers indicate the average molecular weights.

Initial investigations were carried out using the two nut flavors diluted by propylene glycol (1:1) and pure propylene glycol. Those samples with large concentrations of nut flavor were used for increasing the sensor outputs and enhancing the response difference. In total, 38 membranes were tested; however, 12 of the membranes were later eliminated because of the small outputs obtained from them (less than 100 Hz). The discrimination analysis technique was applied to the remaining 26 membranes, and the resulting partial F values are listed in Table 4. The eight membranes with the highest partial F values marked in the table were selected and used in further experiments. These membranes included four GC membranes, one cellulose substance, one pure lipid and two types of lipid mixtures.

After membrane selection, the experiments were performed on the two original flavors and a mixture (ratio 4:1). The dilution ratio of the nut flavors was modified to 5% in propylene glycol since this dilution ratio is widely used in the flavor fabrication spots. The sample data points are plotted in a scatter diagram of principal component analysis, and the results shown in Fig.4 have satisfactory pattern separations among the flavors with slightly different smells, confirming a successful membrane selection.

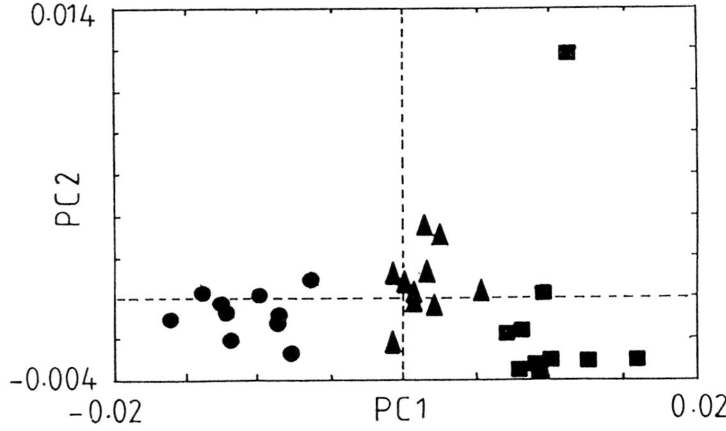

Fig.4 Scatter diagram of principal component analysis after membrane selection. The circles and squares are the data points for the original flavors from different lots, and the triangles are the points corresponding to the mixtures of flavors.

6 Conclusion

The odor sensing system was developed by the authors and applied to whiskey aromas, perfume and flavor identification. Closely similar odors could be discriminated using the system.

It was found during the development of the system that the membrane selection procedure was important for discriminating between closely similar aromas. The appropriate membrane sets for whiskey aromas and nut flavors were selected by multivariate analysis. It can be said that there is a membrane set specific to an odor group. Further efforts in applying the system to other odor groups will reveal some correlation between membrane sets and odor groups.

In addition to identification, an artificial system with the capability to express the human olfactory sense is being developed. Mapping of the sensor output space onto the human sensory space, which can be measured by a sensory test, can be performed by a neural network [20]. Similarities among odors will be measured by the odor sensing system and, furthermore, it will be able to express sensory factors such as pleasantness, freshness, etc. A neural network for nonlinear mapping with high fidelity is indispensable for odor-similarity measurements and the expression of human-like impressions. This kind of development extends the fields of application of the present system especially into comfort science or engineering.

Acknowledgments

The authors wish to thank Mr. J.Ide of T.Hasegawa Co., Ltd. for helpful advice and for providing perfume and flavor samples. They are grateful to A.Fukuda and S.Sasaki for their experimental contributions, Y.Okahata of Tokyo Institute of Technology for offering us the synthesized lipid membrane, Y.Asakura of Suntory Co. Ltd. for providing the whiskey samples and M.Hayashi of Meidensha Corp. for the quartz resonators.

References

1. E.Amoore, Molecular basis of odor (Japanese translation by T.Hara), Koseisha-Koseikaku co., 1972, p.29.
2. K.Kurihara, T.Nomura, M.Kashiwayanagi and T.Kurihara, *Tech. Digest, 4th Int. Conf. Solid-State Sensors and Actuators (Transducers ' 7)*, Tokyo, Japan, June 2-5, 1987, p.569.
3. G.M.Shepherd, Neurobiology (2nd. Ed.), Oxford (1988) 208.
4. K.Persaud and G.Dodd, *Nature*, **299** (1982) 352.
5. W.P.Carey, K.R.Beebe and B.R.Kowalski, *Anal. Chem.*, **58** (1986) 149.

6. D.S.Ballantine, S.L.Rose, J.W.Grante and H.Woltjen, *Anal. Chem.*, **58** (1986) 3058.

7. J.R.Stetter, P.C.Jurs and S.L.Rose, *Anal. Chem.* **58** (1986) 3058.

8. T.Nakamoto, A.Fukuda and T.Moriizumi, *Sens. & Actuators B* **3** (1991)221.

9. T.Nakamoto, A.Fukuda and T.Moriizumi, *Sens. & Actuators B*, **10** (1993) 85.

10. T.Nakamoto, K.Fukunishi and T.Moriizumi, *Sens. & Actuators*, **B1** (1990) 473.

11. T.Nakamoto, A.Fukuda and T.Moriizumi, *Tech. Digest of Transducers'91*, 1991, p.355.

12. T.Nakamoto, S.Sasaki, A.Fukuda and T.Moriizumi, *Sens. & Materials*, 4, **2** (1992) 111.

13. W.H.King, *Anal. Chem.*, **36** (1964) 1735.

14. D.E.Rumelhart, G.E.Hinton and R.J.Williams, *Nature* (London), **323** (1986) 533.

15. Y.Sakuraba, T.Nakamoto and T.Moriizumi, *Systems and Computers in Japan*, **22** (1991) 93.

16. T.Nakamoto and T.Moriizumi, *Proc. IEEE Ultrason. Symp.*, 1988, p.613.

17. W.R.Dillion and M.Goldstein, *Multivariate Analysis*, Wiley, New York, 1984, p.23-349.

18. Y.Tanaka, T.Tarumizu, K.Wakimoto: Handbook of Statistical Analysis by Personal Computer, II:Multivariate Analysis (Kyoritsusyuppan, 1984) p.112 (in Japanese).

19. T.Nomura and K.Kurihara: Am. Chem. Soc. **26** (1987) 6141.

20. Y.Sakuraba, T.Nakamoto and T.Moriizumi, *Tech. Digest Sensor Symposium*, IEE of Japan, 1992, p.197.

Intelligent Sensors
H. Yamasaki (Editor)

Sensor Fusion:The State of the Art

Masatoshi Ishikawa

Faculty of Engineering, University of Tokyo, 7-3-1, Hongo, Bunkyo-ku Tokyo 113, Japan

Abstract

The state of the art of sensor fusion and problems to be solved are described. The basic concept of sensor fusion is classified in contrast to the function of the brain and problems of sensor fusion are defined from the viewpoints of hardware, software and algorithms such as signal processing, artificial intelligence and neural networks. A new concept and an architecture for active sensor fusion are described. Development of the sensor fusion architecture is difficult and many problems remain to be solved in the future.

1. What Is Sensor Fusion?

Man takes in external information through sensitive organs (sensors) referred to as the five senses. Despite some ambiguities, contradictions or defects in such information, external changes and things can be recognized and judged correctly by compensating and integrating sensor data or checking them with the knowledge already acquired. The technique for achieving in an engineering way new recognizing functions not obtained from a single sensor by uniformly processing data sets from a multiple number of sensors in different modes is sensor fusion. It aims to achieve more advanced recognizing and judging functions in a sensing system by taking out highly reliable information flexibly[1, 2].

As the sensors used in the entire system grew in number and kind, mainly in such areas as intelligent robots, manufacturing, aircraft and automobiles, there came up a need to process a large variety and quantity of sensor information efficiently, inviting demand for high quality information to be extracted. Handling a variety of sensor information implies increasing the information to be processed, which is likely to involve inconsistencies between sensor data sets (discrepancy between sensor data sets, disagreements between measuring times, nonuniformity of sensors, etc.), making it essential to develop techniques, including utilizing knowledge, to eliminate such difficulties. Also necessary is a flexible structure for coherently describing and implementing processing procedure different with sensors. The aim of sensor fusion is to achieve a heterogeneous and polymodal processing system for the entire sensing system and to take out highly reliable information flexibly.

In the present age when the information processing technique is changing from centralized processing to decentralized processing, a sensor technique is also required to make a shift to parallel distributed processing. Here the network structure between sensors forms the basis for such developmental trends. It is essential to make refinements of sensor fusion as a means for providing a framework for improving such networks coherently.

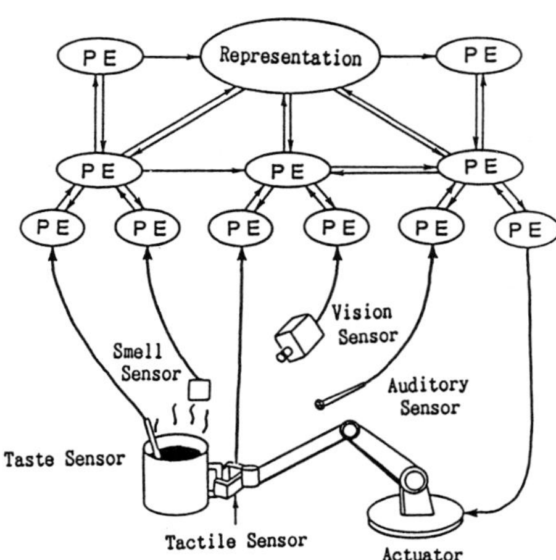

Figure 1: Sensor Fusion System

Figure 1 shows a setup example of actual sensor fusion. Here, a robot, perceiving a coffee cup, is reaching for it.

It uses 6 types of sensors including the five types of sensors. This system involves more information acquired and more redundancy than a system with a single visual or tactile sensor. Studies on sensor fusion aim to develop such functions into a general structure of sensing.

However, the history of such studies is so short that their philosophy has not been refined and, moreover, such concept have not even been defined. At present, despite various practical techniques presented, there is no generally established methodology, while there are still discussions on necessity and abstract discussions in planning phases, leaving much to be studied further. This paper summarizes problems and techniques under consideration, while pointing out other problems involved in them.

2. Intersensory Integration in Human

Before going into sensor fusion as engineering techniques, the human mechanism for processing sensory information should be examined. Man perceives and recognizes changes outside and in himself by using information through visual, aural, tactile, and other senses called the five senses. The coffee cup in the hand is a "coffee cup" as known by visual information obtained from the eyes and at the same time it is also a "coffee cup" as known by tactile information obtained from the touch by the hand. This is also other information related to smell and weight. From these sets of information, man is able to recognize a "coffee cup" as one definite idea of perception, in which different sets of sensory information are fused without inconsistency in this "coffee cup" as representation of

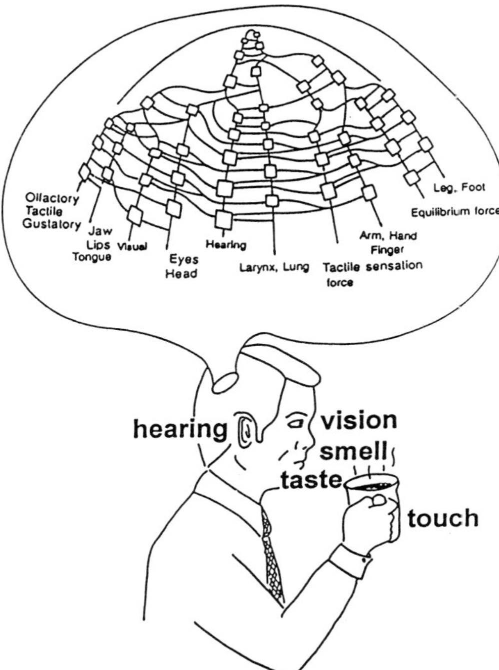

Figure 2: Intersensory Integration in Human

perception. If the coffee cup is moved around his mouth, these sets of sensory information change as expected without inviting any inconsistency with the coffee cup in his memory. If any inconsistency arises, man is able to recognize it as unusual.

Thus, having a higher reliability than a single sets of information, visual or tactile, perception composed of various sets of information including information from memory, and the motor system can provide information having various meanings. It is recognized that the human brain has a hierarchical parallel distributed neural networks as a mechanism for processing various sets of sensory information. It is understood that the association area, in particular, plays an important role and that not only the information of the sensory system but also the control information of the motor system are closely related. **Figure 2** shows a model of the sensory motor control system[3].

This model was presented by Albus, and designed as a layered parallel distributed processing structure. Here, the processing is performed separately by information expressions from the bottom layer to the top layer. Moreover, the upper layers are involved in this process as well as the process right after the sensing. Remarkably, the model also coordinates recognition and behavior together and provides combinations between different sets of sensory information and between sensory and motor information at different layers of processing.

Albus' model shown in **Figure 2** is not directly applicable as a engineering model of sensor fusion but can be taken as an example showing the entire image of sensor fusion correctly. In other words, sensor fusion in a broad sense of the word can be defined as accomplishing the entirety of processing several sets of sensor information in an engineering way.

3. Applications of Sensor Fusion

Sensor fusion can be recognized as a developed variation of conventional sensing technology. Thus, the areas of its application contain all of the applications of conventional sensing technology and in each of them new technological innovation is brought about from sensing engineering. This is not all the effect brought about by sensor fusion but this engineering has the potentialities of creating new application areas by extending the framework of conventional sensing technology.

There are a lot of areas in which sensor fusion is considered as key technology. In the areas of robotics and manufacturing, there can be various applications including handling of objects by using visual and tactile information, and assembly, inspection and defect detection by using several sensors. With mobile robots and underwater working vessels, attempts are being made to automate even operators' recognition and judgment. With aircraft and spaceships, sensor fusion is considered as a principal technique in the processing system for furnishing pilots with highly abstract information, from which viewpoint, application is possible to all human interfaces. Furthermore, sensor fusion is expected to make a great contribution in various other areas beyond the capability of conventional sensor information processing engineering such as advancement of observation machines for remote sensing in outer space, oceans etc., functional improvement of medical and welfare instruments, elevation of diagnostic accuracy, improvement of productivity and safety in plants and automation of work on nature as in agriculture and civil engineering.

As for man-machine interfaces in particular, there are several interface channels, visual, aural, tactile etc. Conventionally these are considered as independent of each other. In reality, man has a function for integrating senses, not treating such channels independently of each other. Thus, on machines, interfaces must be designed with correlations between such channels and human function for integrating senses taken into account. Here, sets of information through channels must be fused in the machine and information should also be fused between man and machine. If a man and a machine are considered as a system, the accomplishment of overall fusion is certainly the requirement for comfortable and highly functional human interfaces. This is particularly necessary with virtual reality systems.

4. The State of the Art and Problem

Thus sensor fusion is utilized in a large number of areas. Depending on objects and levels of processing, various problems have been established. Here we shall consider 3 issues: (1) problems related to hardware focused on intelligent; (2) problems regarding software focusing mainly on descriptive method; and (3) problems related to algorithms realized upon the hardware (refer to **Figure 3**).

4.1. Hardware

In case a number of different sensors are used, each sensor produces information in a different output form peculiar to its attribute. Dealing with a number of different

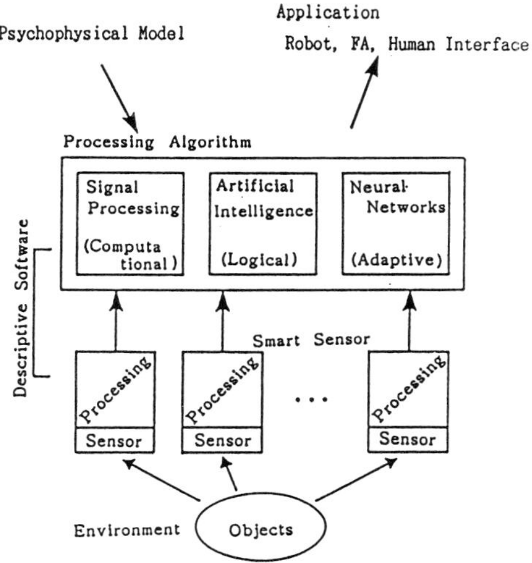

Figure 3: Subjects of Sensor Fusion

sensors with a singe processor would not be efficient. If each sensor can process its own attribute on itself, then the load of the center processor would be greatly reduced, and the speeding up of processing by means of hierarchical parallel distributed can effectively be realized. This type of sensor is called intelligent sensor. There may be a number of ways for intelligent sensor[4].but the important point is how to integrate an effective form while maintaining the computational functions inside sensors simple and general-purpose.

Moreover, since sensors are furnished with a processing function, it has become easier to unify all output signals ; thus the handling of sensors as building block has been facilitated. For this reason, it is now possible to carry out asset building of hardware as well as software, which is to lead to improvements in the developmental efficiency. If this design concept is further advanced, then a network of sensory information processing becomes possible; in fact, a network for sensor fusion has been proposed.

4.2. Software

If the above-mentioned hierarchical distributed processing is assumed to be a prerequisite, then the function of the software necessary for sensor fusion must be able to describe a network and load it in hardware. In order to realize this, it becomes necessary to give a conceptual unification to the intellectualization of sensors, to set a general framework for the sensory information processing format, and thus to rearrange those functions in software which are related to sensory information processing. In particular, some focal points are: (1) description of a flexible but powerful description of network, especially, description of local feedback loops; (2) achievement of efficient distributed processing; (3) realization of realtime processing; (4) description of memorizing and learning functions. These functions are closely related to the structure of hardware, and can be realized by the integration of the functions and the hardware structure.

One approach to these problems may be the thinking in the form of a logical sensor proposed by Henderson et al[5]. This has been developed for the purpose of facilitating the prototyping of a variety of sensors by adopting abstract data forms and then realizing efficient descriptions of networks.

The most important point in creating this type of software is related to the problem of dealing with time. In other words, while there is no problem if each module is operating asynchrously, maintenance of the real-time processing must be focused on in case synchronous processing is required. For solving this problem, there is the method of using time checking in artificial intelligence languages and real-time OSs, but there is a limit in the real-time nature itself. For this reason, there have been discussions on software structures for sensor fusion with the real-time processing maintained[6].

4.3. Algorithms

The above-mentioned problems related to hardware and software are merely problems of arranging tools, and so they do not address the solving of problems in sensor fusion. Instead, a fusion structure must be created with the use of these tools. For that, a number of methods have been employed depending on the form of algorithm, the established environment, knowledge on objects prior to the event, etc.

The case in which the causal relation of a physical phenomenon can be clearly grasped and its inverse problem can, furthermore, be solved by some means and where the computational structure of perception can be described explicitly can be treated as a signal processing problem in a broad sense. The signal processing problem has appeared widely in measuring engineering, and a variety of processing methods have already been established. Accordingly, these methods have only to be rearranged to apply to a multiple number of sensors.

Durrant-Whyte has attempted systematization with regard to fusion of general geometric sensing information. According to his theory, systematization of processing of the evaluation of and constraints for the possibility of sensor information integration can be carried out first by introducing into geometric parameters an uncertainty based on an error distribution and then showing a unified description and evaluation method including this uncertainty[7].

Moreover, when a model of an object is made available by CAD data, then recognition of the model based on multiple pieces of sensory information (model based recognition) is carried out. As examples, there are methods of carrying out recognition according to interpretation trees as shown by Grimson et al.[8] and Gaston et al.[9] These methods have trouble in handling errors, however, and they have the problem that recognition results cannot be identical because of a multiplicity of interpretations. To solve such problem, application of Dempster-Shafer's probability theory[10] or introduction of sensing planning can be considered.

Furthermore, proposal has been made of methods with the introduction of Kalman filters (expanded Kalman filters) into sensing and sensor feedback control[11].

Besides, other proposed ideas include the method of realizing visual-aural integration (cocktail party effect) by means of adaptive filters[12],the method of realizing visual-tactile fusion by using internal model expressing the physical structure of objects as a parallel processing structure[13] or the method of fusing a number of visual sensations by learning of mappings between sensory information[14]. Besides these, there are a large number of possible methods; in particular, many among mathematical ones can be applied for sensor fusion. But basically such methods must explicitly indicate computational structures which can hopefully be connected in parallel in consideration of hardware.

Any system whose logical structure of measurement including the memory structure of knowledge is explicitly expressed can be treated as an application of artificial intelligence or knowledge engineering. What is required then is a logical description of the human process of recognition and judgement based on information from sensors and information previously stored as memory.

However, an expression of all aspects of the structure of an object or an environment requires a massive amount of descriptions. For instance, research on vision and touch fusion carried out by Allen[15] is a typical example of such an attempt. In his system, information on an object from each sensor's integrated as a data base, and this integration is regarded as sensor fusion. This system not only requires a huge data base but also lacks clear discussions on errors.

A distributed blackboard model proposed by Harmon et al.[16] is a software system for simplifying descriptions by distributing the logical structures of sensing. In the study, the present location is found by using the map data base held by a computer and data transmitted by a TV camera, a range finder, or a sonar; then an appropriate root of action is determined. In these cases, each processing module is distributed among a multiple number of computers, which are connected by a network, but this system has been criticized for its problems involving synchronization of processing modules and load distribution.

In addition, proposals have been made of sensor fusion systems by applying production system or expert systems. It should be pointed out further that applications of fuzzy control to sensor fusion are being actively studied.

In cases where the computational structure or logical structure of perception cannot be expressed explicitly, or if the processing takes a long time even when such a structure is available, then the introduction of an adaptive structure such as the neural network theory can be effective[17]. Such a structure is not just confined to neural networks, but can also be found in signal processing systems. These systems are attractive because they can automatically construct a fusion structure by learning, even when an adequate amount of knowledge on an object or the model of an object is not available.

In general, the lower layers of a hierarchical structure are considered more suited for the processing of signals and application of neural networks, while the upper layers are for symbolic and logical processing of artificial intelligence. Consequently, the technique of combining these systems in a skillful way is one of the important topics for research.

5. Intentional Sensing

5.1. Active Sensing

The human hand is an excellent sensing system which is based on the fine fusion of the function as a sensor device for detecting and transforming pressure, temperature, etc. and the function as an actuator. In particular, when cases where the hand makes searching actions (tactile motion) are considered, it is found that the hand as an actuator assumes an important role in sensing. The human eye, on the other hand, does not just limit its function to merely detecting images on the retina but expands its function with the help of an actuator by regulating its focus and accommodation, or changing visual points by moving the head.

In actuality, a number of feature-extracting cells related to the perception associated with this motor system have been found inside the cerebral cortex. In the field of psychology, the importance of active perception has been pointed out, the existence of perception shown which cannot be explained just from combinations of passive receptors, and the assertion has been made that active perception ought to be one perception form which includes man's sensing actions[18].

The most basic of these ideas has been expressed by the phrase "active sensing" in the visual information processing or tactile information processing. In other words, even when there is not enough information on an object, some prediction is made based on the previously obtained information concerning that object, and then an action is started based on the prediction and controlled by sensor feedback. This implies that active motions are part of the sensing[19].

The idea of active sensing is based on the introduction of a structure in which an action prior to recognition (searching movement) or the processing of such action is activated in parallel with the processing of recognition and perception, as a replacement of the previous structure where the processing of "detection" to "recognition" to "action" is carried out. That is, sensing is no longer one-directional but instead perception is carried out cyclically or harmoniously between recognition and action. Because sensor fusion is fed enough information that can be controlled, a proper design of this can provide active sensing with the capability of offering a basic framework common to information of multiple sensors.

5.2. From Active to Intentional

Concepts similar to active sensing are found in the affordance concept of Gibson (an objects exists with actions on it being included)[20] and the perceptive circulation of Neisser (perception is based on the repetition of the cycle: measurement identification of the model prediction and measurement)[21]; the concept of preshaping proposed by Arbib (action prior to measurement)[22] is also a similar concept. What is common to all these concepts is the point that a internal model of an object or its own action is considered and then an action is created by operation of the model. In other words, the purpose of the measurement is obtaining an accurate model and then an action is carried out by operating that internal model. This is equivalent to utilizing models aggressively and in a forward direction as the principle of measurement engineering, and can become a powerful means for sensor fusion.

Let us once again consider the processing of fusion of human sensation with a "coffee cup" as an object. A coffee cup placed on one's hand, for instance, is not only a "coffee cup" based on visual information obtained by the eyes but also a "coffee cup" based on the tactile information of the hand. Moreover, a "coffee cup" is recognized as a perceptive representation through additional information such as aroma or weight. Each piece of sensory information is fused with all the others under the perceptive representation, namely, a "coffee cup," without any contradictions, so that the taste information changes in an expected way when sugar or milk is put into the coffee cup. When the coffee is drunk, then visual information and the tactile information change in an expected manner. Moreover,this information does not contradicts a "coffee cup" stored in his memory. Inversely, if there is a contradiction, then there is a possibility that the cup being held is recognized as a tea cup.

To realize these functions in an engineering way should naturally be a part of the important functions of sensing. At the present time, however, many cases are found in which the necessary judgement is made by man through his vision or by touch, and a suitable sensing system is then designed by him.

Here lies the problem. There always exists in sensing an intention to perceive and recognize. A sensing structure must therefore be constructed to fulfill this intention. The computational processing structure, the generation of searching movement, or the placement and selection of sensors must all be aimed at the implementation of sensing along the line of this intention. Such concept is no longer a problem within the framework of active sensing.

We call this new concept "intentional sensing", which has been proposed in the Japanese national project on sensor fusion. To say that sensing is carried out intentionally means that the intention of sensing (how to measure what type of object and how to utilize the information obtained) is explicitly expressed in the previously supplied information or the knowledge based on environmental conditions or restrictions and that a structure of sensing including active sensing is created based on this intention.

In sensor fusion, establishment of a framework common to all the information of sensors is its essential theme, so that this framework is the most important factor in determining a structure of sensor fusion. In general, the features of an object or internal models are often regarded a priori as a framework; however, the subject of intentional sensing is the rational establishment of such a common framework with respect to an intention. Moreover, when there is a high degree of autonomy, a common framework is hard to obtain, so that fusion of information from different types of sensors, particularly information of mutually highly autonomous sensors, becomes difficult. Therefore, an intention is introduced in order to facilitate the creation of a common framework. In other words, recognition is regarded rather as an intentional process connected with a certain intention than merely as a process of sensory information processing.

A common framework for perception in sensor fusion must be an internal model which generates such an intention. The focal point of this subject is how to describe such an internal model.

6. Problems in the Future

A general view of various subjects involved in sensor fusion has been presented. At the present time, the number of sensors that are incorporated in a system is increasing ever more, and the importance of sensor fusion has accordingly magnified. In spite of this, not a single well-established method exists, and only very few actual examples of practical applications are available. Thus there still are a number of problems yet to be solved, and a number of problems will have to be dealt with in the future. In order to find solutions to these problems, efforts will have to be made not only to merely obtain sensing methods but also to grasp sensing as a system and treat its architecture as a subject for research. In other words, new sensing architectures will have to be developed.

References

[1] M. Ishikawa. The sensor fusion system : Mechanisms for integration of sensory information. *Advanced Robotics*, Vol. 6, No. 3, pp. 335–344, 1991.

[2] R.C. Luo and M.G. Kay. Multisensor integration and fusion in intelligent systems. *IEEE Trans. Sys., Man, Cybern.*, Vol. SMC-19, No. 5, pp. 901–931, 1989.

[3] J.S. Albus. *Brains, Behavior, and Robotics.* McGraw-Hill, 1981.

[4] M. Ishikawa. Parallel processing architecture for sensory information. *Proc. of the 8th Int. Conf. on Solid-State Sensors and Actuators*, pp. 103–106, 1995.

[5] T.C. Henderson and E. Shilcrat. Logical sensor system. *J. Robotics Sys.*, Vol. 1, No. 2, pp. 169–193, 1984.

[6] T. Shimada, K. Nishida, and K. Toda. Real time parallel architecture for sensor fusion. *J. Parallel and Distributed Computing*, Vol. 15, No. 2, pp. 143–152, 1992.

[7] H.F. Durrant-Whyte. Uncertain geometry in robotics. *IEEE Trans. Robotics and Automation*, Vol. 4, No. 1, pp. 23–31, 1988.

[8] W.E.L. Grimson abd T. Lanzano-Perez. Model-based recognition and localization from sparse range or tactile data. *Int. J. Robotics Res.*, Vol. 3, No. 3, pp. 3–35, 1984.

[9] P.C. Gaston and T. Lozano-Perez. Tactile recognition and localization using object models. *IEEE Trans. Pattern Anal. Mach. Intell.*, Vol. PAMI-6, No. 3, pp. 257–265, 1984.

[10] P.L. Bogler. Shafer-dempster reasoning with applications to multisensor target identification system. *IEEE Trans. Syst. Man Cybern.*, Vol. SMC-17, No. 6, pp. 968–977, 1987.

[11] T. Mukai and M. Ishikawa. A sensor fusion system using mapping learning method. *Proc. IEEE Int. Conf. on Multisensor Fusion and Integration for Intelligent Systems*, pp. 288–295, 1994.

[12] K. Takahashi and H. Yamasaki. Self-adapting multiple microphone system. *Sensors and Actuators*, Vol. A21-A23, pp. 610–614, 1990.

[13] A. Takahashi and M. Ishikawa. Signal processing architecture with bidirectional network topology for flexible sensor data integration. *Proc. IEEE/RSJ Int. Conf. on Intelligent Robots and System*, pp. 391–396, 1993.

[14] T. Mukai, T. Mori, and M. Ishikawa. A sensor fusion system using mapping learning method. *Proc. IEEE/RSJ Int. Conf. on Intelligent Robots and System*, pp. 407–413, 1993.

[15] P.K. Allen. Integrating vision and touch for object recognition tasks. *Int. J. Robotics Res.*, Vol. 7, No. 6, pp. 15–33, 1988.

[16] S.Y. Harmon. An architecture for sensor fusion in a mobile robot. *Proc. IEEE Int. Conf. on Robotics and Autmomat.*, pp. 1149–1154, 1986.

[17] Y. Sakaguchi and K. Nakano. Haptic recognition system with sensory integration and attentional perception. *Proc. IEEE Int. Conf. on Multisensor Fusion and Integration for Intelligent Systems*, pp. 288–295, 1994.

[18] J.J. Gibson. Observation on active touch. *Psychol. Rev.*, Vol. 69, No. 6, pp. 477–491, 1962.

[19] A. Blake and A. Yuille. *Active Vision.* MIT Press, 1992.

[20] J.J. Gibson. *The Ecological Approach to Visual Perception.* Houghton Mifflin, 1979.

[21] U. Neisser. *Cognition and Reality.* Freeman, 1976.

[22] M.A. Arbib. Visuomotor coordination : Neural models and perceptual robotics. In J.Ewert and M.A. Arbib, editors, *Visuomotor Coordination.* Plenum, 1989.

Intelligent Sensors
H. Yamasaki (Editor)

Intelligent Visualizing Systems

Yukio Hiranaka

Department of Electrical and Information Engineering, Faculty of Engineering, Yamagata University, 4-3-16 Jonan, Yonezawa, 992 Japan

Visualization is an effective means of teaching because much data can be delivered at one time. Visualization of physical and chemical fields can be realized universally using an array of sensors that can acquire data from the target field. Interpolation, which converts point data at each sensor into a visible image, is an important operation because the density of array sensors is usually sparse in actual visualization situations. Data density at the sensor point is determined by the sensor itself and can be increased easier than the density of sensors. For example, wide-band temporal data can be converted to spatial data by assuming some model of the field. Two visualization examples are shown. One is the visualization of a gas field, which is aimed at effectively replacing smell-sensing (police) dogs. The other is a real-time visualization of a sound field, which is to be used for monitoring and controlling sound fields.

1. INTRODUCTION

To understand is to create an image of the facts in the mind. In literature, words and sentences are the facts used to make an image in the reader's mind. In scientific works, the values measured are the facts of existence used to make an image in the experimenter's mind.

If the object to be understood is not simple and needs many of its values specified, it will not be easily understood. One way to achieve understanding is to draw the object as a visible figure, portraying the main feature of the object. If the object has many detailed features, we should not try to draw it. Photographic presentation is then the best approach. The simpler the image, the easier it is to understand.

Visualization techniques are pursued in many fields. If we can see things that were not be seen before, we will know whether our image is correct. Also something that was not imagined before might appear. There are a variety of methods and instruments now available for visualization.

Recently, visualizations using computed tomography have been used in many scientific areas, for example, X-ray, NMR, ESR, etc. are used for studying the insides of objects. The visualization of essentially unseen objects such as sound fields, flow fields, etc., are also being undertaken. Table 1 shows examples of field visualization accompanied with their objectives. General information can be found in the literature [1,2].

Table 1. Visualizing fields and objectives

Type of field	Visualization objectives
Electromagnetic fields	Evaluation of antenna directivity, check the level and distribution of electromagnetic noise
Electric fields	Conductivity distribution measure through potential field, tracing functionality of silicon circuits, locating functional centers of brain activity
Magnetic fields	Measurement of permeability distribution, flaw detection in magnetic substances, measurement of electric current distribution
Acoustic fields	Detection of noise source, identifying sound path, evaluation of designed sound environments
Mechanical fields	Vibration analysis, viewing stress on each structural member, foreseeing dangerous areas
Gravity fields	Evaluation of earth shell movements, earthquake inference
Flow fields	Analysis of flow field, evaluation of flow design
Chemical fields	Replacement of police dogs, detection of gas leakage
Thermal fields	Detection of heat sources, estimation of various activity distribution
Radiant ray fields	Detection of radio-active pollution

In the following sections, some of these visualizations are described. They are different from the conventional techniques used in processing field images. The computer is the principal means in the visualization. It can process anything the viewer would like to see. Computers make the visualizing system intelligent.

2. VISUALIZING WITH AN ARRAY OF SENSORS

We have to change field data into a two-dimensional or three-dimensional visual scene. Until computers were used, visualizing techniques were mainly dependent on the direct conversion of methods and/or materials from physical characteristics to an optical image. In these cases, human interfaces had a second or later priority. Once computers with their versatile image display could be used, direct conversion was no longer necessary. The most preferable conversion is to change everything into electrical signals. Information presentation in a planar form can be done by computer processing, for example, with appropriate interpolation. In such a situation, information acquisition is separated from the display, and can be modified to increase the performance of the measurements. At the other end of visualizing system, the human interface can be improved to make the system versatile and easy to use. For example, it is not easy to create a brain waves viewer that is directly attached to the human head. There are, however, visualizing systems to see brain waves on computer displays using a helmet of electrodes to show a specific portion of the brain waves, a stimulated response distribution, etc. [3]. It is better to visualize various field patterns using high sensitivity sensing elements than to design an elegant method of visualization.

One simple and universal way for acquiring spatial data is to use a scanning sensor in the field. If we want to visualize in a real-time mode, an array of sensors should be used. Miniaturization of sensors has made it easy to build up a dense array of sensors. High-speed data acquisition can also be attained with rapid-response sensors, another result of miniaturization. The disadvantage of sensor arrays is the variation of sensor characteristics, which should be compensated for to achieve good quality visualization. In the case of sound visualization, variation in frequency characteristics should also be considered. Nevertheless, the performance increase of computers in almost every process has resulted in a shortening of CPU time. We can now consider just the essential capability of sensors without all the error sources.

Once we use the computer in our visualizing system, we can not only to create an image of the field, but also conduct advanced processes, such as adjusting the threshold level and amplitude of information variables for optimal display, automatically eliminating part of the signal which disturbs the display of more essential information in the field, automatically changing the visualizing form into a more easily understandable display form by monitoring the field situation. Such intelligent and dynamic processes make the visualizing system a very convenient tool for scientific and engineering works.

3. INTERPOLATION

This section describes how images are made from the data from sensor array points. We assume the array is equally spaced both horizontally and vertically, in a rectangular form. If we have as many sensor elements as dots on our computer display, we have to do nothing. Usually, however, there are far fewer sensor elements than display dots, for example, 8x8 elements is a better number than we usually have. This number is generally one or two orders lower than the number of display dots. Therefore, we need some interpolation to make the data points large enough to be seen and to give a feeling of spatial distribution.

One simple way to display sparse data is to segment the display area by the number of sensor points. Then, data magnitude is displayed by the gray level of rectangular segments. Such a display image requires the viewer to infer the original data and reconstruct the image of real existence. By using an appropriate display scheme, such as the one described below, displayed images can be directly accepted.

Two-dimensional linear interpolation can be done using linear interpolation in each of the dimensions. One must interpolate the value for the point (i,j) by

$$f(i+\xi,j+\eta)=a+(b-a)\xi+(c-a)\eta+(a+d-b-c)\xi\eta \tag{1}$$

where $a=f(i,j)$, $b=f(i,j+1)$, $c=f(i+1,j)$, $d=f(i+1,j+1)$, $0\leq\xi<1$, $0\leq\eta<1$.

Another interpolation can be done by segmenting the rectangle into four triangles defined by 4 vertexes and the center of the rectangle. The value for the center can be calculated as the mean of the four vertexes. Interpolation for each triangle is to assume the triangle as a flat plane in the three-dimensional space.

Calculation time for linear interpolation is short. However, if there are peaks between sensor elements, they will be inevitably disappear, while some linear effects may appear, so that smoothness of the resultant image is damaged. Third-order spline interpolation is frequently used as a polynomial interpolation to fit a smooth transition on the resulting interpolation. It should be noted, however, that this interpolation sometimes makes false peaks in the interpolated portions.

If we know the the standard model of the distribution under investigation, it is best to use it in the interpolation. For example, when visualizing a gas field, the diffusing process is the main controlling factor determining its spatial distribution. Therefore, the following equation can be used in the first-order interpolating scheme,

$$f(x,y)=\sum_i C_i e^{-\frac{(x-x_i)^2+(y-y_i)^2}{\sigma_i^2}} \tag{2}$$

where (x_i,y_i) is the center of diffusion and C_i and σ_i are the diffusion parameters. Nevertheless, the direct determination of the parameters C_i, x_i, y_i, σ_i demands nonlinear least-square calculation, which requires a lot of time to reach a solution and is not suitable in real-time visualization.

In the case of visualizing sound fields, band-limited assumption is usually set to interpolate spatially and temporally. If we sample instantaneous sound pressure $f(i,j)$ at an equally spaced two-dimensional array of microphones, we can assume that all components of the sound field are completely sampled. The sound pressure everywhere on the sampled plane can be interpolated using the $\sin x/x$ function.

$$f(x,y)=\sum_i\sum_j f(x,y)\frac{\sin\pi(x-i)}{\pi(x-i)}\frac{\sin\pi(y-j)}{\pi(y-j)}. \tag{3}$$

This equation assumes that we can obtain infinite number of sampling points, which corresponds to the situation that we have an infinitely large sensor array. Also, the spatial frequency components should be in the range lower than the Nyquest spatial frequency of the array. If the assumptions are practically satisfied, we can obtain almost real-field images from the interpolation, where the peak disappearances occurring in the linear interpolation will not exist.

With regard to the display, gray scale or contor line representations are the commonest form of presenting computer-generated output to human viewers. What kind of representation format is suitable depends on the image renewal rate and the objectives of the visualization. The gray scale display is better than the contor line display in rapidly changing fields. The contor line display is suitable for magnifying changes in slowly changing fields.

The general way of studying a fast changing field is to slow it down by slow-motion replay. Non-periodical changes in periodically changing fields can be shown by stroboscopic representation. In cases of sound fields, wave-front movement can be simulated by slightly changing the refreshment period of the displayed image with regard to the period of the sound. Some temporal filtering should be applied if noises or fluctuations that are not the object of the visualization are large.

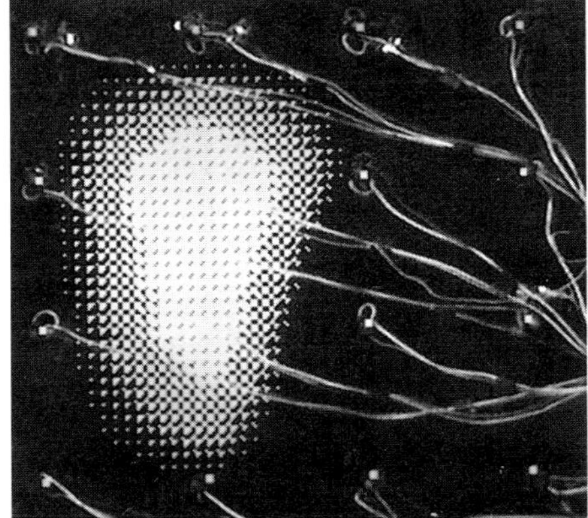

Figure 1. Visualization example of gas distribution[4].

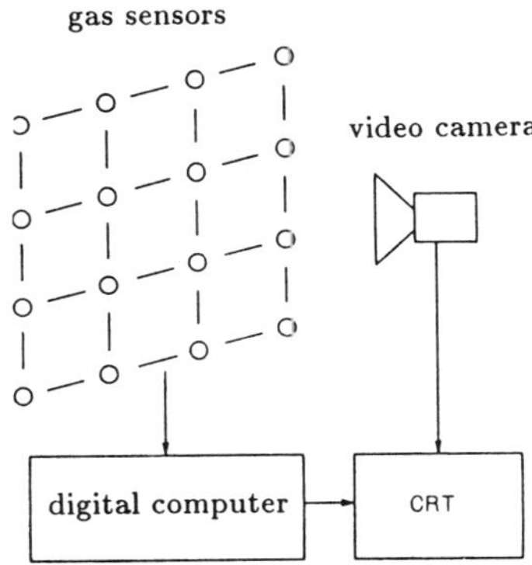

Figure 2. Visualizing system of gas distribution.

4. VISUALIZATION OF SMELL [4-6]

Figure 1 shows an introductory visualization of smell which in reality is a visualization of the concentration distribution of alcoholic gas using a two-dimensional array of gas sensors. The cloudy portion on the left shows the location and concentration of gas components. The CRT display shows the scene in colors, with the gas cloud being red. We used SnO_2 semiconductor gas sensors as the element sensors. This type of gas sensor has a relatively high sensitivity (ppb sensitivity can be attained for some gas species) and is small in size (typically packaged in a case 15 mm in diameter and 15 mm in height). The sensing mechanism detects conductance changes in accordance with the quantity change of surface-adsorbed oxygen.

One weak point of semiconductor gas sensors is their low selectivity. It is not easy to detect particular gas species using a single sensor. In contrast, this low selectivity indicates that we can detect a large variety of gas species using a single sensor. There have been many studies on the identification of gas species using an array of different types of gas sensors and using specific transient response patterns.

Figure 1 shows a two-dimensional 4x4 sensor array with a 15-cm spacing and wirings. The system for this visualization is shown in Figure 2. The interpolation was reached using eq. (1). Gray scale representation was attained using pseudodensity representation (3x3 dots for one pixel). The calculated gas-field image is overlaid on the actual image of the scene using a superpose interface on the CRT display. Thus, we can see a realistic and easily conceivable image of the field.

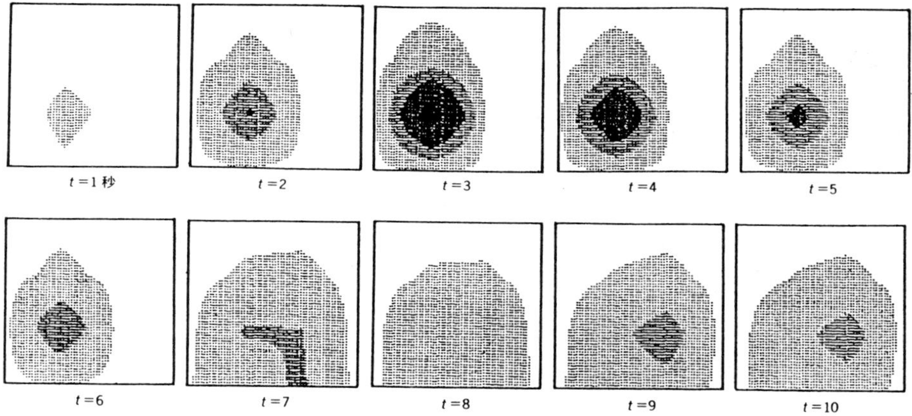

Figure 3. Changes in gas distribution (one second steps)[4].

Figure 3 shows a temporally changing field image of a two-dimensional alcoholic gas concentration. The field is almost limited to two dimensions by a transparent cover plate and base plate with a 10-cm spacing. One second is the time interval between each subfigure. The figure shows that one drop of liquid ethylalcohol rapidly evaporates and diffuses away from the dropped point. Several diffusion patterns after the dropping show an ideal field of circular diffusion. The area of low concentration first expanded rapidly, then gradually increased. In the latter subfigures, evaporation ceased and convection affected the distribution. The center of the distribution moved to the right and the concentration decreased.

In cases of detecting the actual point of gas leakage from road surfaces, we may be faced with more difficult situation: the gas source might be obscure with a low volume of leakage and gas diffusion may be in a three dimensional space where the gas drifts and moves with the wind. Such cases are the most suitable for this type of visualization because human intelligence and experience can be utilized for locating the gas sources. Then, intelligent data processing may be helpful to the human viewers. For example, it is helpful to take the average over time to suppress noisy field fluctuations that may interfere with human perception, to indicate flow direction and velocity from calculations of cross-correlation between neighboring sensors, and to represent the gas distribution image simulated for the case in which air flow does not exist.

The gas sensor has its own sensitivity, differs from others and changes its value with time. Therefore, we need to compensate for these variations. One method to evaluate the sensitivities of element sensors is to estimate sensor sensitivity by monitoring sensor output variance caused by the environmental variation of gas concentration for each sensor element.

One of the common weak point of the visualizing system is the shortage of the display's dynamic range. We cannot convert linearly from ppb order data to percent order data on a common CRT display. Logarithmic conversion is an ordinary means for such situations, but there are rarely changes in a same order of concentration magnitude. Thus, one compromise is to display in a short dynamic range mode, and to increase the quantization step when a great change occurs and to decrease the step when the magnitude decreases. By doing this, we may induce the illusion that the display has a very wide dynamic range for representations.

One important characteristic of the semiconductor gas sensor is that its sensitivity changes with its surface temperature and the change depends on exposed gas species. Figure 4 is an example of such dependence, where gas specificity appears as the temperature of the conductance peak. Since the diffusion process in air is a relatively slow phenomenon, we can measure the peak temperature by periodically changing the surface temperature. Then, we can show peak temperature on the visualized plane by colors or some other means. This is similar to the visualization of smell distributions and an example of temporal data converted into spatial information.

5. VISUALIZATION OF SOUND [4,8]

Sound fields change faster than smell fields. If we could see sound, more specifically sound wave fronts, we could easily identify the location of the sound source and the path of sound propagation. If we could visualize sound in real time, we could effectively devise sound suppression and arrange acoustic spaces. Real acoustic spaces are complicated: not only characteristics of wall materials, but also their

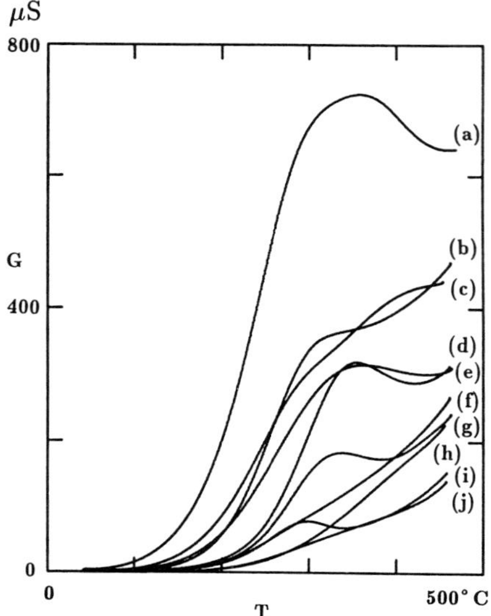

Figure 4. Gas specificity in transient conductance G versus surface temperature. The sample gases are (a)diethyl ether, (b)acetone, (c)methanol, (d)n-pentane, (e)ethanol, (f)isopropyl alcohol, (g)hexane, (h)acetic acid, (i)benzene and (j)butanol[7].

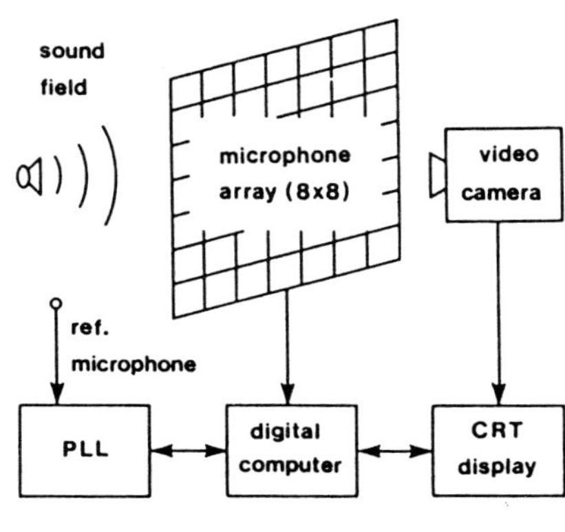

Figure 5. Visualizing system of sound[8].

geometrical factors significantly influence the resultant acoustic field. Perfect sound absorbers, perfect reflectors, or ideal point-sound sources do not really exist; therefore, we should directly measure and treat the real acoustic situation. A direct visualizing means is the most valuable for observing actual sound environment and the effects of any acoustical arrangements.

The Schrieren method is a well-known visualizing one for sound fields. However, its application is limited to a laboratory experimental setup. The use of microphone(s) is the most practical way of visualizing ordinary sound levels in air. By replacing the gas sensors in the smell visualization system with microphones, we can achieve a sound visualization system. One of the differences is that sound propagates much faster than gas diffusion. We need sample holders for each microphone to get sound pressure values at the same moments. By sampling sound pressure, wave fronts can easily be calculated by the interpolating the sound pressure field.

Figure 5 shows the system including the frequency tracking circuit for stroboscopic visualization of sound movements. The number of microphones is 64 and the spacing is 20 cm vertically and horizontally (Figure 6). The frequency tracking circuit comprises a phase lock loop and two frequency dividers (Figure 7). This circuit produces a signal that can be used to trigger sample holders with the period

Figure 6. The microphone array with its supporting frame[4].

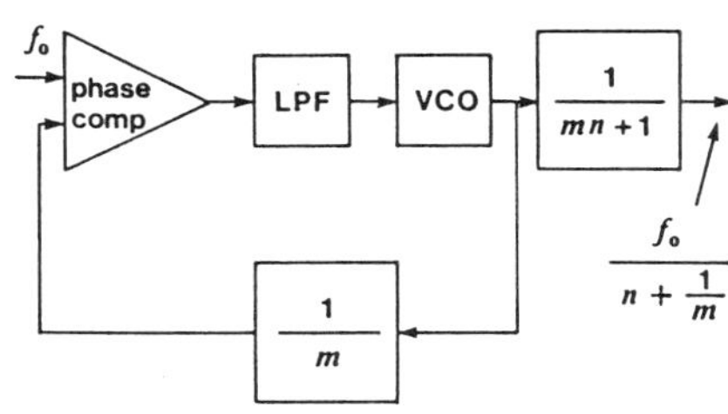

Figure 7. The PLL circuit for stroboscopic representation of sound image[8].

advancing $1/m$ period for each n/f_0 second, where f_0 is the frequency of the field sound and n and m are integers.

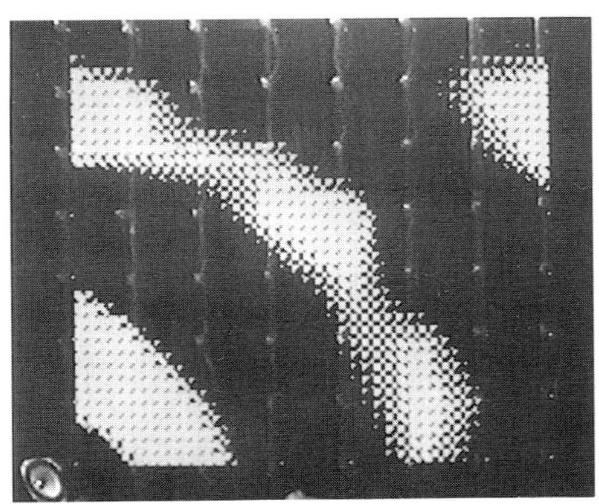

Figure 8. Example of sound visualization. The 500 Hz source is located at the lower left corner[4].

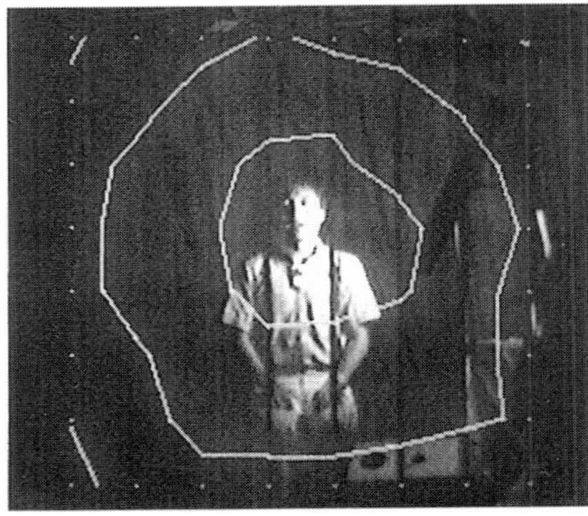

Figure 9. Wavefront of the human voice[4].

An overlaid image shows us floating and moving sound wave fronts (Figures 8 and 9). In the Figure 8 only the positive portion of the sound field is shown, with the pseudo-density display showing the magnitude of sound pressure. The distance between the white belts corresponds to half the wave length. Image calculation time is two image frames per second with a 1 MIPS minicomputer(PDP11/45). Figure 9 shows only the line of zero sound pressure. The distance between the lines is also half the wave length. Calculation of this image is faster than that seen in Figure 8; six image frames per second can be calculated using the same computer.

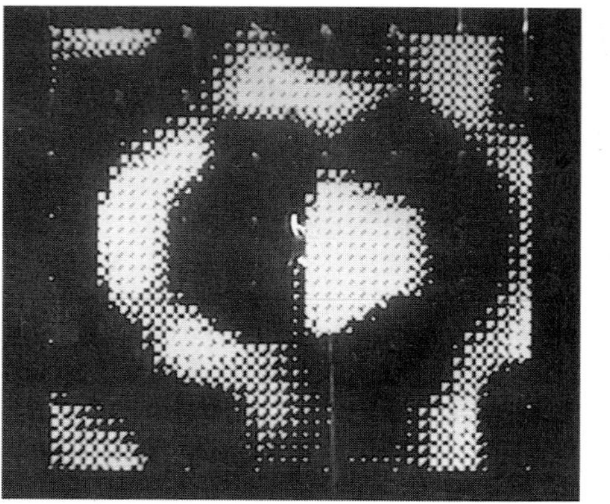

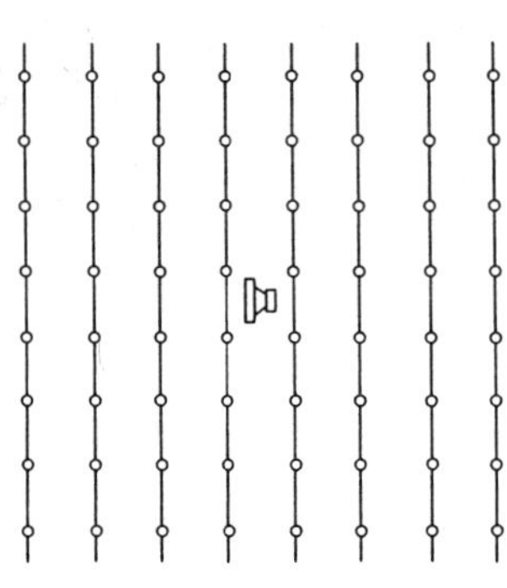

location of loud speaker

Figure 10. Phase distribution of emitted 750 Hz sound from a loudspeaker situated at the center[4].

Figure 10 shows an instantaneous sound pressure field created by a loudspeaker located in the center of the scene and pointing to the left. The view clearly shows the opposite phase emitted from the loudspeaker.

Various signal processes can be applied to make visualization for different field analyses. Figure 11 shows the average sound level created by a loudspeaker, which is calculated by taking first the absolute values and then the average for some period of time at each microphone point. The pattern shows more than directivity; it also includes effects of the sound environment, such as room size and wall reflectance.

If the field includes two or more sound sources, we can lock our stroboscopic displaying interval to one of the sources and see moving wave fronts from the other source(s). If we take the average after locking to one source frequency, we can cancel the remaining sound of different sound frequencies. When the sound from a source is not periodic, we can make cross-correlations between microphone outputs and show correlation delays on the display.

Conversion of temporal information to spatial information can also be done in sound visualization. Spatial sampling is limited by the spacing of microphones, but the temporal band width is only limited by the response of microphones and the AD interface of the computer, which can be set wider than the spatial band width. For

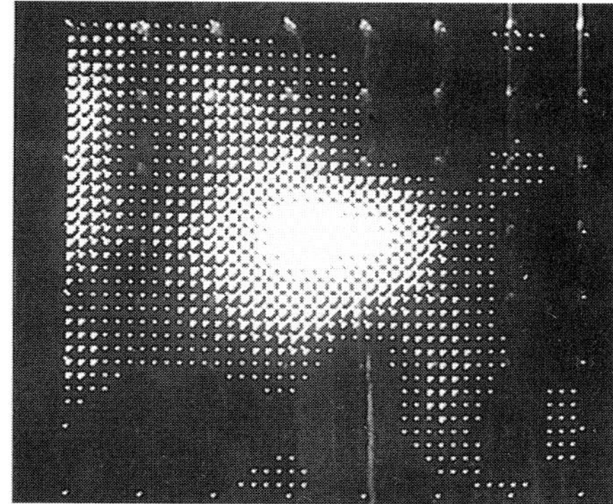

Figure 11. Directivity pattern of a loudspeaker in the same position as Figure 10 [4].

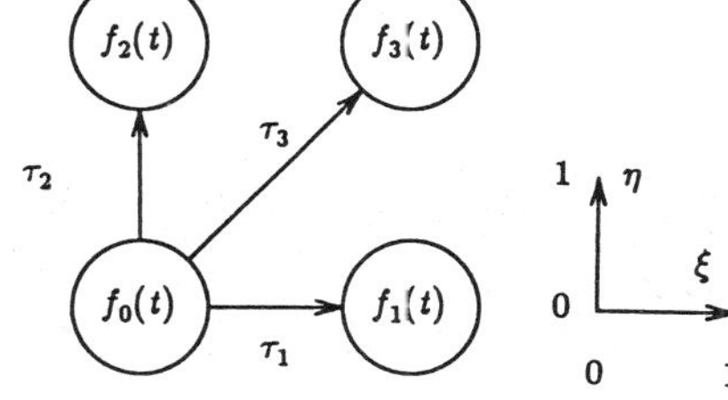

Figure 12. Sound pressures at four neighboring sensor positions and the definition of propagation delays.

example, in our visualizing system the shortest wavelength that can be sampled spatially is 40 cm. On the other hand, the shortest wavelength that can be sampled temporally is about 11 cm (3 kHz sampling at each microphone). Thus, if we know the proper way to estimate spatial information from temporal information, we can get equivalently wider band width than that obtained by spatial sampling only.

Temporal to spatial information conversion needs several assumptions: the signal of each sound source affecting the field must be extracted; propagation delay distribution must be calculated; and propagation delay and attenuation change slowly in the field. The propagation delays can be calculated from a cross-correlation between the data of neighboring microphones. If we get delay times between sensor points (Figure 12), bilinear interpolation can be used for estimating the data for each interpolated point:

$$f(\xi,\eta) = (1-\xi-\eta-\xi\eta)f_0(t-\tau)+\xi(1-\eta)f_1(t-\tau+\tau_1)+\eta(1-\xi)f_2(t-\tau+\tau_2)+\xi\eta f_3(t-\tau+\tau_3) \quad (4)$$

where $\tau = \tau_1\xi+\tau_2\eta+(\tau_3-\tau_1-\tau_2)\xi\eta$.

Figure 13 is an example of the conversion in which a sound propagates from the lower right corner. The cross-correlated component, that is the same as autocorrelation of the sound, is shown in the figure. Magnification of the spatial band width is not very high, but we can see the detailed waveform along the spatial coordinates.

6. SUMMARY

Visualizing systems and visualized examples using an array of sensors have been described. The quality of the visualized images depends mainly on the density of the

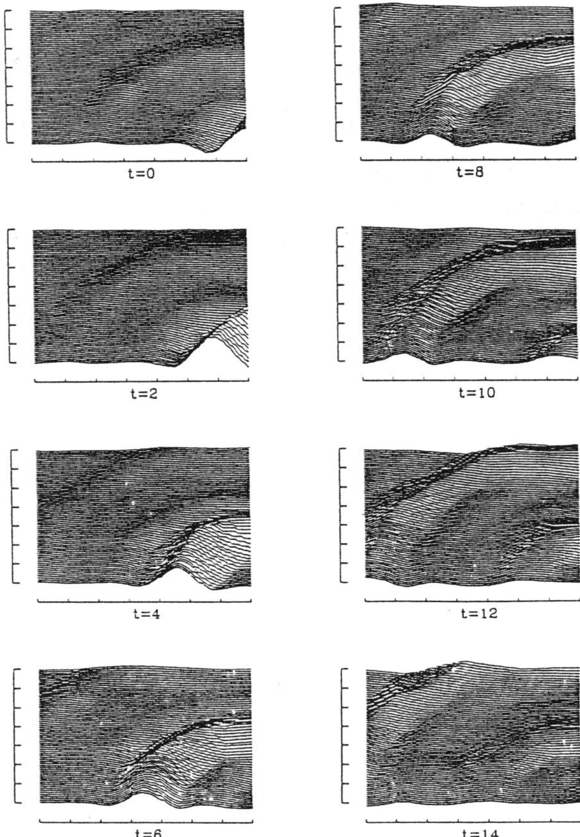

Figure 13. Sound pressure changes with increased spatial band width obtained by temporal to spatial information conversion. Microphones are at 8x8 cross-points. The time interval is 0.35ms.

sensor elements. Moreover, some field model estimations can appropriately interpolate data using not only information from spatially distributed sensors, but also dense temporal data from each sensor. Nevertheless, it should be easy to improve image quality because small-sized sensors and high integration are currently being studied.

Improvement of element sensor functionality, i.e., sensitivity, selectivity, noise rejection, multiple information acquisition, etc., makes it easier to obtain visualized images actually required by the viewer. Some intelligent functions in visualizing systems are just mentioned in this paper. Visualizing systems are extending the means of human functionality, adapting human interfaces as well as improving visualizing performance.

REFERENCES

1. Flow Visualization Soc. ed., Flow visualization handbook, Asakura, Tokyo, 1986 (in Japanese).
2. Television Soc. ed., Visualization of unseen information, Shouko-do, Tokyo, 1979 (in Japanese)
3. D.G. Childers, Proc. IEEE, 65 (1977) 611.
4. Y. Hiranaka, Computrol, 21 (1987) 119 (in Japanese).
5. H. Yamasaki and Y. Hiranaka, Sensors and Actuators A, 35 (1992) 1.
6. Y. Hiranaka and H. Yamasaki, International Congress on Analytical Sciences 1991, (1991) 435.
7. Y. Hiranaka and H. Murata, Sensors and Materials, 4 (1992) 165.
8. Y. Hiranaka, O. Nishii, T. Genma and H. Yamasaki, J. Acoust. Soc. Am., 84 (1988) 1373.